COST CONTROL IN BUILDING DESIGN

Roger Killingsworth

COST CONTROL IN BUILDING DESIGN

Roger Killingsworth

R.S. MEANS COMPANY, INC.
CONSTRUCTION CONSULTANTS & PUBLISHERS
100 Construction Plaza
P.O. Box 800
Kingston, Ma 02364-0800
(617) 585-7880

Printed in the United States of America

10 9 8 7 6 5 4 3 2 1

Library of Congress Cataloging in Publication Data

ISBN 0-87629-103-5

TABLE OF CONTENTS

ACKNOWLEDGMENTS

I would like to express my appreciation to those whose help and encouragement made this book possible. My thanks go to Paul Stephens, Manager, Facilities Planning Division; the Texas A & M University System, for the opportunity and encouragement to develop as a conceptual and cost control estimator; to Paul Brandt, Head, Department of Building Sciences, Auburn University, for a light fall quarter and much patience and understanding; to Mary Greene of R. S. Means Company for her quiet encouragement; to Wade, Carol, Sara, David, and Erin for understanding why Daddy spent so much time sequestered with the computer; and to Mary, my beloved life partner, for her prayers, support, understanding, and love.

INTRODUCTION

Cost Control In Building Design presents the control system and estimating methods that the author has found effective as a conceptual and cost control estimator in construction management. The system enables owners, facilities managers, and construction managers to perform reasonably accurate budgeting and feasibility studies, and to monitor costs during design. It provides guidelines for construction managers, architects, engineers, and design/build contractors for the performance of quick and accurate cost comparison studies, for controlling costs during design, and for running quick checks of change costs during construction. The book also provides an introduction to conceptual and design cost control estimating for students in engineering, architecture, facilities management, construction management, and construction.

The key word in the system is *accuracy*. Estimate accuracy is required for reliable feasibility studies, realistic budgets, or effective design cost control. There are limitations, however, to the accuracy that can be achieved. Some of the limitations are inherent in the nature of the construction industry, others in the planning and estimating process. The factors that the author has found to most strongly affect estimate accuracy are the amount of project scope and design information available, the accuracy of cost information, the accuracy of the bid conditions forecast, and the skill of the estimator.

Complete project scope and design information are required for accurate estimating. However, estimates for feasibility studies (and often for project budgets) are carried out before all of the project requirements are known. Not having this complete information increases the probability that the cost for a necessary item will be omitted. Early design cost control estimates are made based on schematic drawings and partial outline specifications. These estimates are made using assumptions for structural design, finishes, and mechanical or electrical systems. Often these assumptions are incorrect, or another system is substituted at a later date, making these check estimates invalid.

The cost information used in estimating should accurately reflect local labor and equipment rates, labor productivity, and material costs. This data may be from past projects, a proprietary data base, or current local labor and material costs. The historical and proprietary information must be updated and adjusted for location. All of the information must be adjusted for the inflation that may occur between the time that the estimate is created and the beginning of construction.

In estimating, some allowance must be made for the percentage charged by contractors, subcontractors, and suppliers for overhead and profit. This percentage changes as the competition changes, decreasing as competition increases, and increasing as competition decreases. This level of competition is reflected in the *bid conditions*. The author has seen changes in bid conditions which, in a few months' time, affected project costs by as much as 10%.

Through training and experience, an estimator can learn the mechanics and proper application of commonly used estimating methods. Accuracy can be improved by constantly monitoring the local construction market for current bid conditions and inflation rates, and by checking the accuracy of each estimate by comparing it with actual project costs. Nevertheless, the process involves correctly pricing and forecasting so many variables that accuracy is sometimes a matter of luck and intuition as much as it is skill.

The estimating system in *Cost Control in Building Design* provides ways to minimize the effects of the aforementioned limitations. Chapter 1 outlines the phases of project development and describes the amount and kind of information available at each phase. Chapter 2 explains the different estimating methods, the amount of information and time required for each, and their expected accuracy. Chapter 3 reviews various aspects of construction economics, and outlines methods to use in preparing forecasts for bid conditions and inflation. Chapter 4 describes and demonstrates methods to use in preparing estimates for early project planning. Chapter 5 details the steps involved in preparing a *Program of Requirements*. Chapter 6 shows how assemblies and square foot estimating methods can be used to prepare a project budget. Chapter 7 explains the methods used to prepare cost control estimates, and ways that the estimate can be used to suggest design changes that will reduce project costs. Chapters 8, 9, and 10 illustrate a proven system for estimating plumbing, HVAC, and electrical work. Finally, Chapter 11 describes a project report system that allows the estimator to build a project cost data base and to run checks on estimate accuracy; and Chapter 12 covers computer applications appropriate for conceptual and cost control estimating.

1

THE PLANNING PROCESS

1

THE PLANNING PROCESS

The goal of the building process is to construct a facility that meets the needs of the owner's organization and/or provides an adequate return on investment. A structured planning process, or checklist of activities, is needed to ensure that this goal is reached. The actual steps in the process may vary somewhat depending on whether the traditional design-then-build process or the phased construction, "fast-track" method is used. This chapter outlines the phases of the design-then-build process and describes in detail what occurs during each phase. Some of the key points are *when* the different types of estimates are needed, and *what information is available* for each estimate. Finally, we review the similarities and differences between the traditional design-then-build method and phased construction.

Traditional Construction

Traditional construction is a linear process in which each step is completed before the next begins. Planning is completed before the design is started. This means that the plans and specifications for the entire project are complete prior to the start of construction. The steps in the traditional process are as follows.

1. Need Identification
2. Need Analysis
3. Detailed Feasibility Study
4. Funding
5. Conceptual Design
6. Design Development
7. Bid/Negotiation
8. Construction
9. Occupancy

The application of the process is a little different for each project. In some cases, the separate, formal performance of each step may not be necessary, while for others, some steps may have to be repeated or divided into separate activities.

The advantages of the traditional method are better design coordination and lower construction costs. The primary disadvantage is the longer time required for all of the consecutive steps. A large commercial project usually takes three to four years for planning, design, and construction. The process is discussed here in order to show what activities must take place to ensure that the needed facilities are provided and the required design standards are met within the budget.

Need Identification

In the Need Identification phase, the need for new facilities is recognized; then studies are made to determine the best way to meet that need. These studies may include a market analysis, general business conditions forecasts, funding capacity, site selection, optimum size, renovation vs. leasing vs. building, etc. Estimates of probable costs are needed for the analysis of each option. Little is known about the project, however, other than the general type of facility needed. Chapter 4 includes an example of the information available at this early stage, and the appropriate type of estimate.

Need Analysis

In the Need Analysis phase, all of the functional, aesthetic, time, and financial requirements for the facility are identified. Studies are made to further define the purpose of the facility, to identify specific design and budget limitations, and to determine space, equipment, furnishings, and schedule requirements. This information is combined into a written document, the *project program*, which becomes the basis for a design contract. The program includes:

1. A brief history and statement of the goals and purposes of the organization and the reason for the new facility.
2. A brief description of the project, including general building type, location, and desired design concept.
3. General design requirements, including codes to be followed: site and site circulation requirements; and structural, finish, mechanical, and electrical design limitations.
4. Space requirements, including a detailed listing of all the spaces with information about use, occupancy, circulation, finishes, equipment, furnishings, and utilities.
5. A milestone schedule for the project, including dates for the completion of both the conceptual design phase and certain portions of the design development phase, bid negotiation, and occupancy.
6. The project budget providing a breakdown of the total project costs including construction costs, design fees, owner-provided equipment and furnishings, financing, and planning costs.
7. Exhibits showing locations, sites, and special layout requirements.

A detailed estimate based on the program requirements is needed in this phase to set the project budget. The goal of this estimate is to set a budget sufficient to build the facility, while still providing a challenge to the designer. While project requirements are known, the actual design is not. Chapter 5 includes a further explanation of project programming. Chapter 6 contains an example of the appropriate estimating procedure.

Detailed Feasibility Study

This final feasibility study uses the cost information in the program to confirm that the facility as programmed is economically feasible. The program estimate is based on more complete information, and is therefore more accurate than the Need Identification estimates. The business forecasts are also more accurate since this study takes place closer to the time when the facility will be constructed and used. An analysis of these estimates and forecasts may suggest changes in the program and budget that would make the project more profitable and/or practical in meeting the revised needs of the company.

Funding

Using the program and feasibility study, funding is secured. The sources for the funds may be various lending institutions, government appropriations, or bond sales.

Conceptual Design

In the Conceptual Design phase, the designer works through a number of schematic design studies to arrive at several schemes that potentially meet project requirements. These schemes are presented to the owner. The elements of each scheme are:

- a detailed site analysis
- site layout
- building layout and circulation studies
- elevation studies
- massing studies
- outlines of the structural, HVAC, plumbing, and electrical systems
- material/system cost comparison studies
- scheme cost estimate

The scheme chosen by the owner is then developed to the point that the site development, floor plan, and elevations are set, and the engineering systems roughed out. A check estimate is run to confirm that the design is within the budget. While project requirements and the general design scheme are known at this point, the actual details of systems and materials have not yet been determined. Chapter 7 shows examples of the information available in the conceptual design stage and the appropriate estimating methods.

Design Development

The design development phase involves making the final choices for materials, methods, and systems and preparing detailed plans and specifications for construction. Throughout this phase, *check estimates* are needed to monitor the cost of the developing design. Estimates should be run when the documents are approximately 50% complete, 75% complete, 90% complete, and when the documents are ready for bid/negotiation. As this phase progresses, more information becomes available for estimating. As materials and assemblies choices are made and installation details are drawn, the conceptual design estimate is updated to show current project costs. The same estimating methods used for the conceptual design estimate provide a sufficient degree of accuracy at this stage for all of the building systems except plumbing, HVAC, and electrical. Usually, there is not enough information for accurate estimates of these systems until the documents are about 90% complete. When this stage is reached, more detailed and accurate estimating methods can be used. Examples of these more detailed methods can be found in Chapters 8, 9, and 10.

Bid/Negotiation

Using the completed contract documents, the builder prepares an estimate of what he will charge to build the project. This estimate may then be submitted as a bid in competition with other builders, or may serve as the basis for negotiation. When a bid has been accepted, or an agreement reached in negotiation, a construction contract is issued.

Construction

During construction, the site is cleared, the foundation excavated and built, the structure erected and made weather-tight, the interior finishes applied, the engineering systems installed, and the site development completed. Seldom will the project be constructed without the need for changes in the design. These changes may be caused by unknown soil conditions or subsurface utilities, changing requirements of the owner's business, or an oversight in design. Such adjustments usually have some effect on project cost. The designer prepares drawings and specifications illustrating the change, and estimates the cost. Complete information is available at this stage for this type of estimate, which is used to check the builder's quote and/or for negotiating the change price. Estimates for changes can be prepared using the same methods used for the cost control estimates.

Occupancy

At the occupancy phase, construction is complete and the owner occupies the facilities.

Phased Construction

Until the design development phase, the steps in both the traditional process and phased construction are much the same. At the beginning of design development in phased construction, the project is divided into bid packages, with a portion of the project budget assigned for each. These bid packages are broken down according to the sequencing of construction activities. In general, the sequencing is as follows: initial site preparation followed by foundation work, structural frame, exterior closure, interior finishes, and site work. The first package, initial site preparation, is designed, bid, and construction is begun while the other packages are still in design. The remaining packages are developed, bid, and constructed in turn.

The advantage of phased construction is that construction can begin without waiting for the design and bid documents to be completed for the entire project. This allows the project to be finished and occupied by the owner months sooner. The disadvantages of this method are the generally higher construction costs and the coordination problems that sometimes occur between contractors working on separate packages. Changes can be prohibitively expensive because of their effect on work already in place. Generally, the additional cost for phased construction is more than offset by the savings resulting from the project's early completion. These savings include reduced interest on the construction loan and other benefits to the owner's business.

As the design for each portion develops, check estimates are needed at various intervals. Check estimates are performed when the documents are 50% complete, 75% complete, 90% complete, and when they are ready for bid, just as for the traditional construction process. As the actual cost of each package becomes known, its effect on the overall project budget is analyzed and the budgets for remaining packages adjusted accordingly.

Summary

A structured process is needed to ensure that the completed construction project will contain the required facilities built to the desired standards on time and within budget. The steps in this process for design-then-build construction are *Need Identification, Need Analysis, Feasibility Study, Funding, Conceptual Design, Design Development, Bid/Negotiation, Construction, and Occupancy*. The process is modified somewhat in phased construction in that construction is begun on portions of the project prior to the completion of the Design Development phase. Regardless of the method, estimates are needed throughout the process to control the cost of the project.

2

ESTIMATING METHODS

2

ESTIMATING METHODS

In the last chapter we looked at the phases necessary to control the development of a project. We identified which phases required estimates, how much information is available for each estimate, and which estimating method is appropriate for each phase. In this chapter we will explain and demonstrate the different estimating methods, discussing how much information, expertise, and time are needed for each, as well as their expected accuracy ranges. We will also identify sources of cost information.

The four most commonly used estimating methods are *Order of Magnitude*, *Square Foot/Volume*, *Assemblies*, and *Unit Price*. Each has different requirements for information, time, and expertise, and each achieves a different level of accuracy. Figure 2.1 shows a direct correlation between the amount of time required and the accuracy that can be expected for each of the estimating methods. As the time requirements increase, accuracy also increases. Not shown is the fact that as time and accuracy increase, the amount of information and level of expertise required also increase. The four estimating methods are described in the following pages.

Order of Magnitude

The Order of Magnitude method is the quickest and least accurate of the four types of estimates. Its purpose is to provide an approximate, "ballpark" estimate of project cost. It requires only the most basic information, such as the type of facility, takes only a few minutes to prepare, has an expected accuracy range of ±20%, and requires very little expertise on the part of the estimator. The cost information may come from past projects or proprietary cost data, such as the square foot/cubic foot costs in *Means Assemblies Cost Data*.

An example of Order of Magnitude estimate data is shown in Figure 2.2. Order of Magnitude costs take the form of a *per unit of occupancy* cost, like the *per bed* price for a low-rise dormitory in line 141-310-9000. These costs are derived from more than 11,000 projects contained in the Means data bank of construction costs. Contractor's overhead and profit are included in these prices, but architectural fees and land costs are not. The costs are shown in 1/4, median, and 3/4 ranges. The 1/4 column is a low cost range in which one fourth of the reported costs for the particular building type are lower, and three fourths higher than the given amount. The median column is a mid-range cost category; one half of the reported costs are lower and one half higher. The 3/4 column is the high-range cost category; three fourths of the reported costs are lower, and one fourth higher. These ranges reflect different qualities of construction and can be used to adjust a planning estimate to the quality proposed for a particular project. Using the mid-range price from Figure 2.2, line 141-310-9000, an example of a simplified Order of Magnitude for a 250-bed dormitory is as follows:

250 beds at $15,500/bed = $3,875,000 Building Cost

Square Foot/ Cubic Foot

The Square Foot/Cubic Foot method produces a more refined version of a ballpark estimate of project cost. This method requires the approximate building size, can take up to an hour, has an accuracy range of ±15%, and requires a little more expertise on the part of the estimator than the Order of Magnitude estimate. Square foot/cubic foot cost data may also be obtained from past projects or from proprietary cost data bases, such as square foot/cubic foot costs in *Means Assemblies Cost Data*. Examples of Means square foot/cubic foot costs (1988 edition) are shown in Figure 2.2. These costs are also derived from the Means data bank of construction costs. They include the contractor's overhead and profit, exclude architectural fees and land costs, and provide low, median, and high cost ranges. A breakdown of the total square foot cost for each building type is given in *Means Assemblies Cost Data*. This breakdown includes, when appropriate, such items as site work, masonry, finishes, equipment, plumbing, HVAC, and electrical. The breakdown also includes an *order of magnitude per unit of occupancy* where applicable.

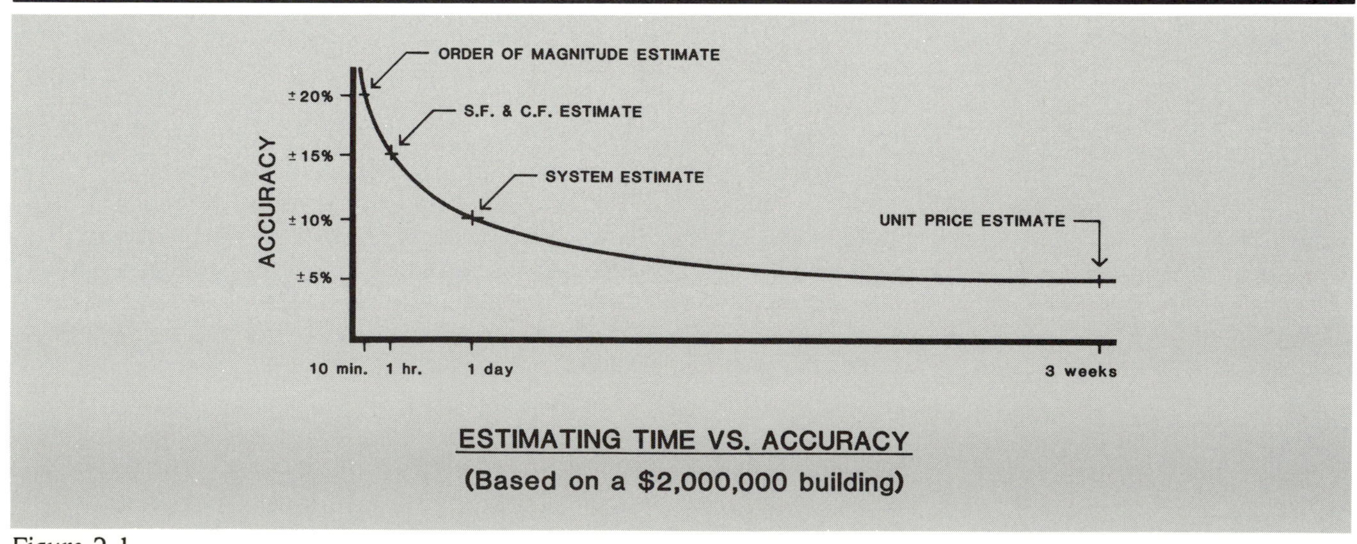

Figure 2.1

141 | S.F. & C.F. and % of Total Costs

	141 200	S.F. & C.F. Costs	UNIT	UNIT COSTS ¼	UNIT COSTS MEDIAN	UNIT COSTS ¾	% OF TOTAL ¼	% OF TOTAL MEDIAN	% OF TOTAL ¾	
250	2770	Heating, ventilating, air conditioning	S.F.	4.40	6.25	8.35	7.60%	10.20%	12.90%	250
	2900	Electrical		4.40	5.65	8.05	7.50%	9.10%	10.90%	
	3100	Total: Mechanical & Electrical		10.75	16	22.75	19.20%	26%	30.90%	
280	0010	**COURT HOUSES**		70.10	86.25	106				280
	0020	Total project costs	C.F.	5.70	6.40	7.45				
	0500	Masonry	S.F.	4.33	10.10	10.25	4.40%	5.40%	8.70%	
	1140	Roofing		.55	1.07	1.58	.70%	.90%	1.20%	
	2720	Plumbing		3.62	4.93	5.75	3.90%	6.30%	7.40%	
	2770	Heating, ventilating, air conditioning		9.60	10.70	15.70	10.30%	11.90%	14.80%	
	2900	Electrical		6.65	8.25	13.30	8.40%	9.80%	12%	
	3100	Total: Mechanical & Electrical		15.35	21.90	31.55	20.10%	25.70%	28.70%	
300	0010	**DEPARTMENT STORES**		28.55	38.10	43.50				300
	0020	Total project costs	C.F.	1.28	1.88	2.49				
	2720	Plumbing	S.F.	.93	1.30	1.68	2.80%	3.90%	5.30%	
	2770	Heating, ventilating, air conditioning		2.97	3.95	5.95	8.30%	11.80%	14.80%	
	2900	Electrical		3.26	4.43	5.45	10.40%	11.90%	13.60%	
	3100	Total: Mechanical & Electrical		6.15	10.80	12.40	22.10%	26.70%	32.30%	
310	0010	**DORMITORIES** Low Rise (1 to 3 story)		46.10	61.45	79.25				310
	0020	Total project costs	C.F.	3.44	5.35	7.25				
	2720	Plumbing	S.F.	3.21	4.03	5.40	8%	8.90%	9.60%	
	2770	Heating, ventilating, air conditioning		3.39	3.90	5.40	4.60%	7.60%	9.90%	
	2900	Electrical		3.29	4.66	5.70	6.40%	8.70%	9.40%	
	3100	Total: Mechanical & Electrical		7.75	12.40	17.30	18.40%	21.80%	28.40%	
	9000	Per bed, total cost	Bed	9,500	15,500	27,600				
320	0010	**DORMITORIES** Mid Rise (4 to 8 story)	S.F.	66.90	79.20	95.45				320
	0020	Total project costs	C.F.	6	7.25	9.15				
	2720	Plumbing	S.F.	5.25	5.30	7.75	6.40%	6.60%	10.30%	
	2900	Electrical		4.73	7.60	9.30	8.20%	9.40%	10.10%	
	3100	Total: Mechanical & Electrical		14.05	17.50	25	17.70%	22.90%	30.60%	
	9000	Per bed, total cost	Bed	9,250	16,000	19,600				
340	0010	**FACTORIES**	S.F.	22.90	33.35	54.60				340
	0020	Total project costs	C.F.	1.76	2.30	3.40				
	0100	Sitework	S.F.	2.83	5.30	8.05	6.90%	9%	16.70%	
	2720	Plumbing		2.52	2.95	4.44	4.60%	4.80%	10.40%	
	2770	Heating, ventilating, air conditioning		2.31	3.73	4.15	3.20%	5%	6.90%	
	2900	Electrical		3.63	7.40	8.25	8.90%	11.30%	14.40%	
	3100	Total: Mechanical & Electrical		8.95	14.65	20.25	17.80%	26.30%	33%	
360	0010	**FIRE STATIONS**		49.55	66.30	80.95				360
	0020	Total project costs	C.F.	3.15	4.24	5.35				
	0500	Masonry	S.F.	8.10	13.20	18.40	10.80%	15.60%	18.90%	
	1140	Roofing		2.22	3.35	4.65	1.80%	3.30%	4.90%	
	1350	Glass & glazing		.57	.78	1.69	.50%	1%	1.40%	
	1570	Floor covering		.32	.40	.58	.20%	.30%	.80%	
	1580	Painting		1.44	1.68	1.89	1.40%	1.60%	2.10%	
	1800	Equipment		.90	1.47	3.53	1.10%	2.50%	4.40%	
	2720	Plumbing		3.20	4.98	6.75	5.90%	7.30%	9.50%	
	2770	Heating, ventilating, air conditioning		2.74	4.41	7	4.80%	7.30%	9.20%	
	2900	Electrical		3.72	6.25	8.90	7.10%	9.70%	11.90%	
	3100	Total: Mechanical & Electrical		9.35	14.70	20.10	17.50%	22.60%	27.50%	
370	0010	**FRATERNITY HOUSES** And Sorority Houses		49.25	58.80	65.60				370
	0020	Total project costs	C.F.	4.71	5.75	6.35				
	2720	Plumbing	S.F.	3.72	4.34	5.75	5.90%	8%	10.80%	
	2900	Electrical		3.24	4.28	7.75	6.50%	8.80%	10.40%	
	3100	Total: Mechanical & Electrical		9.20	13.05	15.75	14.60%	20.70%	24.20%	

For expanded coverage of these items see *Means Square Foot Cost Data 1988*

489

Figure 2.2

To calculate a square foot estimate for the example dormitory, the number of beds must be converted into square feet of floor area. Building size, the floor area in square feet, is calculated in two ways: *gross* square feet and *net* square feet. The gross area of a building is calculated using the outside dimensions of the building. It includes all of the building area enclosed by the exterior walls, and one half of the area under overhangs, as well as one half of unfinished and unheated areas, such as parking basements. The *net* area of a building is the floor area usable for the building's intended function. It is calculated using the inside dimensions of each individual space. It excludes the floor area occupied by walls, corridors, restrooms, stairs, elevators, chases, mechanical rooms, overhangs, etc. The square foot costs from *Means Assemblies Cost Data* are based on gross building areas. Most square foot prices and code occupancy determinations are also based on gross building area. Some prices and code conversion ratios, such as the Life Safety Code occupancy rates, are based on net areas. When preparing estimates or making code applications, it is important to be sure of which unit is used.

Since the Means square foot prices are based on gross building area, the units of occupancy must be converted to gross building area in order to prepare a square foot estimate for the dormitory. The Unit Gross Area Requirements table from *Means Assemblies Cost Data* (Figure 2.3) provides factors to convert occupancy units into gross building area. The factors for most of the building types include low, medium, and high size ranges. Just as the cost ranges from the square foot/cubic foot cost tables, these ranges allow an estimate to be adjusted to more accurately reflect the conditions of a particular project. Assuming that our dormitory will fall in the medium size range, a conversion factor of 230 S.F./bed is used. The calculation to determine the approximate building size of our dormitory is as follows:

250 beds × 230 S.F./bed = 57,500 S.F.

Next, the building square footage is multiplied by a square foot cost. Assuming that the dormitory will be of medium quality, the mid-range cost from line 141-310-0010 in Figure 2.2 is used as follows:

57,500 S.F. × $61.45/S.F. = $3,533,375

It may be noted that the square foot estimate is nearly $350,000 less than the Order of Magnitude estimate. The reasons for this difference are the expected accuracy range of the estimating methods and the amount of information about the scope of the project that is used in each type of estimate. Both methods provide "ballpark" estimates. The size of the "ballpark", or the expected accuracy range, for the order of magnitude estimate, is ±20%. The accuracy range for the Square Foot estimate is ±15%. Since the difference between the two is less than 10%, both estimates are within their expected accuracy range, or "ballpark". Further, the information used for the Order of Magnitude estimate is limited to a particular building type, occupancy rate, and cost range. A wide variety of designs with greater or lesser floor areas and more or less expensive building assemblies could be used to match these parameters. The scope of the project would be defined in more detail for the Square Foot estimate, where a size range per occupancy is used to determine the building floor area. The inclusion of additional information further restricts the possible design solutions and allows the estimate to more closely "zero in" on the actual cost of the project.

S.F. & C.F. COSTS	C14.1-000	General

(91) Square Foot and Cubic Foot Building Costs

The cost figures in division 14.1 were derived from more than 11,200 projects contained in the Means Data Bank of Construction Costs and include the contractor's overhead and profit, but do not include architectural fees or land costs. The figures have been adjusted to January 1, 1988. New projects are added to our files each year and projects over ten years old are discarded. For this reason, certain costs may not show a uniform annual progression. In no case are all subdivisions of a project listed.

These projects were located throughout the U.S. and reflect tremendous differences in S.F. and C.F. costs. This is due to both differences in labor and material costs, plus differences in the owner's requirements. For instance, a bank in a large city would have different features than one in a rural area. This is true of all different types of buildings analyzed. As a general rule, the projects on the low side did not include any site work or equipment, but the projects on the high side may include both equipment and site work. The median figures do not generally include site work.

None of the figures "go with" any others. All individual cost items were computed and tabulated separately. Thus the sum of the median figures for Plumbing, HVAC and Electrical will not normally total up to the total Mechanical and Electrical costs arrived at by separate analysis and tabulation of the projects.

Each building was analyzed as to total and component costs and percentages. The figures were arranged in ascending order with the results tabulated as shown. The 1/4 column shows that 25% of the projects had lower costs, 75% higher. The 3/4 column shows that 75% of the projects had lower costs, 25% had higher. The median column shows that 50% of the projects had lower costs, 50% had higher.

There are two times when square foot costs are useful. The first is in the conceptual stage when no details are available. Then square foot costs make a useful starting point. The second is after the bids are in and the costs can be worked back into their appropriate units for information purposes. As soon as details become available in the project design, the square foot approach should be discontinued and the project priced as to its particular components. When more precision is required or for estimating the replacement cost of specific buildings, the "Means Square Foot Costs 1987" should be used.

In using the figures in division 14.1, it is recommended that the median column be used for preliminary figures if no additional information is available. The median figures, when multiplied by the total city construction cost index figures and then multiplied by the project size modifier, should present a fairly accurate base figure, which would then have to be adjusted in view of the estimator's experience, local economic conditions, code requirements and the owner's particular requirements. There is no need to factor the percentage figures as these should remain constant from city to city. All tabulations mentioning air conditioning had at least partial air conditioning.

The editors of this book would greatly appreciate receiving cost figures on one or more of your recent projects which would then be included in the averages for next year. All cost figures received will be kept confidential except that they will be averaged with other similar projects to arrive at S.F. and C.F. cost figures for next year's book. See advertising section in back of this publication for Means Project Cost Report for details and the discount available for submitting one or more of your projects.

S.F. & C.F. COSTS	C14.1-030	Space Planning

The figures in the table below indicate typical ranges in square feet as a function of the "occupant" unit. This table is best used in the preliminary design stages to help determine the probable size requirement for the total project. See next page for the typical total size ranges for various types of buildings.

Table 14.1-101 Unit Gross Area Requirements

Building Type	Unit	Gross Area in S.F.		
		1/4	Median	3/4
Apartments	Unit	660	860	1100
Auditorium & Play Theaters	Seat	18	25	38
Bowling Alleys	Lane		940	
Churches & Synagogues	Seat	20	28	39
Dormitories	Bed	200	230	275
Fraternity & Sorority Houses	Bed	220	315	370
Garages, Parking	Car	325	355	385
Hospitals	Bed	685	850	1075
Hotels	Rental Unit	475	600	710
Housing for the Elderly	Unit	515	635	755
Housing, Public	Unit	700	875	1030
Ice Skating Rinks	Total	27,000	30,000	36,000
Motels	Rental Unit	360	465	620
Nursing Homes	Bed	290	350	450
Restaurants	Seat	23	29	39
Schools, Elementary	Pupil	65	77	90
Junior High & Middle	↓	85	110	129
Senior High	↓	102	130	145
Vocational	↓	110	135	195
Shooting Ranges	Point		450	
Theaters & Movies	Seat		15	

485

Figure 2.3

The Cubic Foot estimating method requires that the total building volume be known. Using the dormitory as an example, the floor area, 57,500 S.F., must be converted into cubic feet for building volume. For the relatively small rooms of a dormitory, an eight foot ceiling height is normal. Space must be allowed above the ceiling for the floor or roof structure, lighting, and HVAC duct work. This dimension may be as little as six inches or as much as four feet, depending on the structural system. Allowing for the eight foot ceiling height and four feet for structure, lighting and HVAC duct work, a twelve foot floor-to-floor height will be used. The building volume is:

57,500 S.F. × 12′ = 690,000 C.F.

This building volume is then multiplied by a cubic foot price. Assuming medium quality, the mid-range cost from line 141-310-0020 in Figure 2.2 is used as follows:

690,000 C.F. × $5.35/C.F. = $3,691,500

This estimate falls between the Order of Magnitude estimate at $3,875,000 and the Square Foot estimate at $3,533,375. It is in the same "ballpark" of ±15%. However, by defining the volume contained in the building, it further limits design choices and more closely estimates project costs.

Assemblies

The Assemblies, or Systems, method gives a more detailed estimate of project cost. It requires a detailed listing of project requirements, several hours to several days to complete, has an accuracy range of around ±10%, and requires that the estimator have some knowledge of building systems and assemblies estimating methods. The information may come from job cost records or from proprietary data bases such as *Means Assemblies Cost Data*. The information used takes the form of a price for a complete construction assembly such as a structural floor system (shown in Figure 2.4) or a bathroom fixture system (shown in Figure 2.5). The cost for a floor system—a cast-in-place multispan joist slab—includes the costs for metal pans, beam forms, slab edge forms, reinforcing, and concrete, as well as concrete placing, finishing, and curing. Costs per square foot of floor area are given for different superimposed loads and bay spacings. The cost for the bathroom assembly includes the installed prices of a water closet, lavatory, bathtub or shower, and rough-in piping. Costs for each assembly are given for different fixture groups and arrangements. The Assemblies method can be used in modeling, comparison estimates, and in monitoring costs during design. Examples are given in Chapters 7, 8, 9, and 10.

Unit Cost

A unit cost estimate provides a complete listing of the quantities and prices of the labor, equipment, and materials required for the project. This method requires complete project plans and specifications, takes weeks or months to complete, has an accuracy range of about ±5%, and requires a great deal of expertise to prepare. It is used by contractors to prepare a bid for a project. Data is usually obtained from company cost records, and from current wage and materials prices. Unit price information can also be obtained from proprietary data bases such as Means' annually updated *Building Construction Cost Data*.

FLOORS	B3.5-160	C.I.P. Multispan Joist Slab

General: Combination of thin concrete slab and monolithic ribs at uniform spacing to reduce dead weight and increase rigidity. The ribs (or joists) are arranged parallel in one direction between supports.

Square end joists simplify forming. Tapered ends can increase span or provide for heavy load.

Costs for multiple span joists are provided in this section. Single span joist costs are not provided here.

Design and Pricing Assumptions:
Concrete f'c = 4 KSI, normal weight placed by concrete pump.
Reinforcement, fy = 60 KSI.
Forms, four use.
4-1/2" slab.
30" pans, sq. ends (except for shear req.).
6" rib thickness.
Distribution ribs as required.
Finish, steel trowel.
Curing, spray on membrane.
Based on 4 bay x 4 bay structure.

System Components	QUANTITY	UNIT	COST PER S.F. MAT.	INST.	TOTAL
SYSTEM 03.5-160-2000					
15'X15' BAY, 40 PSF, S. LOAD 12" MIN. COLUMN					
Forms in place, floor slab with 30" metal pans, 4 use	.905	S.F.	.91	2.73	3.64
Forms in place, exterior spandrel, 12" wide, 4 uses	.170	SFCA	.13	.83	.96
Forms in place, interior beam. 12" wide, 4 uses	.095	SFCA	.07	.38	.45
Edge forms, 7"-12" high on elevated slab, 4 uses	.010	L.F.		.03	.03
Reinforcing in place, elevated slabs #4 to #7	.628	Lb.	.19	.13	.32
Concrete ready mix, regular weight, 4000 psi	.555	C.F.	1.19		1.19
Place and vibrate concrete, elevated slab, 6" to 10" pump	.555	C.F.		.35	.35
Finish floor, monolithic steel trowel finish for finish floor	1.000	S.F.		.48	.48
Cure with sprayed membrane curing compound	.010	S.F.	.02	.04	.06
TOTAL			2.51	4.97	7.48

3.5-160 Cast In Place Multispan Joist Slab

	BAY SIZE (FT.)	SUPERIMPOSED LOAD (P.S.F.)	MINIMUM COLUMN SIZE (IN.)	RIB DEPTH (IN.)	TOTAL LOAD (P.S.F.)	COST PER S.F. MAT.	INST.	TOTAL
2000	15 x 15	40	12	8	115	2.51	4.97	7.48
2100	(17)(18)	75	12	8	150	2.53	4.98	7.51
2200		125	12	8	200	2.60	5.05	7.65
2300		200	14	8	275	2.69	5.20	7.89
2600	15 x 20	40	12	8	115	2.56	4.97	7.53
2800		75	12	8	150	2.63	5.05	7.68
3000		125	14	8	200	2.77	5.35	8.12
3300		200	16	8	275	2.92	5.40	8.32
3600	20 x 20	40	12	10	120	2.63	4.85	7.48
3900		75	14	10	155	2.79	5.15	7.94
4000		125	16	10	205	2.81	5.25	8.06
4100		200	18	10	280	2.99	5.45	8.44
4300	20 x 25	40	12	10	120	2.67	4.92	7.59
4400		75	14	10	155	2.80	5.15	7.95
4500		125	16	10	205	2.98	5.45	8.43
4600		200	18	12	280	3.16	5.70	8.86
4700	25 x 25	40	12	12	125	2.65	4.80	7.45
4800		75	16	12	160	2.86	5.05	7.91
4900		125	18	12	210	3.25	5.60	8.85
5000		200	20	14	291	3.49	5.70	9.19

64

For expanded coverage of these items see *Means Concrete Cost Data 1988*

Figure 2.4

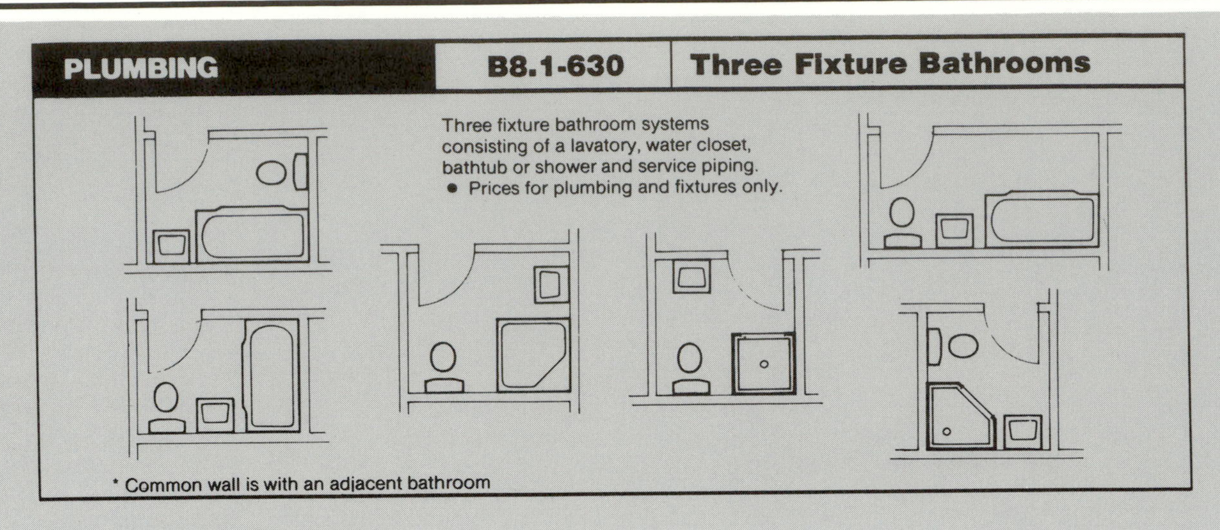

System Components	QUANTITY	UNIT	COST EACH MAT.	COST EACH INST.	COST EACH TOTAL
SYSTEM 08.1-630-1170					
BATHROOM, LAVATORY, WATER CLOSET & BATHTUB					
ONE WALL PLUMBING, STAND ALONE					
Wtr closet, 2 pc close cpld vit china flr mntd w/seat supply & stop	1.000	Ea.	126.50	93.50	220
Water closet, rough-in waste & vent	1.000	Set	91.45	258.55	350
Lavatory w/ftngs, wall hung, white, PE on CI, 20" x 18"	1.000	Ea.	139.70	65.30	205
Lavatory, rough-in waste & vent	1.000	Set	119.60	305.40	425
Bathtub, white PE on CI, w/ftgs, mat bottom, recessed, 5' long	1.000	Ea.	288.20	116.80	405
Baths, rough-in waste and vent	1.000	Set	74.58	262.92	337.50
TOTAL			840.03	1,102.47	1,942.50

8.1-630	Three Fixture Bathroom, One Wall Plumbing	COST EACH MAT.	COST EACH INST.	COST EACH TOTAL
1150	Bathroom, three fixture, one wall plumbing			
1160	Lavatory, water closet & bathtub			
1170	Stand alone	840	1,100	1,940
1180	Share common plumbing wall *	735	835	1,570

8.1-630	Three Fixture Bathroom, Two Wall Plumbing	COST EACH MAT.	COST EACH INST.	COST EACH TOTAL
2130	Bathroom, three fixture, two wall plumbing			
2140	Lavatory, water closet & bathtub			
2160	Stand alone	845	1,125	1,970
2180	Long plumbing wall common *	765	920	1,685
3610	Lavatory, bathtub & water closet			
3620	Stand alone	905	1,275	2,180
3640	Long plumbing wall common *	855	1,150	2,005
4660	Water closet, corner bathtub & lavatory			
4680	Stand alone	1,550	1,125	2,675
4700	Long plumbing wall common *	1,450	840	2,290
6100	Water closet, stall shower & lavatory			
6120	Stand alone	895	1,425	2,320
6140	Long plumbing wall common *	855	1,325	2,180
7060	Lavatory, corner stall shower & water closet			
7080	Stand alone	1,050	1,300	2,350
7100	Short plumbing wall common *	980	930	1,910

268

For expanded coverage of these items see *Means Mechanical Cost Data 1988*

Figure 2.5

To prepare a unit cost estimate for the cast-in-place multispan joist slab floor system, such as shown in Figure 2.4, the estimator would take off and price units, square feet of forms, tons of reinforcing, cubic yards of concrete, and square feet of concrete finishes and curing. While this system would be priced in just a few minutes using the Assemblies method, it could take several weeks for a large project using the Unit Price method.

Additional Considerations

A common problem in providing estimates for clients occurs when the client reviews the costs from another project and, in comparison, the estimator's costs appear high. A close check usually reveals that the client's price is incomplete, that it is out of date, and/or that it is based on another location where costs are different. To provide accurate estimates, one must know what the cost data includes, the size range and location of the project on which the data is based, and the date of the information.

The items included in cost data vary considerably, depending on the source. Cost data may include the complete project costs, it may be limited to the construction contract costs, or it may include only building costs. The Means data, for example, includes only the cost to construct the building. The costs for site development, design fees, construction contingencies, furniture, movable equipment, construction financing, and planning costs must be added to get a complete project cost estimate.

Inflation, the increase in prices over time, is a fact of life. Though short term price decreases are occasionally experienced, the long term trends are always up. Cost data must be adjusted to reflect this change.

Project Size

Generally speaking, building costs vary according to the size of the project. As the building increases in size, the square foot/cubic foot cost decreases. This is due primarily to the decreasing contribution of exterior wall costs to the square foot/cubic foot prices as the project increases in size, and secondarily, to economies of scale. Square foot and cubic foot prices must be adjusted to the size of the project that is being estimated.

Project Location

Prices also change from city to city. The price information from a project in one city must be adjusted for use in another city because of different materials and labor prices, different labor practices, different sales tax rates, etc.

Cost Adjustment

To obtain an accurate estimate of complete project costs, one must adjust the cost data for size, location, and inflation. One must also add estimates of design fees, construction contingencies, furniture costs, equipment costs, financing costs, and planning costs. Following is an example of the adjustments and additions needed to prepare a complete project cost estimate for the dormitory. First, the building cost will be adjusted for size. Next, the estimated costs of design fees, construction contingencies, furniture, equipment, financing, and planning will be added. Then, the total price will be adjusted for location and inflation.

The Means square foot/cubic foot cost data is based on the information given in the *Square Foot Base Size* table in Figure 2.6. Shown are the median square foot costs, the typical square foot size, and the typical size ranges for each type of building listed in *Means Assemblies Cost Data*, 1988. The typical size given for low-rise dormitories is 24,500 square feet.

A factor to adjust the Means cost data to the size planned for the dormitory is determined by dividing the proposed building size by the typical size from the chart. The proposed dormitory size divided by the typical low-rise dormitory size is :

$$\frac{57{,}500 \text{ S.F.}}{24{,}500 \text{ S.F.}} = 2.35$$

The cost multiplier is determined by plotting the size factor on the graph given with the *Square Foot Base Size* table. 2.35 on the horizontal scale yields a cost multiplier of approximately .93 from the vertical scale. The building cost adjustment is as follows:

Building Cost	$3,533,375
Cost Multiplier	× .93
Adjusted Building Cost	$3,286,039

Next, the costs for site development, design fees, construction contingencies, furniture, movable equipment, and planning must be added to get a complete project cost estimate. A detailed study is required for an accurate estimate of site development costs and is normally done in the Need Analysis Phase. For preliminary planning, an allowance of 10% of the building cost can be used.

Design fees are usually a percentage of the construction contract costs. These percentages vary according to the type of construction and the cost of the project. The fee percentages for different types of projects with different cost ranges are shown in Figure 2.7. Falling in the *Apartments, banks, schools* category with a cost between $2,500,000 and $5,000,000, the design fee for the dormitory is in the 6.4% to 6.7% range. 6.7% will be used in the example estimate.

Construction Contingency is an allowance for changes during construction. The amount varies according to the phase and size of the project. A project cost range based on the expected accuracy of the estimating method suffices for early planning. The construction contingency amount for the dormitory estimate will be assumed to be included in the cost ranges that are calculated later in the chapter.

The amount needed for furniture and movable equipment varies according to the type of project and the amount of existing equipment to be reused. It is assumed that new furniture will be purchased for the dormitory. A *per student* price for dormitory furniture can be obtained from the 1988 edition of *Means Building Construction Cost Data*. Lines 126-206-8000 and 126-206-8050 in Figure 2.8 provide a cost range for dormitory furniture of $1850/student to $3600/student. Because a mid-range cost is assumed for the building cost, a mid-range cost of $2500/student will be used to price the furniture in the example. No moveable equipment is needed for the dormitory.

Funding is obtained for most construction projects through financing. This financing may take the form of an interim financing/permanent financing package, capital improvement bonds, or, in special cases, a government subsidized loan. Financing costs are the interest and handling charges for the loans or bonds. These costs vary according to current interest rates and handling charge fees. An expert in financing should be consulted to aid in forecasting future rates and calculating the resulting costs. Occasionally a project such as our dormitory will be financed by government appropriation. In such cases, financing costs are not a factor. Therefore, no financing costs will be added to the dormitory estimate.

S.F & C.F. COSTS | **C14.1-000** | **Project Size Modifier**

(92) Square Foot Project Size Modifier

One factor that affects the S.F. cost of a particular building is the size. In general, for buildings built to the same specifications in the same locality, the larger building will have the lower S.F. Cost. This is due mainly to the decreasing contribution of the exterior walls plus the economy of scale usually achievable in larger buildings. The Area Conversion Scale shown below will give a factor to convert costs for the typical size building to an adjusted cost for the particular project.

The Square Foot Base Size lists the median costs, most typical project size in our accumulated data and the range in size of the projects.

The Size Factor for your project is determined by dividing your project area in S.F. by the typical project size for the particular Building Type. With this factor, enter the Area Conversion Scale at the appropriate Size Factor and determine the appropriate cost multiplier for your building size.

Example: Determine the cost per S.F. for a 100,000 S.F. Mid-rise apartment building.

$$\frac{\text{Proposed building area} = 100{,}000 \text{ S.F.}}{\text{Typical size from below} = 50{,}000 \text{ S.F.}} = 2.00$$

Enter Area Conversion scale at 2.0, intersect curve, read horizontally the appropriate cost multiplier of 0.94. Size adjusted cost becomes 0.94x $51.15=$48.08 based on national average costs.

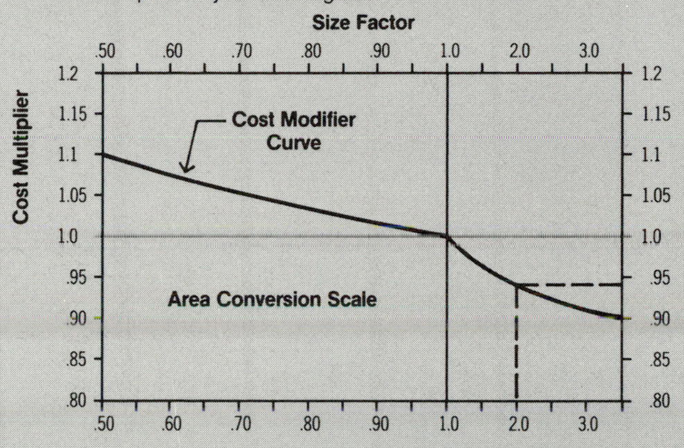

Building Type	Median Cost per S.F.	Typical Size Gross S.F.	Typical Range Gross S.F.	Building Type	Median Cost per S.F.	Typical Size Gross S.F.	Typical Range Gross S.F.
	Square Foot Base Size						
Apartments, Low Rise	$ 40.45	21,000	9,700 - 37,200	Jails	$119.00	13,700	7,500 - 28,000
Apartments, Mid Rise	51.15	50,000	32,000 - 100,000	Libraries	72.60	12,000	7,000 - 31,000
Apartments, High Rise	56.75	310,000	100,000 - 650,000	Medical Clinics	69.90	7,200	4,200 - 15,700
Auditoriums	68.25	25,000	7,600 - 39,000	Medical Offices	66.05	6,000	4,000 - 15,000
Auto Sales	43.15	20,000	10,800 - 28,600	Motels	51.55	27,000	15,800 - 51,000
Banks	94.20	4,200	2,500 - 7,500	Nursing Homes	69.75	23,000	15,000 - 37,000
Churches	62.60	9,000	5,300 - 13,200	Offices, Low Rise	55.40	8,600	4,700 - 19,000
Clubs, Country	60.55	6,500	4,500 - 15,000	Offices, Mid Rise	59.05	52,000	31,300 - 83,100
Clubs, Social	59.60	10,000	6,000 - 13,500	Offices, High Rise	73.70	260,000	151,000 - 468,000
Clubs, YMCA	64.00	28,300	12,800 - 39,400	Police Stations	92.30	10,500	4,000 - 19,000
Colleges (Class)	82.70	50,000	23,500 - 98,500	Post Offices	69.75	12,400	6,800 - 30,000
Colleges (Science Lab)	96.55	45,600	16,600 - 80,000	Power Plants	465.00	7,500	1,000 - 20,000
College (Student Union)	88.95	33,400	16,000 - 85,000	Religious Education	51.80	9,000	6,000 - 12,000
Community Center	64.55	9,400	5,300 - 16,700	Research	90.75	19,000	6,300 - 45,000
Court Houses	86.25	32,400	17,800 - 106,000	Restaurants	82.30	4,400	2,800 - 6,000
Dept. Stores	38.10	90,000	44,000 - 122,000	Retail Stores	40.00	7,200	4,000 - 17,600
Dormitories, Low Rise	61.45	24,500	13,400 - 40,000	Schools, Elementary	60.70	41,000	24,500 - 55,000
Dormitories, Mid Rise	79.20	55,600	36,100 - 90,000	Schools, Jr. High	60.50	92,000	52,000 - 119,000
Factories	33.35	26,400	12,900 - 50,000	Schools, Sr. High	60.05	101,000	50,500 - 175,000
Fire Stations	66.30	5,800	4,000 - 8,700	Schools, Vocational	57.60	37,000	20,500 - 82,000
Fraternity Houses	58.80	12,500	8,200 - 14,800	Sports Arenas	47.15	15,000	5,000 - 40,000
Funeral Homes	58.95	7,800	4,500 - 11,000	Supermarkets	39.40	20,000	12,000 - 30,000
Garages, Commercial	44.65	9,300	5,000 - 13,600	Swimming Pools	68.00	13,000	7,800 - 22,000
Garages, Municipal	47.25	8,300	4,500 - 12,600	Telephone Exchange	104.00	4,500	1,200 - 10,600
Garages, Parking	20.40	163,000	76,400 - 225,300	Terminals, Bus	52.50	11,400	6,300 - 16,500
Gymnasiums	56.35	19,200	11,600 - 41,000	Theaters	55.55	10,500	8,800 - 17,500
Hospitals	114.00	55,000	27,200 - 125,000	Town Halls	66.50	10,800	4,800 - 23,400
House (Elderly)	57.45	37,000	21,000 - 66,000	Warehouses	26.40	25,000	8,000 - 72,000
Housing (Public)	48.20	36,000	14,400 - 74,400	Warehouse & Office	30.10	25,000	8,000 - 72,000
Ice Rinks	45.80	29,000	27,200 - 33,600				

486

Figure 2.6

GENERAL CONDITIONS | C10.1-100 | Design & Engineering

Table 10.1-101 Architectural Fees

Tabulated below are typical percentage fees, below which adequate service cannot be expected. Fees may vary from those listed due to economic conditions.

Rates can be interpolated horizontally and vertically. Various portions of the same project requiring different rates should be adjusted proportionately. For alterations, add 50% to the fee for the first $500,000 of project cost and add 25% to the fee for project cost over $500,000.

Architectural fees tabulated below include Engineering Fees.

Building Type	Total Project Size in Thousands of Dollars						
	100	250	500	1,000	2,500	5,000	10,000
Factories, garages, warehouses repetitive housing	9.0%	8.0%	7.0%	6.2%	5.6%	5.3%	4.9%
Apartments, banks, schools, libraries, offices, municipal buildings	11.7	10.8	8.5	7.3	6.7	6.4	6.0
Churches, hospitals, homes, laboratories, museums, research	14.0	12.8	11.9	10.9	9.5	8.5	7.8
Memorials, monumental work, decorative furnishings	—	16.0	14.5	13.1	11.3	10.0	9.0

(75) Table 10.1-102 Engineering Fees

Typical **Structural Engineering Fees** based on type of construction and total project size. These fees are included in Architectural Fees.

Type of Construction	Total Project Size			
	To $250,000	$250,000-$500,000	$1,000,000	$5,000,000 & Over
Industrial buildings, factories & warehouses	Technical payroll times 2.0 to 2.5	1.60%	1.25%	1.00%
Hotels, apartments, offices, dormitories, hospitals, public buildings, food stores	↓	2.00%	1.70%	1.20%
Museums, banks, churches and cathedrals	↓	2.00%	1.75%	1.25%
Thin shells, prestressed concrete, earthquake resistive	↓	2.00%	1.75%	1.50%
Parking ramps, auditoriums, stadiums, convention halls, hangars & boiler houses	↓	2.50%	2.00%	1.75%
Special buildings, major alterations, underpinning & future expansion	↓	Add to above 0.5%	Add to above 0.5%	Add to above 0.5%

For complex reinforced concrete or unusually complicated structures, add 20% to 50%.

Table 10.1-103 Mechanical and Electrical Fees

Typical **Mechanical and Electrical Engineering Fees** based on the size of the subcontract. These fees are included in Architectural Fees.

Type of Construction	Subcontract Size							
	$25,000	$50,000	$100,000	$225,000	$350,000	$500,000	$750,000	$1,000,000
Simple structures	6.4%	5.7%	4.8%	4.5%	4.4%	4.3%	4.2%	4.1%
Intermediate structures	8.0	7.3	6.5	5.6	5.1	5.0	4.9	4.8
Complex structures	12.0	9.0	9.0	8.0	7.5	7.5	7.0	7.0

For renovations, add 15% to 25% to applicable fee.

465

Figure 2.7

126 | Furniture and Accessories

		126 100 Landscape Partitions	CREW	DAILY OUTPUT	MAN-HOURS	UNIT	BARE COSTS MAT.	LABOR	EQUIP.	TOTAL	TOTAL INCL O&P	
107	0010	POSTS Portable for pedestrian traffic control, standard, minimum				Ea.	54			54	59	107
	0100	Maximum					95			95	105	
	0300	Deluxe posts, minimum					87			87	96	
	0400	Maximum				↓	237			237	260	
	0600	Ropes for above posts, plastic covered, 1-½" diameter				L.F.	5			5	5.50	
	0700	Chain core				"	6			6	6.60	
		126 200 Furniture										
203	0010	BANK FURNITURE See division 126-222										203
204	0010	CHURCH FURNITURE See division 110-400										204
206	0010	FURNITURE, DORMITORY										206
	1000	Chest, four drawer, minimum				Ea.	260			260	285	
	1020	Maximum				"	403			403	445	
	1050	Built in, minimum	2 Carp	13	1.230	L.F.	72	26		98	120	
	1150	Maximum		10	1.600		143	34		177	210	
	1200	Desk top, built-in, laminated plastic, 24" deep, minimum		50	.320		16	6.80		22.80	28	
	1300	Maximum		40	.400		47	8.50		55.50	64	
	1450	30" deep, minimum		50	.320		16	6.80		22.80	28	
	1550	Maximum		40	.400		69	8.50		77.50	88	
	1750	Dressing unit, built-in, minimum		12	1.330		116	28		144	170	
	1850	Maximum	↓	8	2	↓	345	42		387	440	
	4001											
	8000	Rule of thumb: total cost for furniture, minimum				Student					1,850	
	8050	Maximum				"					3,600	
210	0010	FURNITURE, HOSPITAL Beds, manual, minimum				Ea.	536			536	590	210
	0100	Maximum					932			932	1,025	
	0300	Manual and electric beds, minimum					670			670	735	
	0400	Maximum					1,442			1,442	1,575	
	0600	All electric hospital beds, minimum					891			891	980	
	0700	Maximum					2,215			2,215	2,425	
	0900	Manual, nursing home beds, minimum					422			422	465	
	1000	Maximum					876			876	965	
	1020	Overbed table, laminated top, minimum					180			180	200	
	1040	Maximum				↓	448			448	495	
	1100	Patient wall systems, not incl. plumbing, minimum				Room	536			536	590	
	1200	Maximum				"	989			989	1,100	
	2000	Geriatric chairs, minimum				Ea.	196			196	215	
	2020	Maximum				"	366			366	405	
214	0010	FURNITURE, HOTEL Standard quality set, minimum				Room	1,725			1,725	1,900	214
	0200	Maximum				"	5,300			5,300	5,825	
218	0010	FURNITURE, LIBRARY										218
	0100	Attendant desk, 36" x 62" x 29" high	1 Carp	16	.500	Ea.	1,487	10.60		1,497	1,650	
	0200	Book display, "A" frame display, both sides		16	.500		1,676	10.60		1,686	1,850	
	0220	Table with bulletin board		16	.500		887	10.60		897.60	990	
	0800	Card catalogue, 30 tray unit		16	.500		1,785	10.60		1,795	1,975	
	0840	60 tray unit	↓	16	.500		3,028	10.60		3,038	3,350	
	0880	72 tray unit	2 Carp	16	1	↓	3,137	21		3,158	3,475	
	0900											
	1000	Carrels, single face, initial unit	1 Carp	16	.500	Ea.	438	10.60		448.60	500	
	1500	Double face, initial unit	2 Carp	16	1		649	21		670	745	
	4000	Dictionary stand, stationary	1 Carp	16	.500		497	10.60		507.60	560	
	4020	Revolving		16	.500		162	10.60		172.60	195	
	4200	Exhibit case, table style, 60" x 28" x 36"		11	.727		2,812	15.40		2,827	3,125	
	7000	Tables, card catalog reference, 24" x 60" x 42"	↓	16	.500	↓	562	10.60		572.60	635	
222	0010	FURNITURE, OFFICE										222
	0020	Desks, 29" high, double pedestal, 30" x 60", metal, minimum				Ea.	224			224	245	

For expanded coverage of these items see *Means Interior Cost Data 1988*

263

Figure 2.8

Planning costs vary depending on project needs and the size of the project. This cost can be approximated for planning purposes by adding a percentage of the construction costs. Usually, an allowance of 1% to 3% is sufficient. This price varies inversely with the size of the project. For projects having a construction cost less than $1,000,000, use 3%; for those having a construction cost between $1,000,000 and $3,000,000, use 2%; and for those having a construction cost greater than $3,000,000, use 1%. The building cost of the dormitory, plus 10% for site development, is greater than $3,000,000. Therefore, 1% will be used in the example estimate.

Adding for these items modifies the dormitory estimate as follows:

Adjusted Building Cost	$3,286,039
Site Development @ 10%	328,604
Construction Contract Cost	$3,614,643
Design Fees @ 6.7%	242,181
Total Cost of Construction	**$3,856,824**
Furniture—250 students @ $2,500	625,000
Planning	38,568
Total Project Cost	$4,520,392

Next, adjustments for **location and inflation** must be made. Location adjustments can be calculated using the City Cost Indexes from *Means Assemblies Cost Data* shown in Figure 2.9. The City Cost Indexes include values for material, installation, and total costs for the major UNIFORMAT divisions. Also included is a weighted average of the material, installation, and total costs for the complete project. Converting an estimate based on Means national average cost data to a specific location is done by multiplying the estimated cost by the total weighted average cost index for the location.

We will assume that the dormitory is to be built in Atlanta, Georgia. The total weighted cost index from Figure 2.9 for Atlanta is 89.2. This means that the total weighted construction costs in Atlanta are 89.2% of the national average. Adjusting the Total Project Cost for the dormitory to construction costs in Atlanta is as follows:

$4,520,392 × .892 = $4,032,190

The estimate must also be also adjusted for the inflation that occurs between the effective date of the cost data and the scheduled project bid date. The cost data used to prepare the dormitory estimate is from the 1988 editions of *Means Assemblies Cost Data* and *Building Construction Cost Data*. These data sources have an effective date of January 1, 1988. Inflation must be added to bring the costs up to the scheduled bid date. Assuming the date of the estimate is June 1, 1988 and allowing 1-1/2 years for planning and design, the price must be escalated to January, 1990. Thus, the price must be escalated by two years, from January 1, 1988 to January 1, 1990. Allowing 4% inflation per year yields:

$4,032,190 × 1.04 × 1.04 = $4,361,216

Finally, it must be remembered that each estimating method has an accuracy range. The Square Foot method used in the dormitory estimate has an accuracy range of ±15%. This means that the actual project costs can be any value in the range of ±15% of the total estimated cost. This cost range is calculated as follows:

$4,361,216 + 15% to $4,361,216 − 15%

or

$5,015,398 to $3,707,034

CITY COST INDEXES | C13.3-000 | Building Systems

DIV. NO.	BUILDING SYSTEMS	NEW HAVEN, CT MAT.	INST.	TOTAL	NORWALK, CT MAT.	INST.	TOTAL	STAMFORD, CT MAT.	INST.	TOTAL	WATERBURY, CT MAT.	INST.	TOTAL	WILMINGTON, DE MAT.	INST.	TOTAL
1-2	FOUND/SUBSTRUCTURES	101.8	103.3	102.7	106.4	103.3	104.5	118.9	103.9	109.6	112.4	101.5	105.6	102.5	109.0	106.6
3	SUPERSTRUCTURES	95.1	103.7	99.6	96.3	103.6	100.1	101.5	104.3	103.0	99.8	101.6	100.8	94.8	108.4	101.9
4	EXTERIOR CLOSURE	106.9	106.0	106.3	99.2	104.5	102.6	103.1	104.4	103.9	97.3	103.4	101.2	91.4	96.8	94.9
5	ROOFING	88.8	105.4	94.5	89.9	105.9	95.4	89.0	106.9	95.1	90.0	102.4	94.3	91.2	103.1	95.3
6	INTERIOR CONSTRUCTION	106.4	105.3	105.8	109.5	105.2	107.2	109.6	106.1	107.7	105.4	102.4	103.8	97.4	101.0	99.4
7	CONVEYING	100.0	105.9	101.9	100.0	105.9	101.9	100.0	105.9	101.9	100.0	105.9	101.9	100.0	105.1	101.6
8	MECHANICAL	101.9	103.2	102.6	102.4	103.2	102.8	101.3	107.8	104.6	100.6	97.9	99.2	100.0	105.3	102.7
9	ELECTRICAL	93.7	105.4	101.9	97.7	104.9	102.7	95.5	106.2	103.0	92.5	102.5	99.5	106.9	105.2	105.7
11	EQUIPMENT	100.0	105.3	101.4	100.0	104.8	101.3	100.0	106.1	101.6	100.0	102.4	100.6	100.0	105.1	101.3
12	SITEWORK	120.1	99.5	110.7	126.0	101.6	114.9	126.9	102.2	115.6	108.8	100.5	105.1	118.7	101.7	111.0
1-12	WEIGHTED AVERAGE	101.4	104.4	103.0	102.1	104.1	103.2	104.2	105.4	104.9	100.9	101.6	101.3	98.9	103.9	101.6

DIV. NO.	BUILDING SYSTEMS	WASHINGTON, D.C. MAT.	INST.	TOTAL	FT LAUDERDALE, FL MAT.	INST.	TOTAL	JACKSONVILLE, FL MAT.	INST.	TOTAL	MIAMI, FL MAT.	INST.	TOTAL	ORLANDO, FL MAT.	INST.	TOTAL
1-2	FOUND/SUBSTRUCTURES	105.3	87.9	94.6	96.8	92.7	94.3	96.6	82.0	87.5	94.7	94.1	94.3	97.4	83.2	88.6
3	SUPERSTRUCTURES	104.1	87.8	95.6	92.7	91.3	92.0	95.8	81.0	88.1	92.2	92.4	92.3	92.4	81.6	86.8
4	EXTERIOR CLOSURE	99.3	94.5	96.2	90.7	88.8	89.5	88.0	73.5	78.7	89.6	83.2	85.5	88.8	64.2	73.0
5	ROOFING	105.9	90.3	100.6	89.2	88.2	88.9	88.3	71.4	82.5	87.9	83.1	86.3	88.1	67.7	81.1
6	INTERIOR CONSTRUCTION	108.9	91.7	99.5	101.2	82.7	91.1	103.0	74.3	87.3	103.2	80.5	90.8	100.8	71.9	85.0
7	CONVEYING	100.0	91.9	97.4	100.0	85.7	95.4	100.0	74.5	91.8	100.0	84.6	95.0	100.0	73.6	91.5
8	MECHANICAL	101.6	84.7	92.9	100.7	85.8	93.0	100.0	78.8	89.1	97.6	89.4	93.4	97.0	78.3	87.4
9	ELECTRICAL	97.4	87.8	90.6	99.3	85.3	89.5	100.1	73.6	81.5	100.5	95.4	96.9	93.4	73.7	79.5
11	EQUIPMENT	100.0	90.5	97.4	100.0	85.7	96.1	100.0	74.5	93.1	100.0	84.6	95.8	100.0	73.6	92.9
12	SITEWORK	88.5	93.0	90.6	110.4	91.7	101.9	121.7	81.7	103.5	98.8	86.3	93.1	97.7	89.6	94.0
1-12	WEIGHTED AVERAGE	102.0	89.4	95.2	97.5	87.9	92.3	98.4	77.1	86.8	96.1	88.4	92.0	95.4	75.3	84.5

DIV. NO.	BUILDING SYSTEMS	TALLAHASSEE, FL MAT.	INST.	TOTAL	TAMPA, FL MAT.	INST.	TOTAL	ALBANY, GA MAT.	INST.	TOTAL	ATLANTA, GA MAT.	INST.	TOTAL	COLUMBUS, GA MAT.	INST.	TOTAL
1-2	FOUND/SUBSTRUCTURES	106.2	75.0	86.9	99.4	96.3	97.5	107.0	81.6	91.3	88.8	84.5	86.1	105.2	79.3	89.2
3	SUPERSTRUCTURES	97.4	73.2	84.8	99.2	93.3	96.1	104.2	79.6	91.3	99.6	82.3	90.6	99.9	77.8	88.4
4	EXTERIOR CLOSURE	88.2	66.3	74.1	97.7	78.6	85.5	85.8	70.0	75.6	91.8	77.6	82.7	89.6	54.4	67.0
5	ROOFING	87.7	59.9	78.2	104.3	69.2	92.3	92.8	68.9	84.6	97.1	71.1	88.2	93.3	66.6	84.1
6	INTERIOR CONSTRUCTION	99.4	60.0	77.9	98.5	79.4	88.1	92.9	70.1	80.5	106.1	82.3	93.1	93.9	62.9	77.0
7	CONVEYING	100.0	67.9	89.7	100.0	79.3	93.3	100.0	78.9	93.2	100.0	78.9	93.2	100.0	78.9	93.2
8	MECHANICAL	99.5	69.2	83.9	97.2	80.1	88.4	100.7	72.4	86.1	102.6	78.7	90.3	98.3	68.5	83.0
9	ELECTRICAL	92.1	69.2	76.0	93.4	80.9	84.6	102.4	70.8	80.2	95.7	80.3	84.9	97.7	61.6	72.4
11	EQUIPMENT	100.0	67.9	91.4	100.0	79.3	94.4	100.0	70.0	91.9	100.0	78.1	94.1	100.0	64.9	90.6
12	SITEWORK	121.0	85.7	104.9	112.3	91.9	103.0	120.1	88.9	105.9	107.0	92.2	100.3	126.2	88.3	108.9
1-12	WEIGHTED AVERAGE	98.4	69.2	82.6	99.2	84.0	91.0	99.7	74.4	86.0	99.2	80.7	89.2	98.9	68.0	82.2

DIV. NO.	BUILDING SYSTEMS	MACON, GA MAT.	INST.	TOTAL	SAVANNAH, GA MAT.	INST.	TOTAL	HONOLULU, HI MAT.	INST.	TOTAL	CEDAR RAPIDS, IA MAT.	INST.	TOTAL	DES MOINES, IA MAT.	INST.	TOTAL
1-2	FOUND/SUBSTRUCTURES	98.3	85.8	90.6	98.6	88.0	92.0	118.3	107.4	111.6	106.7	87.7	94.9	111.8	88.2	97.2
3	SUPERSTRUCTURES	93.7	83.5	88.4	99.7	85.1	92.1	114.8	107.2	110.9	98.9	85.8	92.1	99.6	87.1	93.1
4	EXTERIOR CLOSURE	89.8	69.3	76.6	89.3	70.7	77.4	117.7	115.5	116.3	95.8	79.4	85.2	93.5	82.0	86.1
5	ROOFING	93.0	68.1	84.4	90.7	71.3	84.0	109.5	114.5	111.2	89.9	80.4	86.6	89.1	82.0	86.6
6	INTERIOR CONSTRUCTION	93.5	69.4	80.4	106.9	70.7	87.1	134.9	116.8	125.1	106.1	80.3	92.1	105.4	82.0	92.6
7	CONVEYING	100.0	78.9	93.2	100.0	76.2	92.3	100.0	116.9	105.4	100.0	78.3	93.0	100.0	84.2	94.9
8	MECHANICAL	98.6	71.8	84.8	98.4	72.2	84.9	111.3	114.7	113.0	99.6	80.4	89.7	96.6	82.1	89.1
9	ELECTRICAL	108.5	70.8	82.0	101.8	76.7	84.2	106.2	105.5	105.7	97.6	80.4	85.5	99.4	82.0	87.2
11	EQUIPMENT	100.0	69.3	91.7	100.0	70.6	92.1	100.0	115.8	104.2	100.0	80.3	94.7	100.0	81.9	95.1
12	SITEWORK	119.2	88.1	105.1	121.9	87.2	106.1	135.6	105.9	122.1	103.3	92.6	98.4	102.5	92.4	97.9
1-12	WEIGHTED AVERAGE	97.6	75.1	85.4	99.8	76.8	87.4	115.6	111.7	113.5	100.0	82.3	90.4	99.6	84.0	91.1

DIV. NO.	BUILDING SYSTEMS	DAVENPORT, IA MAT.	INST.	TOTAL	SIOUX CITY, IA MAT.	INST.	TOTAL	WATERLOO, IA MAT.	INST.	TOTAL	BOISE, ID MAT.	INST.	TOTAL	POCATELLO, ID MAT.	INST.	TOTAL
1-2	FOUND/SUBSTRUCTURES	95.5	94.4	94.8	100.3	82.2	89.1	103.8	84.1	91.6	102.9	92.8	96.7	109.7	92.7	99.2
3	SUPERSTRUCTURES	93.1	93.9	93.5	95.5	80.2	87.5	106.3	82.6	94.0	98.0	92.0	94.8	100.6	91.9	96.1
4	EXTERIOR CLOSURE	101.2	90.1	94.0	98.2	70.8	80.6	102.5	73.3	83.8	104.6	86.8	93.2	108.3	88.1	95.4
5	ROOFING	90.4	90.1	90.3	87.4	70.8	81.7	87.7	73.2	82.7	105.5	88.3	99.6	103.6	88.1	98.3
6	INTERIOR CONSTRUCTION	102.5	90.0	95.6	98.3	70.8	83.3	110.5	73.2	90.1	98.8	88.3	93.0	103.5	88.2	95.2
7	CONVEYING	100.0	78.3	93.0	100.0	73.3	91.4	100.0	78.3	93.0	100.0	87.5	96.0	100.0	87.5	96.0
8	MECHANICAL	99.8	90.1	94.8	97.9	71.0	84.0	101.1	73.3	86.7	96.6	88.4	92.4	101.2	88.3	94.6
9	ELECTRICAL	98.1	90.1	92.4	99.7	81.6	87.0	93.0	73.2	79.1	96.7	88.3	90.8	93.7	88.2	89.8
11	EQUIPMENT	100.0	90.0	97.3	100.0	70.7	92.1	100.0	73.1	92.8	100.0	88.2	96.8	100.0	88.1	96.8
12	SITEWORK	110.4	88.4	100.4	115.9	89.9	104.1	113.1	90.2	102.7	92.8	95.8	94.2	102.8	95.0	99.3
1-12	WEIGHTED AVERAGE	98.8	90.8	94.4	98.5	75.8	86.2	102.9	76.8	88.8	99.3	89.4	93.9	102.5	89.6	95.5

DIV. NO.	BUILDING SYSTEMS	CHICAGO, IL MAT.	INST.	TOTAL	DECATUR, IL MAT.	INST.	TOTAL	JOLIET, IL MAT.	INST.	TOTAL	PEORIA, IL MAT.	INST.	TOTAL	ROCKFORD, IL MAT.	INST.	TOTAL
1-2	FOUND/SUBSTRUCTURES	98.8	102.4	101.0	103.4	95.2	98.3	100.8	102.1	101.6	97.5	88.1	91.7	105.2	95.5	99.2
3	SUPERSTRUCTURES	93.8	103.7	99.0	101.0	94.7	97.7	100.5	102.6	101.6	94.6	88.7	91.5	106.8	95.1	100.7
4	EXTERIOR CLOSURE	101.3	104.3	103.2	101.2	92.9	95.9	105.5	100.8	102.5	94.5	94.5	94.5	98.3	98.5	98.4
5	ROOFING	97.5	109.1	101.5	100.8	94.5	98.6	102.0	106.2	103.5	95.6	100.8	97.4	107.3	96.4	103.6
6	INTERIOR CONSTRUCTION	99.3	100.5	99.9	107.3	92.0	99.0	104.2	102.7	103.4	105.9	94.3	99.6	99.8	91.2	95.1
7	CONVEYING	100.0	103.9	101.2	100.0	88.7	96.3	100.0	103.9	101.2	100.0	86.3	95.6	100.0	100.2	100.0
8	MECHANICAL	96.4	96.5	96.4	95.6	93.1	94.3	97.2	95.0	96.1	96.1	92.4	94.2	101.7	99.7	100.7
9	ELECTRICAL	93.9	101.3	99.1	94.2	92.5	93.0	102.5	101.2	101.6	94.6	94.3	94.4	93.4	100.3	98.2
11	EQUIPMENT	100.0	102.6	100.7	100.0	92.4	97.9	100.0	101.1	100.3	100.0	94.2	98.4	100.0	100.2	100.0
12	SITEWORK	106.6	104.8	105.7	112.4	95.3	104.6	113.4	102.2	108.3	119.3	98.9	110.0	115.5	101.8	109.3
1-12	WEIGHTED AVERAGE	98.0	101.8	100.0	100.9	93.3	96.8	101.8	100.7	101.2	98.5	92.4	95.2	102.5	97.3	99.7

479

Figure 2.9

This means that the actual total project costs for the dormitory could be any value between $3,707,034 and $5,015,398. Need identification feasibility studies should be made using this range of possible costs.

Summary

There are four methods commonly used in construction estimating. These are Order of Magnitude, Square Foot/Cubic Foot, Assemblies, and Unit Cost. Each of these methods has different information, time, and expertise requirements, and each has a different accuracy range. Care should be taken to choose the appropriate method based on the amount of information available, the estimator's level of expertise, and the time available for the estimate.

To obtain a complete estimate of project costs, costs for design, construction contingency, equipment, furniture, and planning must be included. An estimate using cost data from past projects or from a proprietary cost data base must be adjusted for project size, location, and inflation. The cost range used in planning is based on the accuracy of the estimating method.

3

FORECASTING BID CONDITIONS AND INFLATION

3

FORECASTING BID CONDITIONS AND INFLATION

The two factors that often have the most effect on cost control estimate accuracy are bid conditions and inflation rates. The author has seen inflation rates vary from over 10% per year to less than 4% per year within the span of time required for planning and designing a major construction project. Changes in bid conditions have also been seen to lower the bid price of a major construction project by as much as 10% within months. The combined effect of these two factors could increase or decrease the cost of a construction project by 20% or more. An effective cost control system requires some method of forecasting these very significant factors.

Even under the best of circumstances, forecast accuracy is difficult to obtain. The most widely used methods are based on general business and manufacturing conditions. These conditions reflect the mass production of identical products at a central plant in a controlled environment, for an accurately predetermined cost, using a relatively stable labor force. Studies have determined that the demand for these products is affected by price, consumer income, the cost of alternate or complementary goods, and taste.

The construction industry cannot be analyzed in quite the same way. Rather than producing one easily definable group of products, construction is several industries which produce a variety of products, but are grouped together because of similar methods and technology. The industry as a whole consists of three broad categories, each of which has distinct sub-categories. These are:

- **Heavy and Highway:** All utilities and power including water, gas, electrical, and nuclear; all transportation facilities including ground, sea and air; and military.
- **Residential:** All single and multi-unit housing.
- **Nonresidential:** Manufacturing, commercial, government and private offices; educational, health care, worship, and recreational.

For each of these categories, every project is effectively unique. Even using the same basic design, the differences in site development, utility service availability, and foundation design caused by different site conditions are enough to make each project unique. Projects are built not in controlled environments in a central location, but out in the weather at widely separated, unique sites. Weather delays are expected and project management facilities and personnel are often duplicated at each site. Because workmen move from project to project, the work force in any given area is unstable. Further, the unknown effects of weather, and labor availability and productivity make it difficult for contractors to accurately predetermine what their costs will be.

Demand for construction is presented to the industry in terms of the number of projects out for bid. This assessment is also our measure of bid conditions. The factors that affect the demand are somewhat different for each category. All of the categories are affected by population, credit, and economic shocks such as the oil embargo of the 1970's. Per capita income, price, and amenities are significant factors affecting residential construction. Nonresidential construction is affected primarily by economic activity and image and, secondarily, by price. Heavy and highway construction, financed largely by government appropriations, is significantly affected by government policy.

Another challenge in predicting competition in construction is the fact that there is little uniformity among the great number of companies involved. Each is different in terms of the types, sizes, and locations of projects that they are willing to attempt. Some contractors bid on any nonresidential project in a metropolitan area, while others operating in the same area specialize in certain types of projects. Still others bidding the same work may expand their work area to include the entire state or region.

Under the best of circumstances, forecasting is more of an art than a science. The special conditions described above make forecasting in the construction industry even more of a challenge. While it may not be possible to achieve an acceptable degree of accuracy using quantitative forecasts for construction, success can be achieved using cycles, trends, leading indicators, and a general rule of thumb. The application of these methods involves collecting quantitative historical data for a local construction market and the factors that affect it; transforming the data to remove the effects of inflation and seasonal and random fluctuations; and analyzing the transformed data to identify trends, cycles, and leading indicators.

Data Sources

There are a number of data sources that can be used in construction forecasting. These sources include newsletters and magazines published by agencies of national, state, and local governments; business and trade journals; and private data bases. Fifteen years of data (or as close to fifteen years as possible) is needed to identify cycles, trends, and leading indicators. The data must specifically represent the forecast market. The market is defined by the geographical area served by most of the contractors bidding on the project. For small projects, the market may be limited to a portion of a city or county. Generally, as the project size increases, the market size also increases.

Useful national government data sources include the *Federal Reserve Bulletin* for interest rates, the *Consumer Price Index* for general inflation data, and various reports of the Department of Commerce for business and construction volume data. (Examples are the *Statistical Abstract of the United States* and *Construction Reports—The Value of New Construction Put In Place.*) Many of the data series given in these publications are broken down for selected local markets. Much of the data is, however, nearly two years old by the time it is published.

State agencies also produce publications on business, gross sales, and the labor market. Such reports contain information on general business and construction activity, and demographic data for counties and metropolitan areas. Some data can also be obtained directly from state agencies, such as the local or state Employment Commissions, the State Comptroller of Public Accounts, and the Centers for Business and Economic Research at leading universities. Local sources of data include city agencies and Chambers of Commerce.

Many business and trade journals also publish useful data. Some, such as *Texas Contractor*, furnish a monthly report of contract awards by categories for the state. *Engineering News Record* publishes weekly reports of national construction volume by category and sub-category, and a construction cost index for a few selected cities in the United States.

Private data bases often contain the most recent data. Two good sources of up-to-date construction volume data are *Dodge Construction Potentials* and *Dodge Local Construction Potentials*. These reports give construction volume by sub-category for regions, states, counties, and metropolitan areas. *Means Construction Cost Indexes* provide construction inflation data on a quarterly basis for 209 United States and Canadian cities and for a 30-city United States average.

Often data to quantify the exact factor being investigated is not available. Approximations can be substituted or calculated. For example, if a series providing the dollar volume of local business activity is not available, a report of sales tax receipts could be consulted. Such a report reflects the changes in, if not the actual values of local business volume and can, therefore, be used as a substitute. Per capita income figures may not be available, but sales tax receipts, divided by local labor force estimates, can provide an approximation.

Much of this data can be found at local libraries, particularly those that are designated repositories of government documents.

Data Transformation

After it is collected, most data must be transformed before it can be analyzed. Data expressed in dollars must be adjusted to remove the effects of inflation and all data must be checked for random and seasonal fluctuations.

The following example shows the transformation process using the dollar volume of national nonresidential construction from *Construction Reports—Value of New Construction Put in Place*, published by the U.S. Department of Commerce. Figure 3.1 illustrates the reason for carrying out this transformation. The curve shows that the unadjusted dollar volume of nonresidential construction in the United States has a very strong cyclical pattern with an upward trend. The cyclical pattern shows construction volume beginning relatively low in the first quarter of the year, increasing during the year to highs in the second and third quarters, only to drop in the fourth quarter of the year. The yearly repetition of this pattern indicates the presence of seasonal fluctuations.

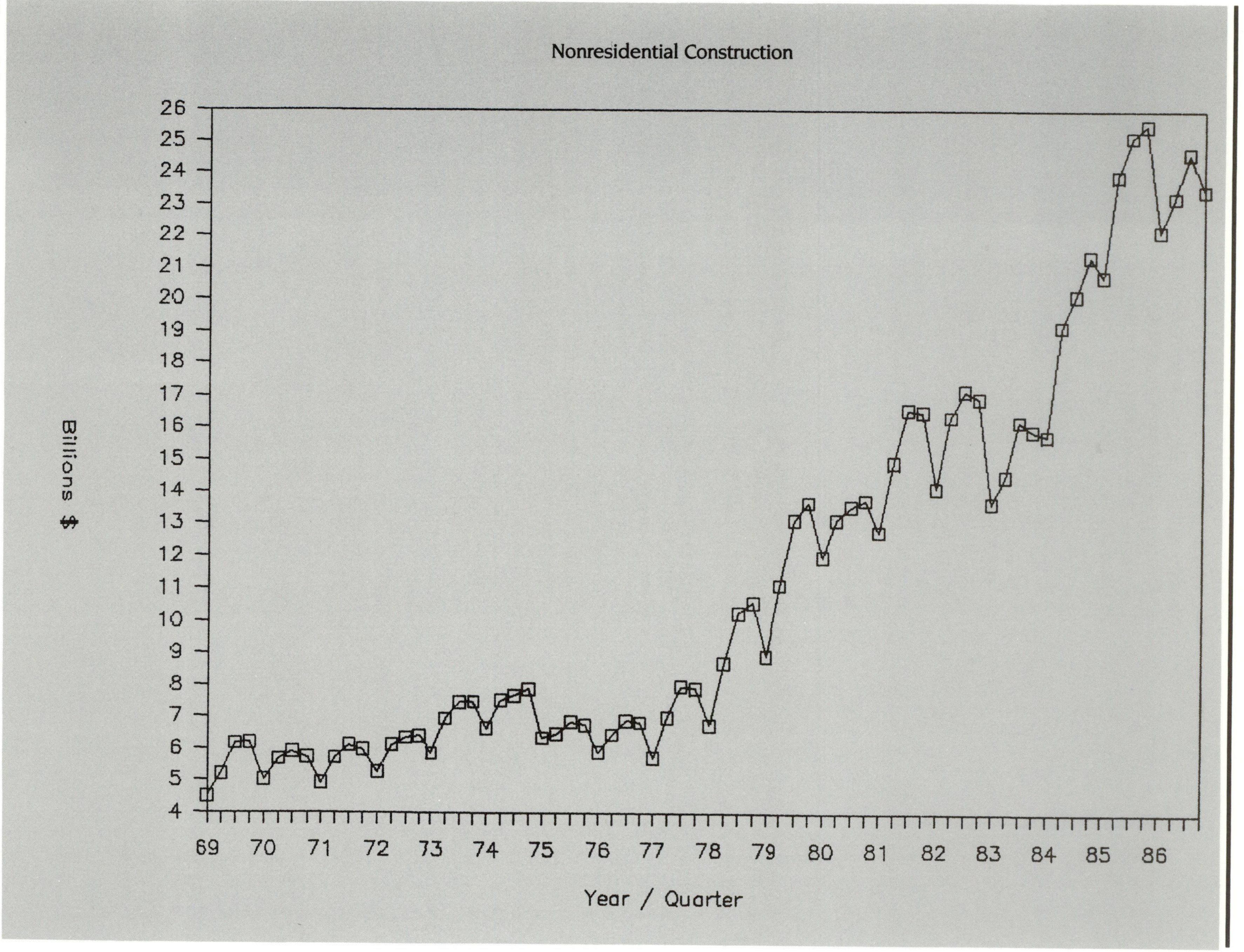

Nonresidential Construction
Billions $
26
25
24
23
22
21
20
19
18
17
16
15
14
13
12
11
10
9
8
7
6
5
4
69
70
71
72
73
74
75
76
77
78
79
80
81
82
83
84
85
86
Year / Quarter

Figure 3.1

The general rising trend in the dollar volume suggests the presence of inflation.

The effects of inflation must be removed first. This is accomplished by using a local index for the industry to adjust the data. The national average *Means Historical Construction Cost Index*, as shown in Figure 3.2, is an appropriate source of national construction data. This information is transformed using the following formula.

$$\text{Period Volume} \times \frac{\text{Target Index}}{\text{Period Index}} = \text{Adjusted Volume}$$

As an example, we will adjust the dollar volume of construction for the first quarter of 1984 into first quarter 1987 dollars. The period volume, the volume for the first quarter of 1984, is $15.75 billion. The period cost index is 184.4 and the target index, the index for the first quarter of 1987, is 196.2. Therefore:

$$\$15.75 \times (196.2/184.4) = \$16.76$$

The cost data for each time period must be transformed in this manner.

Note that the dollar data is expressed in billions of dollars. This is done to keep the data entry from becoming too unwieldy.

The data with the inflation factor removed is shown in Figure 3.3. The continuous upward trend has been replaced by a rough curve showing long term fluctuations. Also shown is a recurring four-quarter cycle indicating the continued presence of seasonal fluctuations. These fluctuations may be removed by a calculated seasonal index; or, seasonal as well as random fluctuations may be removed by a centered moving average. For our purposes, to identify trends and leading indicators, the centered moving average works just as well as a seasonal index, and is easier to calculate. A centered moving average smooths out seasonal and random fluctuations by replacing a time period value with an average value for the year for which that time period (T) is the center. The formula for calculating the centered moving average for quarterly data is as follows:

$$\text{Centered Moving Average} = \frac{(Q - 2/2) + (Q - 1) + Q + (Q + 1)(Q + 2/2)}{4}$$

Where: Q = the value of the quarter without the seasonal factor.
$Q - 2$ = the value for two quarters prior to T.
$Q - 1$ = the value for the quarter prior to T.
$Q + 1$ = the value for the quarter after T.
$Q + 2$ = the value for two quarters after T.

Year	**Quarter**	**Total** Nonresidential Construction for 1987 in Millions
1984	1	$15.75
	2	19.17
	3	20.14
	4	21.37
1985	1	20.74
	2	23.92

Historical Cost Indexes

Year	National 30 City Average	Alabama					Alaska	Arizona		Arkansas		California				
		Birmingham	Huntsville	Mobile	Montgomery	Tuscaloosa	Anchorage	Phoenix	Tuscon	Fort Smith	Little Rock	Anaheim	Bakersfield	Fresno	Los Angeles	Oxnard
Jan. 1988	202.5	171.4	172.9	178.4	167.4	166.1	263.4	190.8	191.0	169.8	173.2	227.7	213.5	219.8	228.8	223.6
1987	196.2	167.6	170.9	174.5	165.5	164.7	258.7	180.1	182.5	165.3	168.2	223.6	210.2	217.2	222.9	220.9
1986	191.6	164.9	166.0	171.8	160.6	159.6	254.4	178.4	178.4	161.8	167.0	219.8	206.1	207.8	218.2	219.5
1985	187.3	161.1	163.1	162.8	157.5	156.0	251.5	176.6	175.9	159.4	163.9	211.5	201.6	205.1	211.0	211.1
1984	184.4	157.8	159.8	168.3	154.1	152.3	246.8	179.4	179.5	158.2	162.3	207.0	196.4	200.6	205.5	205.4
1983	179.0	152.6	159.2	163.2	154.1	150.5	226.7	175.3	176.5	153.0	158.6	201.4	194.1	195.7	200.1	200.6
1982	165.1	143.7	146.0	150.8	143.0	142.0	208.4	163.4	162.3	142.3	149.1	184.2	181.6	182.8	180.5	182.3
1981	151.3	135.9	134.5	139.2	136.1	133.6	198.5	155.9	152.8	132.2	139.1	167.7	167.3	166.6	166.8	168.4
1980	139.0	124.9	125.1	127.1	126.2	124.2	198.2	143.9	141.3	121.5	128.8	152.3	151.7	152.2	150.3	152.7
1979	125.7	113.1	113.2	117.6	112.8	112.0	179.0	126.3	128.0	109.5	114.3	139.4	135.4	138.5	137.7	137.5
1978	117.5	105.7	105.4	109.2	104.2	103.5	163.4	117.4	118.4	102.5	104.8	127.3	124.9	127.8	127.1	128.0
1977	109.7	97.5	100.0	101.1	94.2	93.1	153.1	110.1	109.9	95.2	97.5	118.0	115.4	117.8	119.8	116.8
1976	103.7	92.5	94.3	94.8	89.5	88.7	138.9	105.2	105.0	89.7	91.0	109.9	107.8	111.7	112.0	105.3
1975	100.0	90.5	94.6	93.7	89.1	86.9	124.2	100.6	102.3	89.2	89.4	105.4	101.8	105.7	107.7	102.7
1974	90.5	78.8	82.4	82.0	80.0	79.0	121.2	90.7	91.7	79.1	74.8	97.7	95.6	97.9	94.0	96.4
1973	83.0	70.6	75.6	74.3	73.4	72.5	113.7	82.6	84.1	72.5	66.8	89.6	87.7	89.8	85.1	86.0
1972	76.6	64.3	69.8	69.3	67.7	66.9	106.5	74.8	77.6	66.9	61.7	82.7	81.0	82.9	77.4	81.6
1971	69.7	58.8	63.5	63.4	61.6	60.9	99.2	66.9	70.6	60.9	56.1	75.2	73.7	75.4	70.0	74.2
1970	63.7	54.5	58.0	57.7	56.3	55.6	93.3	61.5	64.5	55.7	51.6	68.8	67.3	68.9	64.7	67.8
1969	59.2	51.3	53.9	52.5	52.3	51.7	88.3	57.8	60.0	51.7	48.6	63.9	62.6	64.1	60.4	63.0
1968	55.4	48.5	50.5	48.5	49.0	48.4	83.4	54.6	56.1	48.4	46.0	59.8	58.5	59.9	57.1	59.0
1967	52.9	47.1	48.2	46.8	46.8	46.2	80.2	52.7	53.6	46.2	44.7	57.1	55.9	57.2	55.1	56.3
1966	50.8	46.0	46.3	45.2	44.9	44.4	77.7	50.8	51.5	44.4	43.9	54.8	53.7	55.0	52.9	54.1
1965	49.1	44.3	44.7	43.8	43.4	42.9	75.7	49.2	49.8	42.9	42.8	53.0	51.9	53.1	50.7	52.3
1964	47.9	43.0	43.6	42.7	42.4	41.8	73.8	48.4	48.5	41.9	41.2	51.7	50.6	51.8	49.4	51.0
1963	46.7	41.9	42.5	41.7	41.3	40.8	72.0	47.2	47.3	40.8	40.7	50.4	49.4	50.5	48.7	49.7
1960	44.6	40.6	40.6	39.8	39.4	38.9	68.7	45.1	45.2	39.0	38.8	48.1	47.1	48.3	46.0	47.5
1955	37.4	33.6	34.1	33.4	33.1	32.7	57.6	37.8	37.9	32.7	33.6	40.4	39.5	40.5	38.6	39.8
1950	30.9	27.7	28.1	27.6	27.3	27.0	47.6	31.2	31.3	27.0	26.9	33.4	32.7	33.4	31.9	32.9
1945	19.7	17.7	17.9	17.5	17.4	17.2	30.4	19.9	20.0	17.2	17.1	21.3	20.8	21.3	20.3	21.0
1940	15.2	13.6	13.8	13.6	13.4	13.3	23.4	15.4	15.4	13.3	13.2	16.4	16.1	16.4	15.7	16.2

Year	National 30 City Average	California							Colorado			Connecticut				
		Riverside	Sacramento	San Diego	San Francisco	Santa Barbara	Stockton	Vallejo	Colorado Springs	Denver	Pueblo	Bridgeport	Bristol	Hartford	New Britain	New Haven
Jan. 1988	202.5	226.4	222.0	223.0	252.8	230.2	221.6	238.1	186.0	194.5	190.1	205.5	201.6	203.1	204.3	203.4
1987	196.2	221.5	218.0	218.7	243.5	225.5	218.2	229.2	181.2	189.6	186.1	204.7	202.5	203.0	205.2	202.5
1986	191.6	217.6	208.5	213.9	236.9	220.2	210.4	220.1	181.9	185.6	183.7	196.0	193.1	194.5	195.5	194.2
1985	187.3	209.2	203.7	208.1	233.0	211.7	205.2	215.3	176.8	181.1	178.6	190.6	187.7	189.7	190.0	188.7
1984	184.4	202.6	207.2	203.1	225.6	205.6	201.2	212.3	177.2	188.0	178.9	185.0	181.5	183.6	183.2	181.9
1983	179.0	197.3	203.4	199.4	221.6	201.6	199.2	206.6	171.6	179.8	171.9	176.3	173.9	176.4	175.4	173.3
1982	165.1	181.3	187.6	185.1	201.7	185.8	185.9	190.3	158.0	164.1	166.6	159.2	155.0	159.4	157.6	157.4
1981	151.3	166.7	171.7	164.5	181.6	172.3	169.5	174.2	145.5	148.1	145.2	148.3	144.4	146.5	146.4	146.5
1980	139.0	151.7	157.5	150.3	165.1	161.5	156.0	159.6	134.8	136.0	132.1	136.2	132.2	134.6	134.0	134.2
1979	125.7	137.3	141.5	136.0	148.1	147.5	140.5	144.8	121.8	125.1	120.0	123.9	120.5	122.7	122.3	122.1
1978	117.5	127.3	132.6	128.7	137.4	134.3	130.3	133.1	113.5	116.3	112.4	116.2	110.9	112.9	114.3	113.8
1977	109.7	117.7	123.1	117.9	127.5	120.4	120.1	120.3	104.9	108.0	104.4	109.1	105.8	105.7	106.5	107.1
1976	103.7	109.8	113.7	110.7	118.2	111.3	112.0	112.3	98.7	102.0	98.7	104.2	101.4	101.9	102.9	104.5
1975	100.0	104.4	108.6	105.4	109.3	104.6	104.3	103.2	95.6	95.5	94.5	100.1	99.0	100.1	99.7	100.5
1974	90.5	97.2	100.9	95.1	98.0	101.0	99.0	100.2	87.0	83.4	86.1	92.3	87.3	91.0	88.5	91.4
1973	83.0	89.1	92.6	86.8	89.3	92.6	90.8	91.9	79.8	74.7	79.0	84.7	80.0	83.8	81.2	83.9
1972	76.6	82.3	85.4	79.7	82.0	85.5	83.8	84.8	73.6	69.3	72.9	77.8	73.9	78.5	74.9	76.3
1971	69.7	74.8	77.7	73.1	75.2	77.8	76.3	77.2	67.0	63.7	66.3	70.4	67.2	71.4	68.2	69.3
1970	63.7	68.4	71.0	67.8	69.4	71.1	69.7	70.5	61.2	58.3	60.6	64.7	61.4	64.4	62.3	64.1
1969	59.2	63.6	66.0	63.1	64.3	66.1	64.8	65.6	56.9	54.2	56.3	60.2	57.1	59.6	57.9	59.6
1968	55.4	59.5	61.8	58.7	59.6	61.8	60.6	61.4	53.3	51.2	52.7	56.0	53.4	55.7	54.2	56.3
1967	52.9	56.8	59.0	56.6	56.6	59.0	57.9	58.6	50.9	49.5	50.3	53.4	51.0	52.9	51.8	54.3
1966	50.8	54.5	56.7	54.9	54.4	56.7	55.6	56.3	48.8	48.0	48.3	51.3	49.0	50.8	49.7	52.4
1965	49.1	52.7	54.8	53.1	51.9	54.8	53.7	54.4	47.2	46.7	46.7	49.7	47.3	49.2	48.0	50.7
1964	47.9	51.4	53.4	51.5	50.6	53.4	52.4	53.0	46.1	45.6	45.6	47.5	46.2	46.7	46.9	47.1
1963	46.7	50.1	52.1	50.2	49.4	52.1	51.1	51.7	44.9	44.4	44.4	46.3	45.0	45.6	45.7	45.9
1960	44.6	47.9	49.7	48.0	47.1	49.8	48.8	49.4	42.9	42.4	42.4	44.3	43.0	43.5	43.6	43.8
1955	37.4	40.2	41.7	40.2	39.5	41.7	40.9	41.4	36.0	35.6	35.6	37.1	36.1	36.5	36.6	36.7
1950	30.9	33.2	34.5	33.2	32.7	34.5	33.8	34.2	29.7	29.4	29.4	30.7	29.8	30.1	30.2	30.4
1945	19.7	21.2	22.0	21.2	20.8	22.0	21.6	21.8	18.9	18.7	18.7	19.5	19.0	19.2	19.3	19.4
1940	15.2	16.3	17.0	16.4	16.1	17.0	16.6	16.8	14.6	14.5	14.5	15.1	14.7	14.8	14.9	14.9

12

Figure 3.2

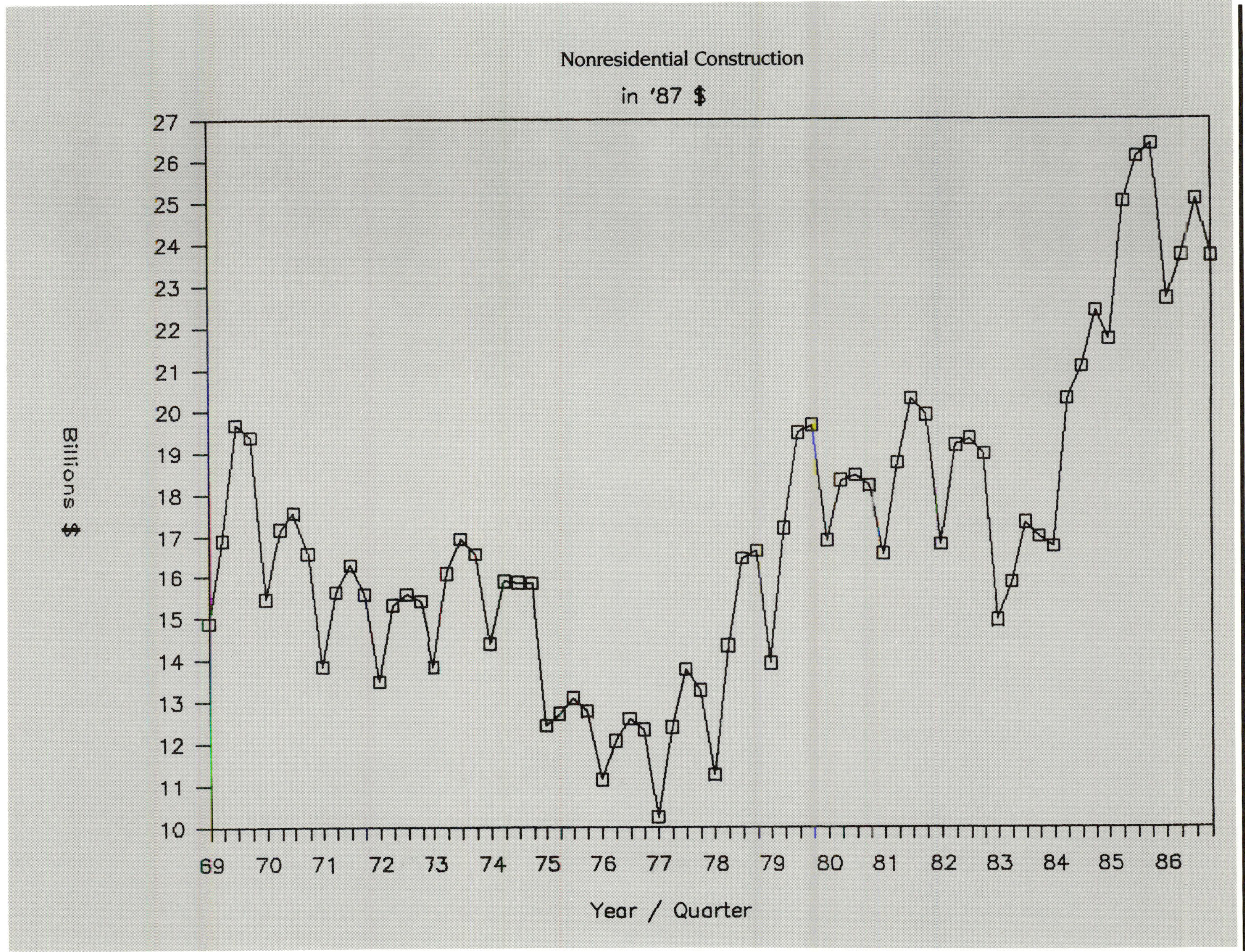

Nonresidential Construction
in '87 $
Billions $
27
26
25
24
23
22
21
20
19
18
17
16
15
14
13
12
11
10
69
70
71
72
73
74
75
76
77
78
79
80
81
82
83
84
85
86
Year / Quarter

Figure 3.3

Using the data acquired thus far, the centered moving average for the third quarter of 1984 is calculated as follows:

$$\frac{(15.75/2) + 19.17 + 20.14 + 21.37 + (20.74/2)}{4} = 19.73$$

The centered moving average for the fourth quarter of 1984 is calculated as:

$$\frac{(19.17/2) + 20.14 + 21.37 + 20.74 + (23{,}92/2)}{4} = 20.95$$

Because there are no preceding or following values available, the averages for the first two quarters and the last two quarters cannot be calculated. The averaging process is repeated for the remainder of the available data. A graph of the transformed data is shown in Figure 3.4.

Cycles and Trends

A graph of the transformed data is used to identify cycles and trends. Figure 3.4 shows a six to seven year cycle and a long term rising trend in U.S. nonresidential construction. The limiting factor in following cycles is that the intervals are not consistent. The intervals between highs and lows, and from high to high and low to low vary. Nevertheless, cycles can be useful in suggesting that a change is due. In the case of my own early estimates, knowing that there was a cyclical pattern would have been sufficient warning that a change in bid conditions was likely. Assuming a fairly consistent six to seven year cycle, the trend starting downward in 1986 can be expected to bottom out in 1989 or 1990. An upward trend with a peak in 1993 or 1994 can be expected to follow.

Trends are useful for tracking construction inflation. A graph of the percent increase in national average construction costs based on *Means Construction Cost Index* is shown in Figure 3.5. Construction inflation appears to be affected primarily by government policy and economic shocks. There seems to be a lag time of about two years between the creation of new administration policy and evidence of its effects on construction inflation. This assumption suggests that the relatively low inflation rates of 1986 and 1987 could be expected to continue until mid-1989 or 1990. It is interesting to note in Figure 3.5 that in four of the five presidential election years in which the incumbent was up for reelection, construction inflation levels decreased, only to rebound the following year.

Leading Indicators

The use of leading indicators is based on the observation that with reasonable consistency, changes in certain elements of the economy precede corresponding changes in other elements of the economy. The relationship between the two may be direct (as one increases, the other increases). Or, there may be an *inverse* relationship between the two (as one increases, the other decreases). By knowing the lag, or interval, between these two factors, one may use the leading indicator in forecasting.

Factors to investigate as leading indicators of construction volume are the determinants of demand, the factors affecting demand, for the different categories of construction. To continue with the example using U.S. nonresidential construction, the factors to investigate as leading indicators are population, credit, shocks to the economy, economic activity, image, and price.

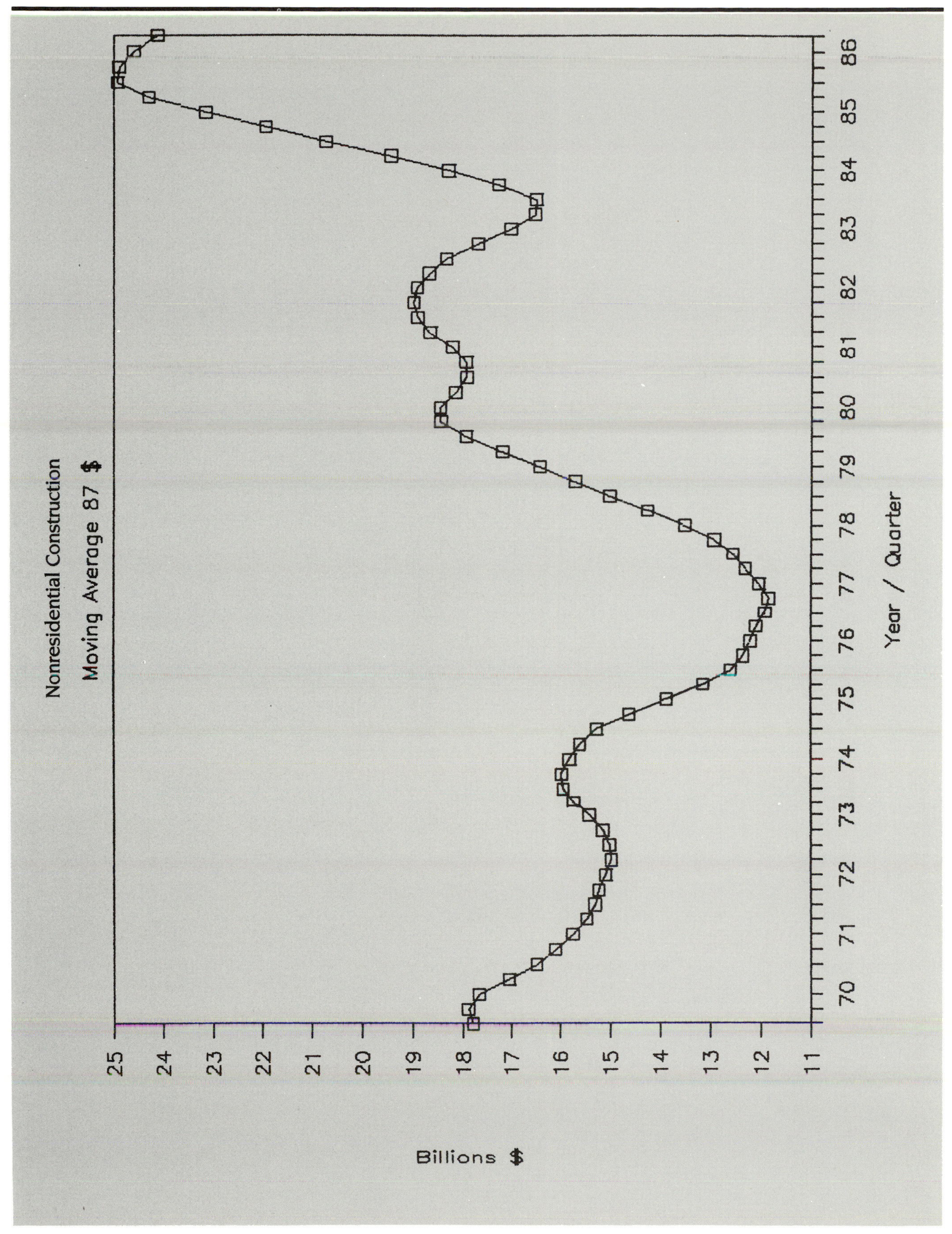
Nonresidential Construction
Moving Average 87 $
Billions $
25
24
23
22
21
20
19
18
17
16
15
14
13
12
11
70
71
72
73
74
75
76
77
78
79
80
81
82
83
84
85
86
Year / Quarter

Figure 3.4

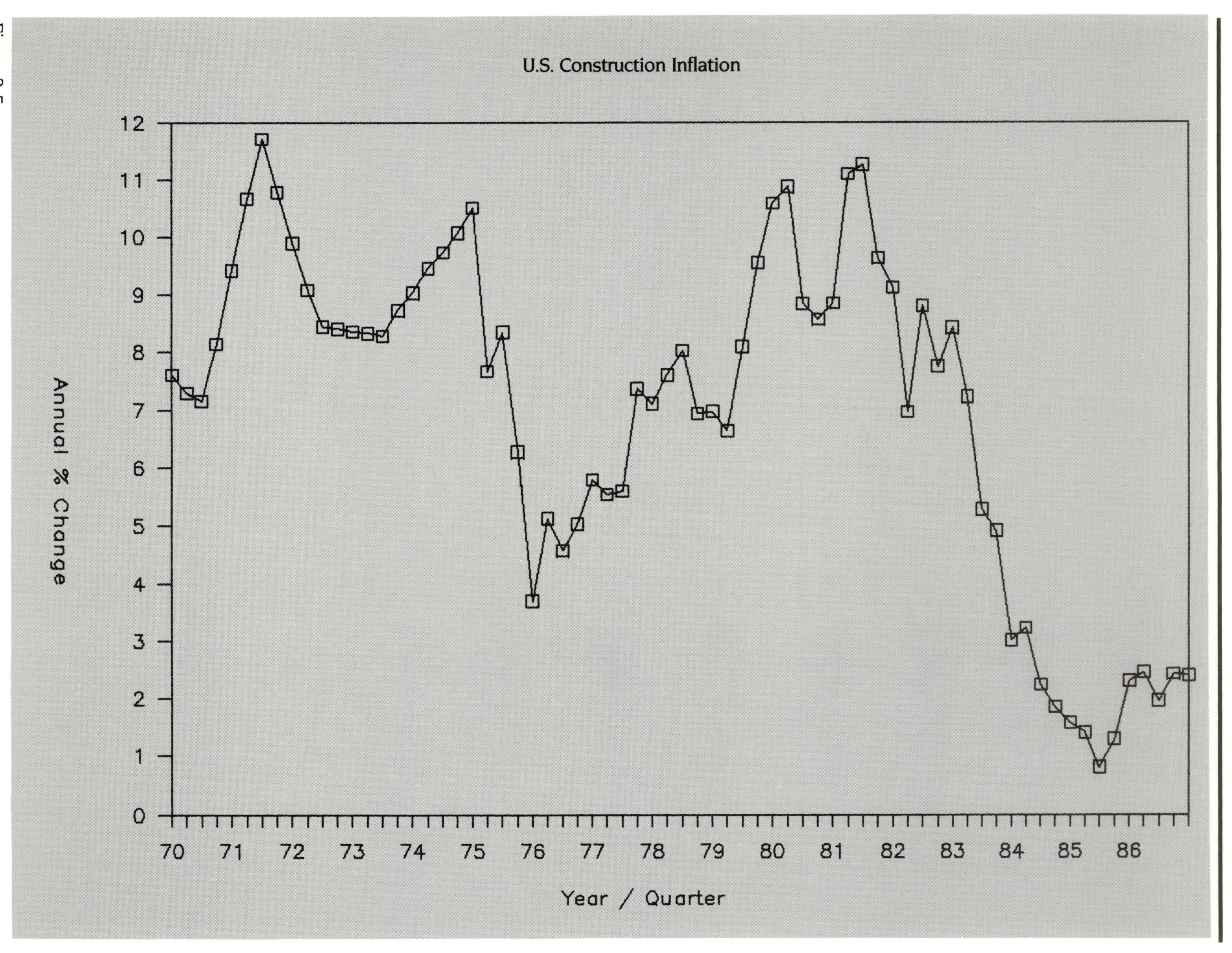
U.S. Construction Inflation
Annual % Change
Year / Quarter
12
11
10
9
8
7
6
5
4
3
2
1
0
70
71
72
73
74
75
76
77
78
79
80
81
82
83
84
85
86

Figure 3.5

Leading indicators can be identified by graphing or calculating correlation coefficients. Graphs are drawn, or correlation coefficients calculated for construction volume, and each of the factors lagged for various time periods. For this application, XY graphs are used. Though test statistics are not available, graphs are easily understood and, with the use of a spreadsheet program on a computer, are quick and easy to generate. A series of XY graphs showing the relationship between U.S. nonresidential construction and national economic activity are given in Figures 3.6 through 3.18. The report of national economic activity used is the Gross National Product from the *Business Conditions Digest* published by the U.S. Department of Commerce. Inflation and cyclical and random fluctuations were removed from the GNP dollar data using the same methods previously demonstrated.

The graphs lag national economic activity for 0 to 12 quarters. A graph in which the grouping of points falls fairly closely around a line shows correlation and a possible leading indicator. Figures 3.9 through 3.12 show fairly tight groupings of points, suggesting that this measure of economic activity is a leading indicator of construction activity by a period of around five quarters.

There are limitations inherent in the use of leading indicators. Whereas data is usually available to quantify the different categories of construction for a local market, that data is not differentiated by size of project. The greatest percentage of construction dollar volume is in the larger projects. Because the planning and design phases of these projects typically require several years, it takes several years for a change in a leading indicator to be reflected in the volume of construction. Because the greatest dollar volume is in larger projects, the leading indicator interval is biased towards larger projects. Since the planning and design phases for smaller projects take less time, the leading indicator interval is shorter. Some adjustment is needed, which leads to the following rule of thumb.

Rule of Thumb

Nonresidential construction volume tends to trail economic activity by the average length of the planning and design phases for the project size range. The average length of these phases for a four million dollar job is 18 to 24 months. Changes in economic activity can be expected to occur 18 to 24 months before corresponding changes in nonresidential construction volume for projects of that size range. For example, if economic activity in Atlanta decreases in the third quarter of 1987, a decrease in the number of projects out for bid in the four million dollar range can be expected 18 to 24 months later, in the first half of 1989. The decrease in the number of projects would signal an increase in bid competition.

Summary

Because of the unique character of the construction industry, many of the commonly used business forecasting techniques are not as reliable as they are for other industries. The study of trends, cycles, and leading indicators does, however, give some indication of future changes in construction and can be used as a "rule of thumb" to provide a rough forecast of bid conditions. A continuing study helps the estimator develop an intuitive feel for the market. This kind of familiarity may be the most reliable forecasting tool of all.

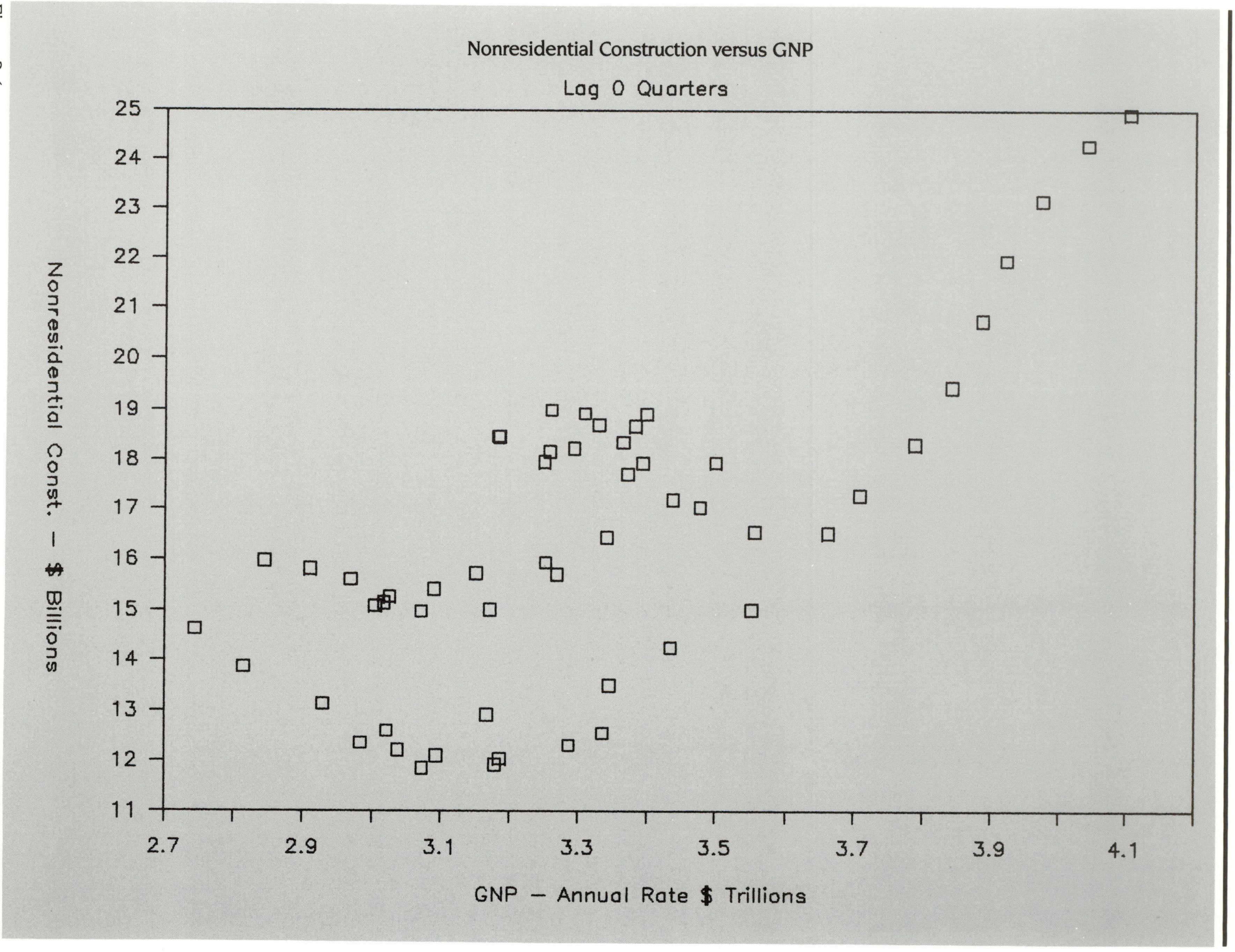
Nonresidential Construction versus GNP
Lag 0 Quarters
Nonresidential Const. — $ Billions
25
24
23
22
21
20
19
18
17
16
15
14
13
12
11
2.7
2.9
3.1
3.3
3.5
3.7
3.9
4.1
GNP — Annual Rate $ Trillions

Figure 3.6

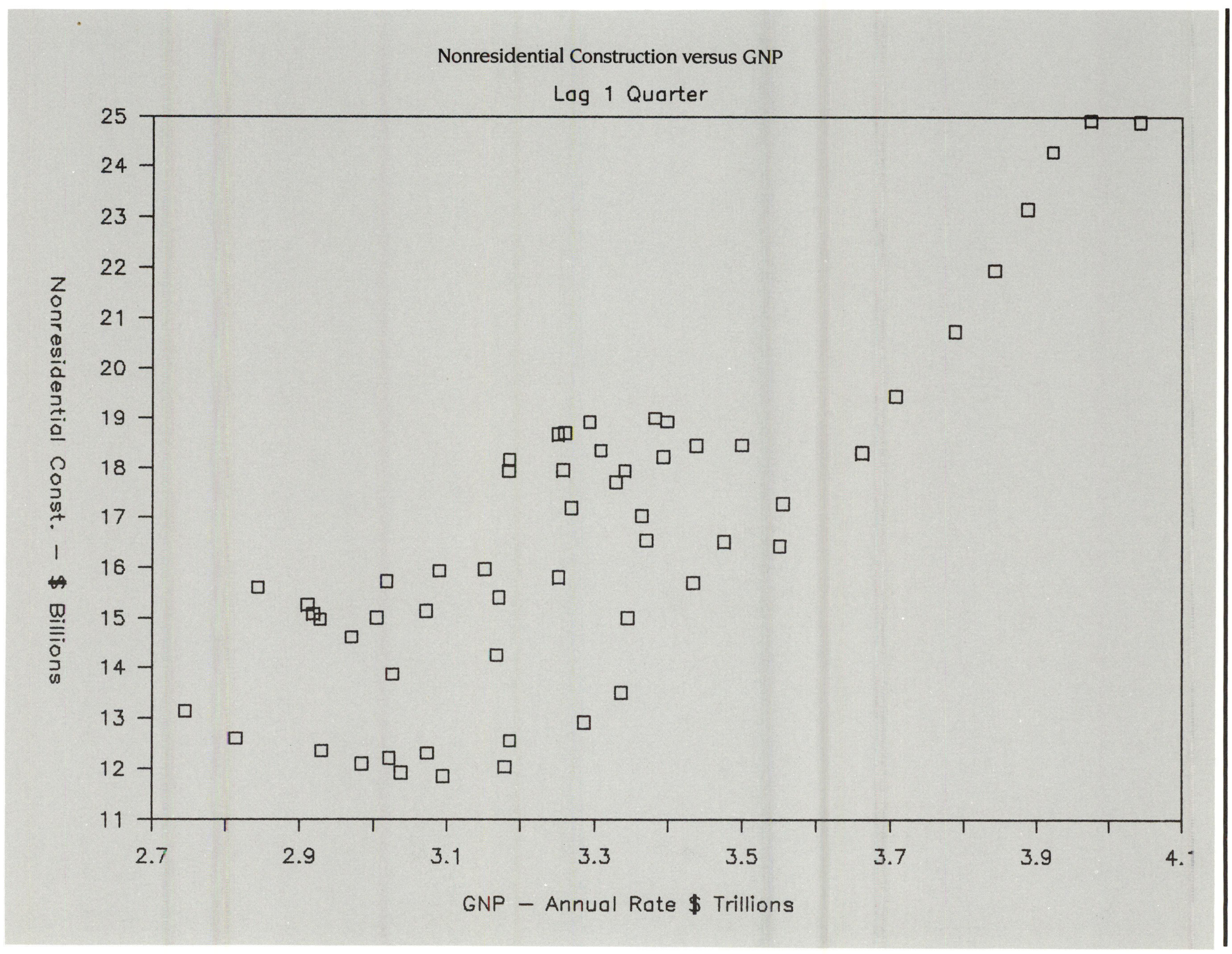
Nonresidential Construction versus GNP
Lag 1 Quarter
25
24
23
22
21
20
19
18
17
16
15
14
13
12
11
Nonresidential Const. — $ Billions
2.7
2.9
3.1
3.3
3.5
3.7
3.9
GNP — Annual Rate $ Trillions

Figure 3.7

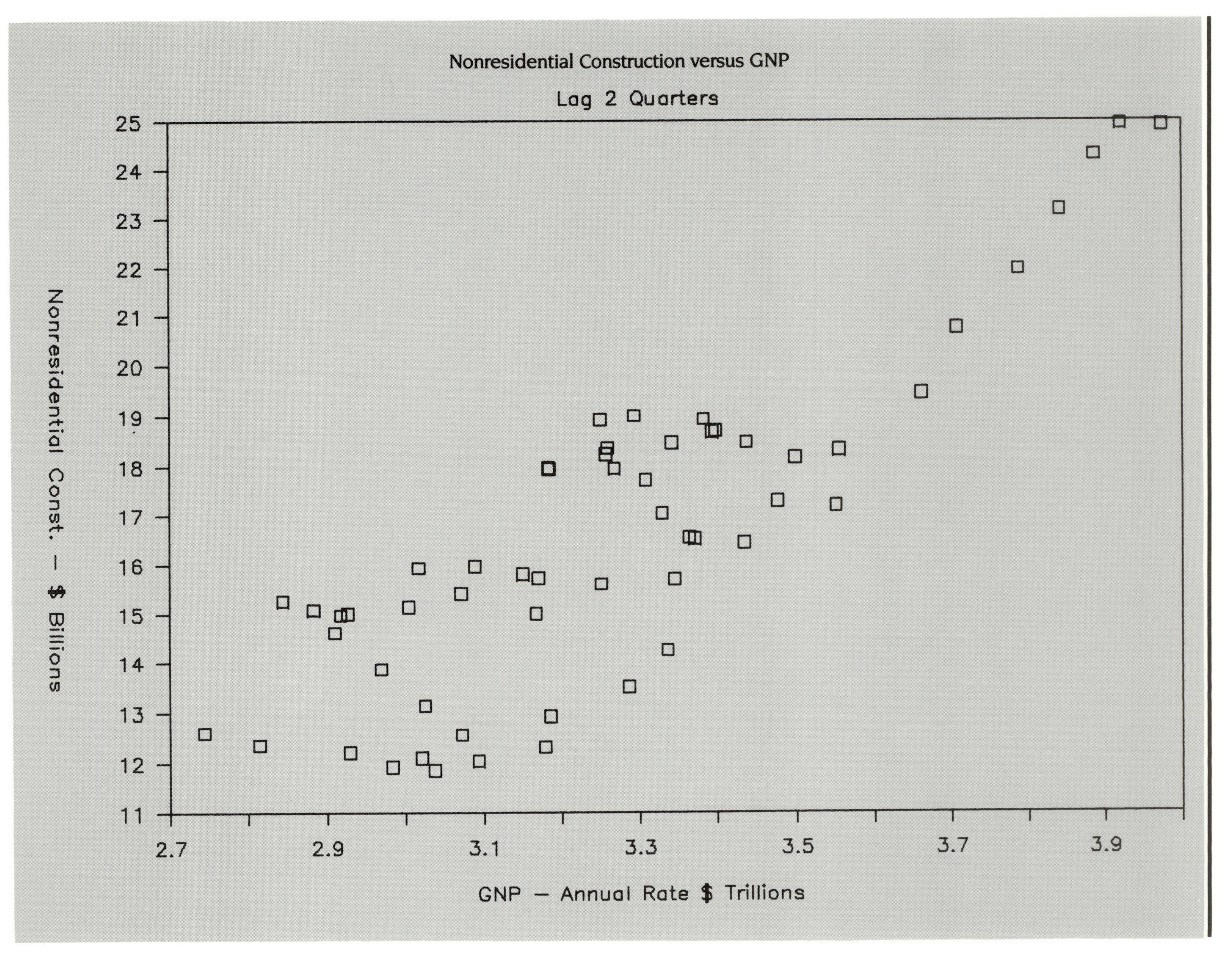
Nonresidential Construction versus GNP
Lag 2 Quarters
Nonresidential Const. — $ Billions
25
24
23
22
21
20
19
18
17
16
15
14
13
12
11
2.7
2.9
3.1
3.3
3.5
3.7
3.9
GNP — Annual Rate $ Trillions

Figure 3.8

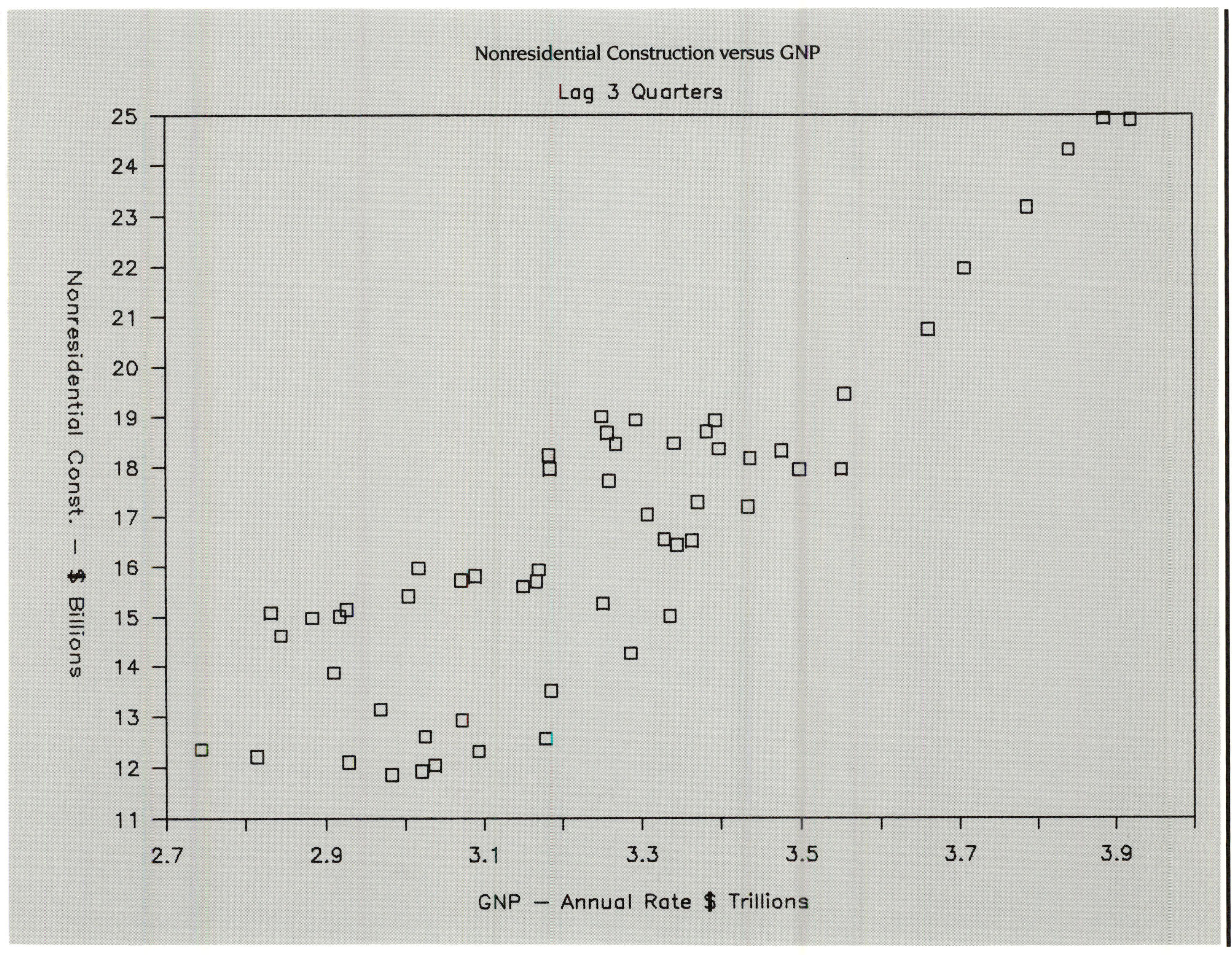

Nonresidential Construction versus GNP
Lag 3 Quarters
Nonresidential Const. — $ Billions
25
24
23
22
21
20
19
18
17
16
15
14
13
12
11
2.7
2.9
3.1
3.3
3.5
3.7
3.9
GNP — Annual Rate $ Trillions

Figure 3.9

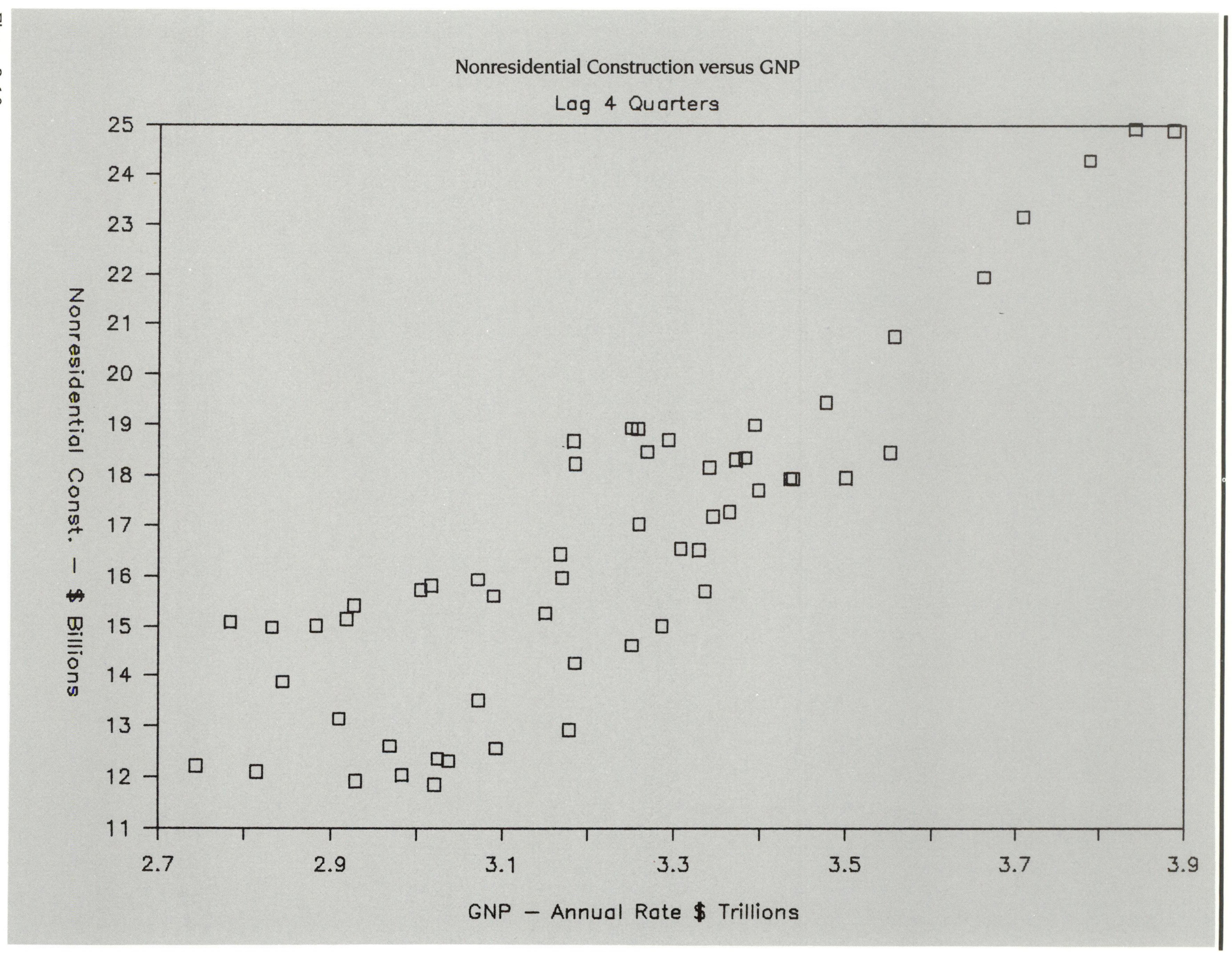
Nonresidential Construction versus GNP
Lag 4 Quarters
Nonresidential Const. — $ Billions
25
24
23
22
21
20
19
18
17
16
15
14
13
12
11
2.7
2.9
3.1
3.3
3.5
3.7
3.9
GNP — Annual Rate $ Trillions

Figure 3.10

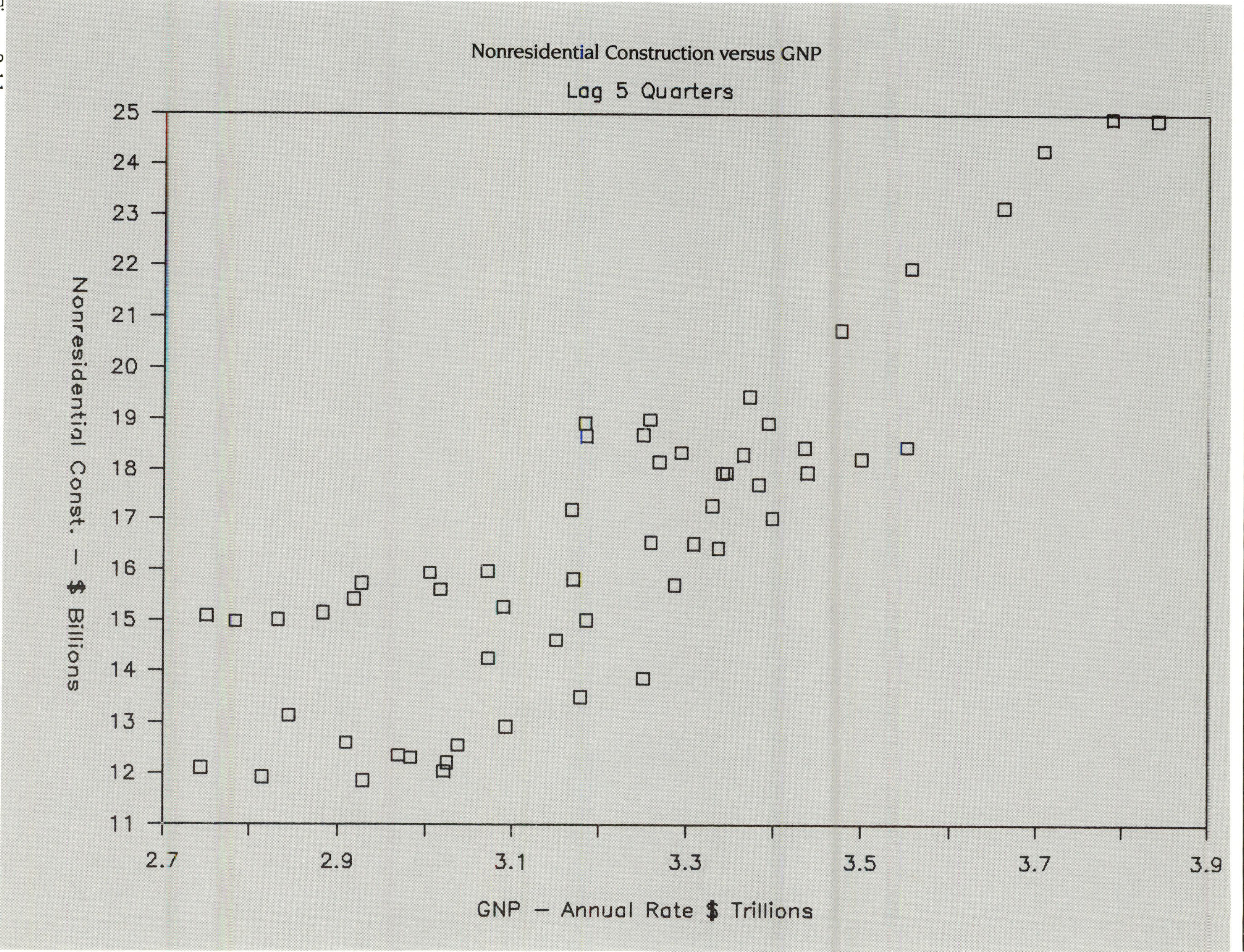
Nonresidential Construction versus GNP
Lag 5 Quarters
Nonresidential Const. – $ Billions
25
24
23
22
21
20
19
18
17
16
15
14
13
12
11
2.7
2.9
3.1
3.3
3.5
3.7
3.9
GNP – Annual Rate $ Trillions

Figure 3.11

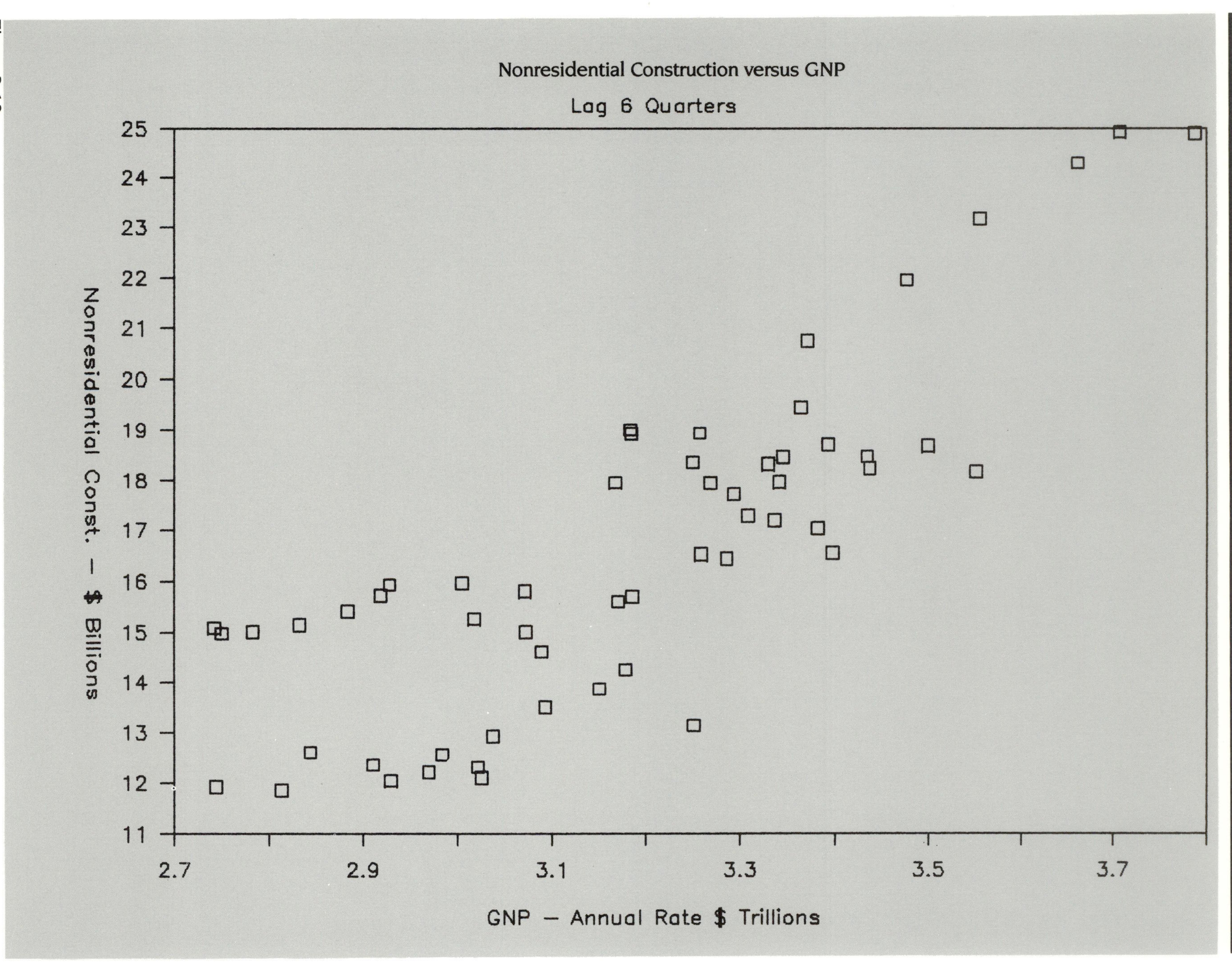

Nonresidential Construction versus GNP
Lag 6 Quarters
Nonresidential Const. — $ Billions
25
24
23
22
21
20
19
18
17
16
15
14
13
12
11
2.7
2.9
3.1
3.3
3.5
3.7
GNP — Annual Rate $ Trillions

Figure 3.12

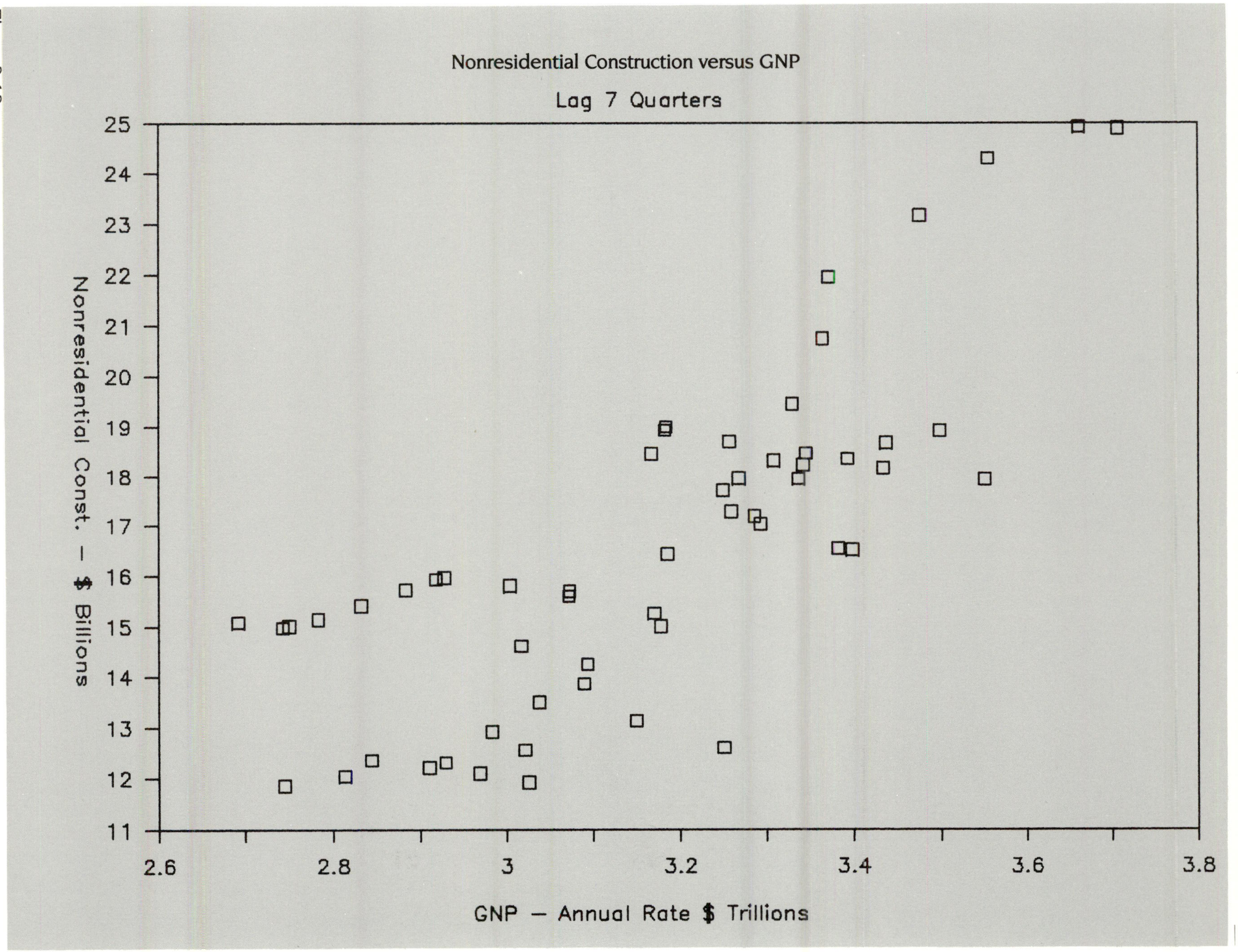

Nonresidential Construction versus GNP
Lag 7 Quarters
Nonresidential Const. — $ Billions
25
24
23
22
21
20
19
18
17
16
15
14
13
12
11
2.6
2.8
3
3.2
3.4
3.6
3.8
GNP — Annual Rate $ Trillions

Figure 3.13

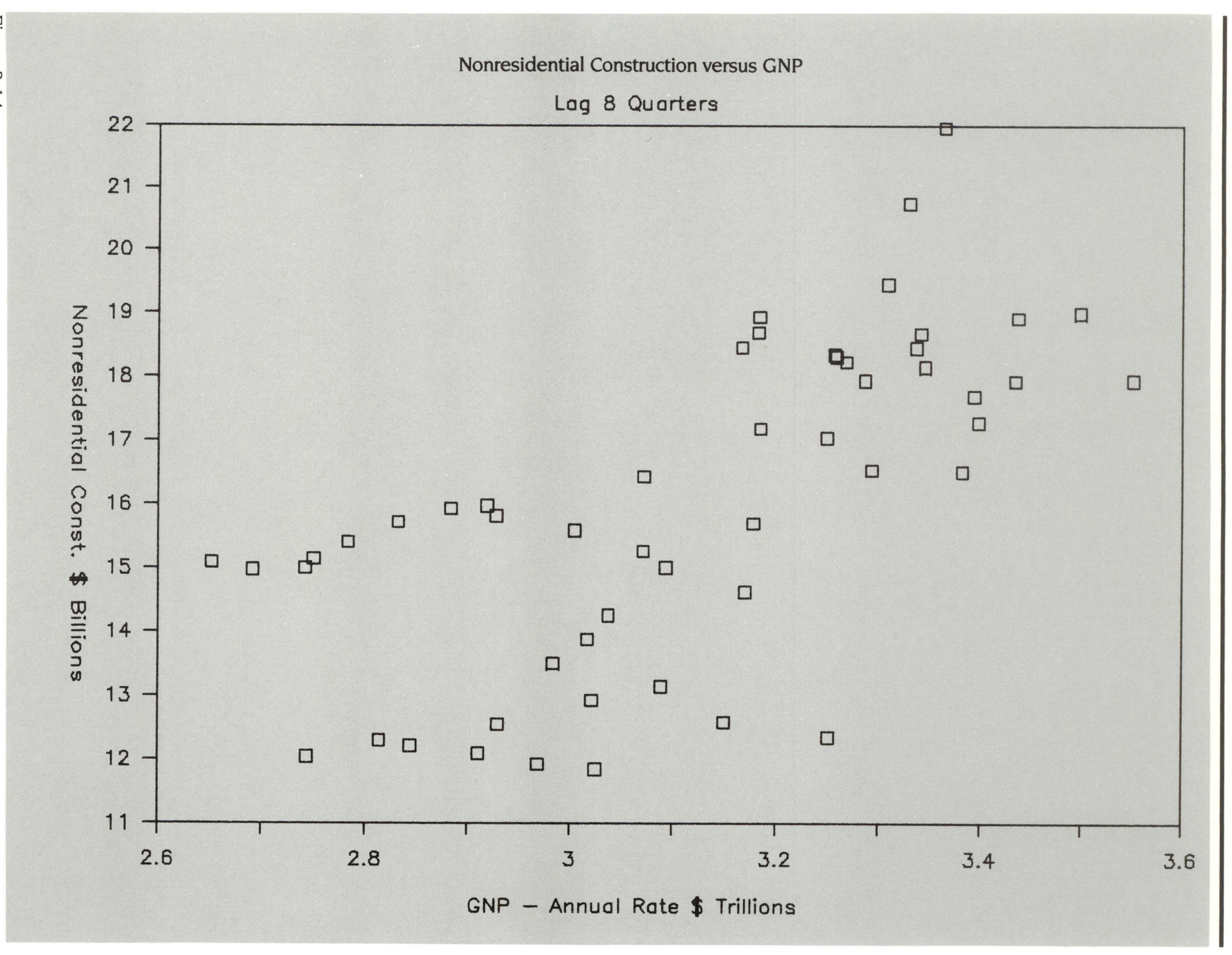

Nonresidential Construction versus GNP
Lag 8 Quarters
Nonresidential Const. $ Billions
22
21
20
19
18
17
16
15
14
13
12
11
2.6
2.8
3
3.2
3.4
3.6
GNP — Annual Rate $ Trillions

Figure 3.14

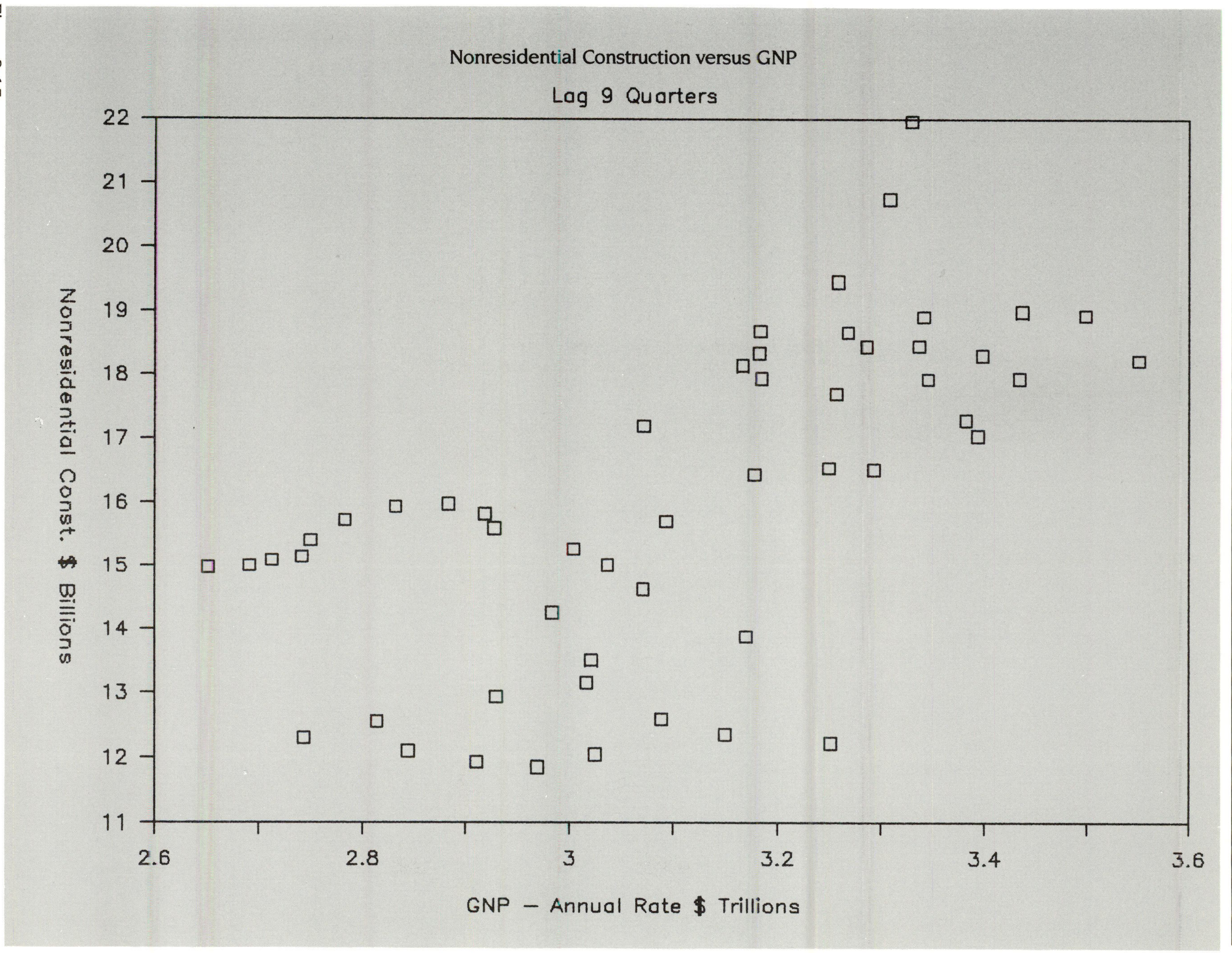
Nonresidential Construction versus GNP
Lag 9 Quarters
Nonresidential Const. $ Billions
22
21
20
19
18
17
16
15
14
13
12
11
2.6
2.8
3
3.2
3.4
3.6
GNP — Annual Rate $ Trillions

Figure 3.15

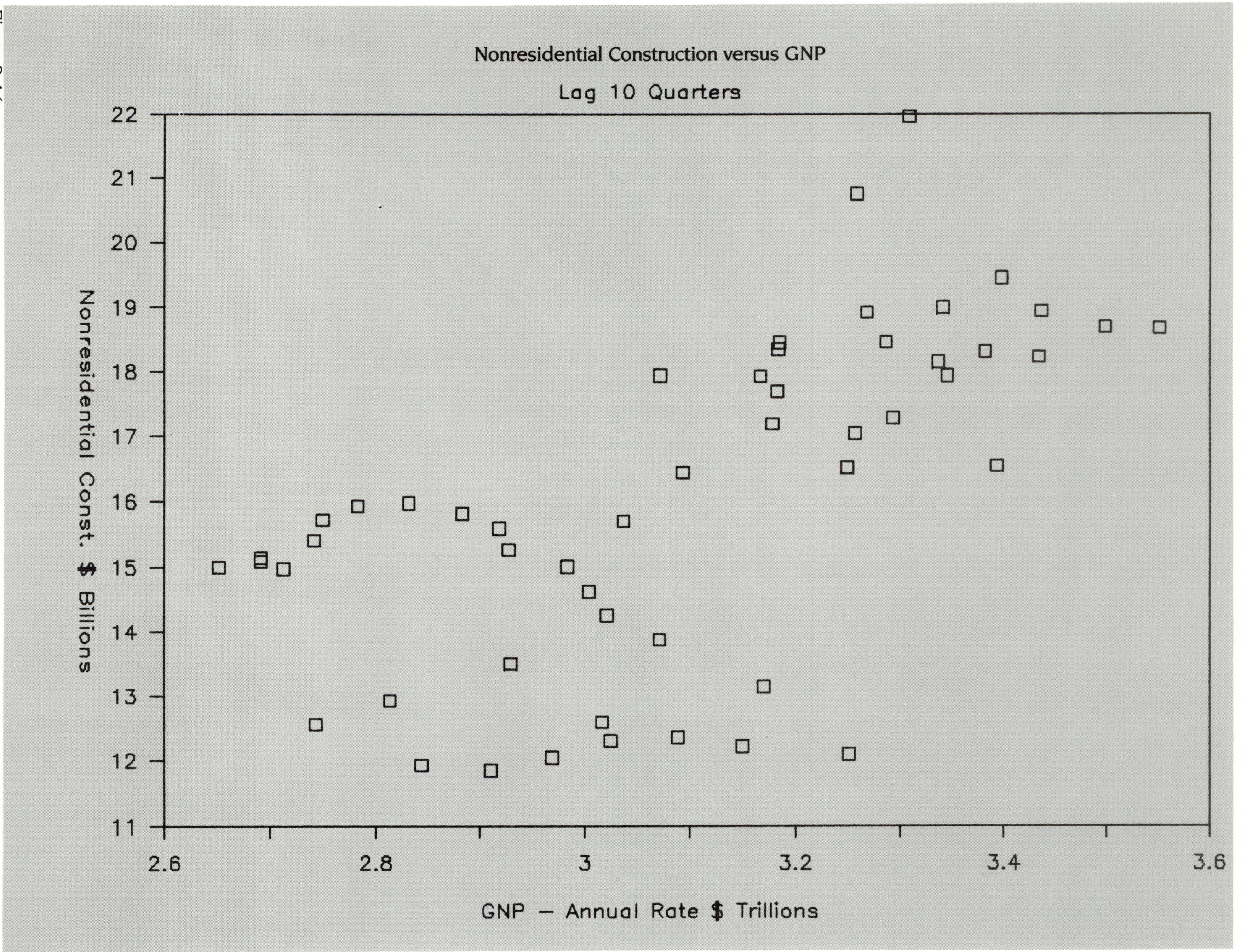

Nonresidential Construction versus GNP
Lag 10 Quarters
Nonresidential Const. $ Billions
22
21
20
19
18
17
16
15
14
13
12
11
2.6
2.8
3
3.2
3.4
3.6
GNP – Annual Rate $ Trillions

Figure 3.16

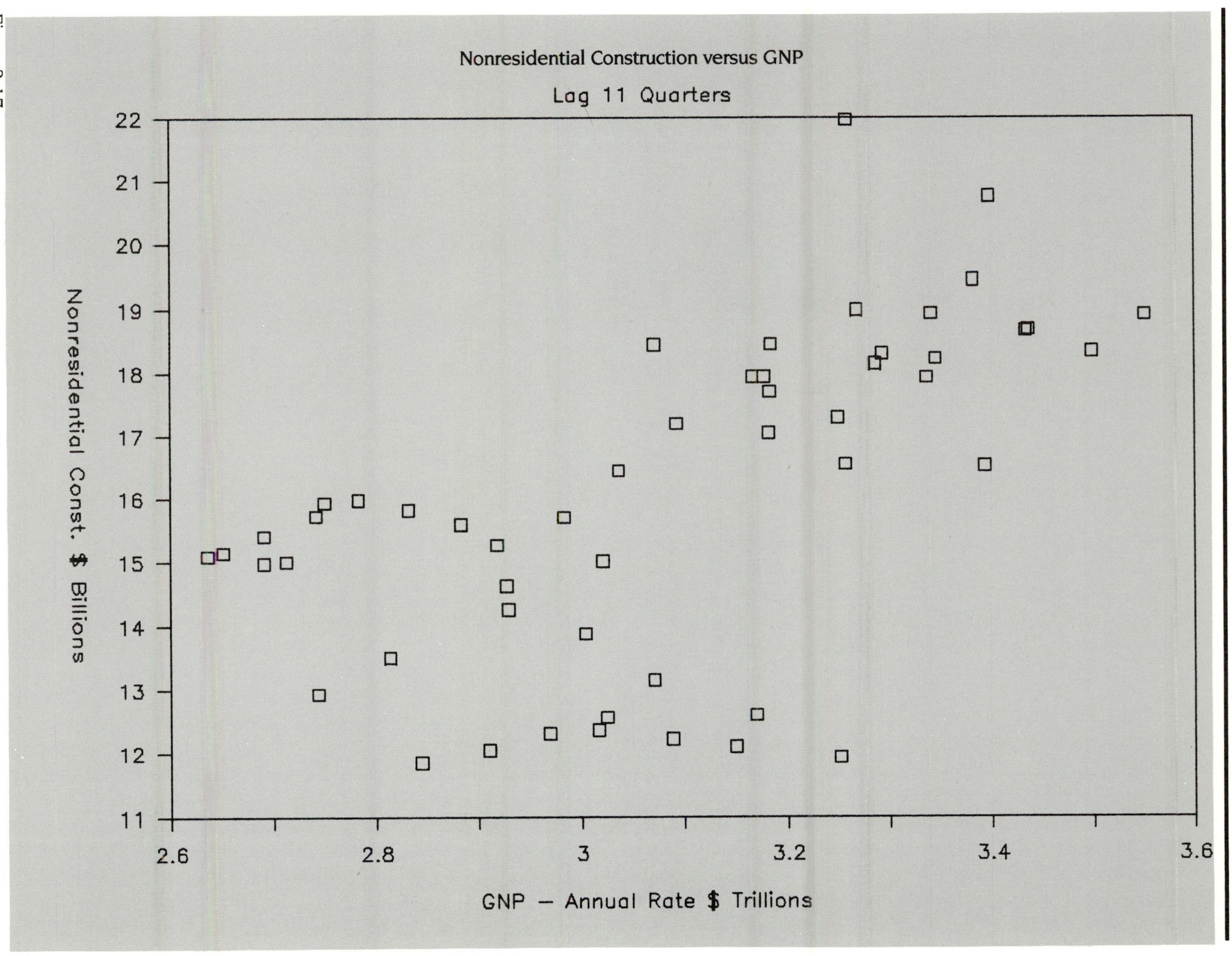

Nonresidential Construction versus GNP
Lag 11 Quarters
Nonresidential Const. $ Billions
22
21
20
19
18
17
16
15
14
13
12
11
2.6
2.8
3
3.2
3.4
3.6
GNP – Annual Rate $ Trillions

Figure 3.17

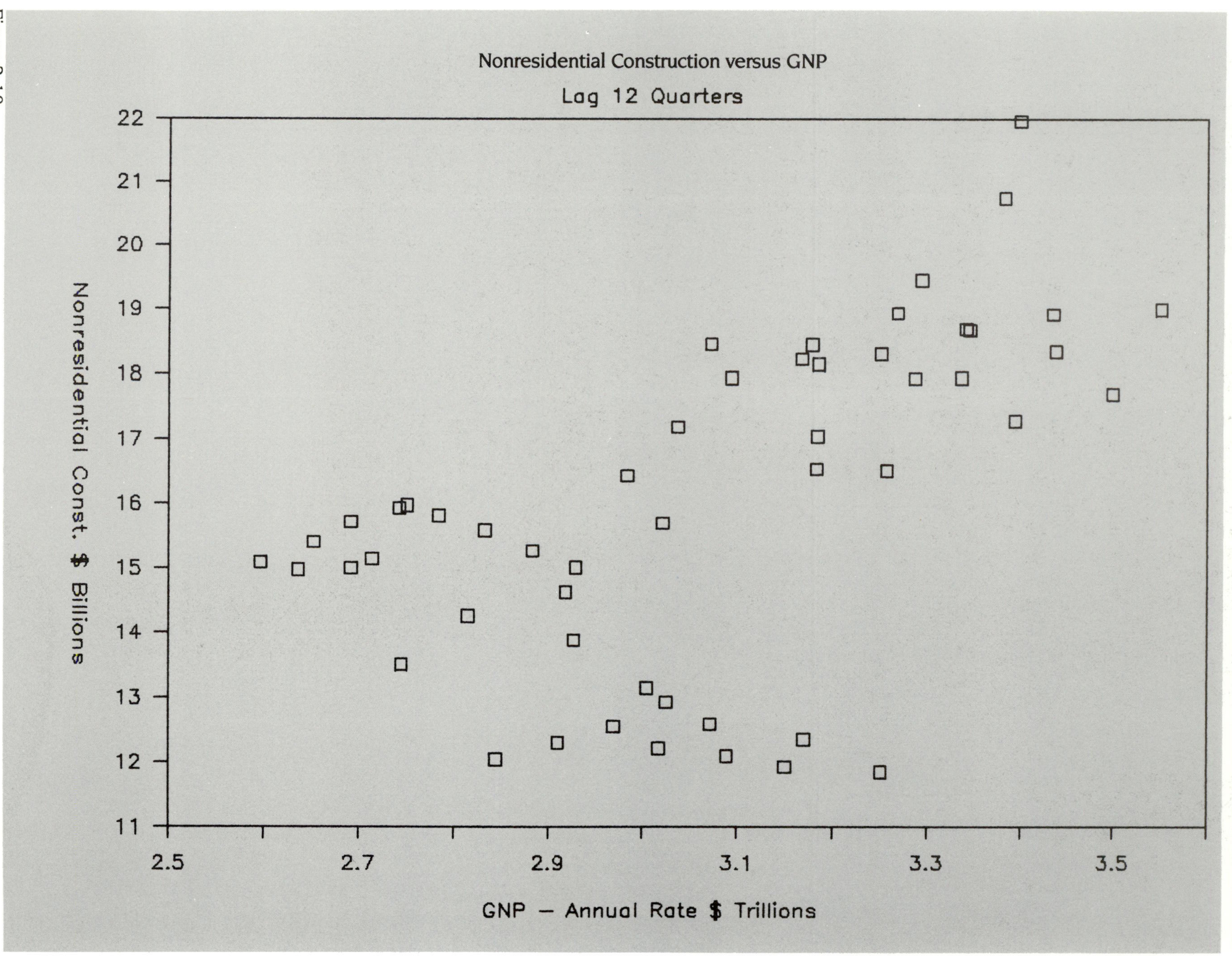

Nonresidential Construction versus GNP
Lag 12 Quarters
Nonresidential Const. $ Billions
22
21
20
19
18
17
16
15
14
13
12
11
2.5
2.7
2.9
3.1
3.3
3.5
GNP – Annual Rate $ Trillions

Figure 3.18

4

THE NEED IDENTIFICATION ESTIMATE

4

THE NEED IDENTIFICATION ESTIMATE

An effective cost control system cannot wait until the project design phases, but must begin with project planning. Realistic building cost estimates must be used in the Need Identification Phase for feasibility studies, and in the Need Analysis Phase to set realistic budgets. The effective cost control system, then, consists of a series of estimates beginning in the Need Identification Phase and continuing through need analysis, conceptual design, design development, and construction.

In the Need Identification Phase, a building need is recognized and studies are made to determine the feasibility of various building options. These studies require "ballpark" estimates of probable project cost ranges. Only the building type is known at this point, and those who prepare the rough estimates may have little expertise in construction estimating. The appropriate estimating methods to use under these conditions are the *Order of Magnitude* method and the *Square Foot/Cubic Foot* method.

During the Need Identification Phase, the estimate requirements take one of two forms. When the approximate size and type of facility is known, an estimate of *total project cost* is needed. When the type of facility is known, but the size must be determined from a set budget, or an optimum size must be determined to maximize return on investment, an estimate of total project cost *per unit of occupancy* or *per unit of building* is needed. Both types of preliminary estimates are demonstrated in this chapter, using a hypothetical project. A *per unit* cost estimate is also shown.

The Project

Chambers, Georgia is located on Interstate 85, less than 45 minutes from downtown Atlanta. Because of its proximity to Atlanta, to nearby high tech industries, educational facilities, and recreational areas; and because of its quality public school system, Chambers is expected to continue to grow as a residential community. A group of local businessmen believe that Chambers can attract several of the businesses and high tech industries that have expressed an interest in moving to the Atlanta area. They are investigating the possibility of constructing a low-rise office building (shell only) for commercial lease space. The group needs a rough cost estimate for a preliminary feasibility study.

Total Project Cost

The costs for low-rise office buildings from the Square Foot/Cubic Foot Costs tables in Means 1988 *Building Construction Cost Data* are shown in Figure 4.1. These figures include square foot and cubic foot costs for the total project, and square foot costs for site work, masonry, equipment, plumbing, HVAC, electrical, and total mechanical and electrical. A ***per unit of occupancy*** price is not given. The ***total project costs*** represent the total cost to construct the building including structure, exterior closure, interior finishes, mechanical and electrical systems and, in this case, site work. Excluded are design, land, furniture, equipment, financing, and planning costs.

Because Means does not specify a *per unit of occupancy* price for low-rise offices, the square foot price will have to be used. The total project cost for this category is shown in line 171-610-0010 from Figure 4.1. This price includes the cost of interior finishes. Since the Chambers project is to be built as a shell, the cost of the interior finishes must be removed. The Means breakdown does not include a square foot price for interior finishes, but the following rule of thumb can be applied to remove this factor from the total project cost. Approximately one half of the general construction portion of the project (the building less mechanical, electrical, plumbing, equipment, and site work costs) is for the structure, one fourth is for exterior closure, and the remaining one fourth is for interior finishes. Using the median price from line 171-610-0010, the adjustment to remove interior finishes from the total project square foot cost is as follows:

Total building costs	$55.40/S.F.
Less total Mechanical, Electrical, and Plumbing	11.15
Less equipment	1.01
Less site work	5.35
General Construction Cost	**$37.89/S.F.**

Of the $37.89, approximately one fourth is for interior finishes. Therefore:

Total building costs	$55.40/S.F.
Less interior finishes ($37.89 × .25)	9.47
Adjusted Building Costs	**$45.93/S.F.**

The adjusted building cost of $45.93/S.F. will be used to estimate the cost of the building shell and site work. Assuming that the group has determined that the desired project size is 56,700 square feet, the cost for the building shell and site work is as follows:

56,700 S.F. × $45.93/S.F. = $2,604,231

To provide a total project cost estimate, costs for design, furniture, equipment, financing, and planning will have to be added. The estimate must also be adjusted for size, location, inflation, and bid conditions. As shown in Chapter 2, the procedure is to first adjust the price for building size. Then, the costs of design fees, furniture, equipment, financing, and planning are estimated. Next, the location, inflation, and bid conditions adjustments are made. Finally, the cost range is calculated.

Project size multiplier information is obtained from Figure 4.2 (1988 *Means Assemblies Cost Data*). The size multiplier is obtained by dividing the proposed building size by the typical size for that building category from the table shown in Figure 4.2. The product is plotted on the Area Conversion Scale to determine the cost multiplier. The typical size for low-rise offices is 8,600 square feet; the proposed building size is 56,700 square feet. Therefore:

56,700 S.F./8,600 S.F. = 6.59

171 | S.F., C.F., and % of Total Costs

	171 000	S.F. & C.F. Costs	UNIT	UNIT COSTS ¼	UNIT COSTS MEDIAN	UNIT COSTS ¾	% OF TOTAL ¼	% OF TOTAL MEDIAN	% OF TOTAL ¾	
600	3100	Total: Mechanical & Electrical	S.F.	12.30	16.40	24.15	22%	28.10%	33.20%	600
	3200									
	9000	Per bed or person, total cost	Bed	20,900	27,100	36,300				
610	0010	OFFICES Low-Rise (1 to 4 story)	S.F.	43.15	55.40	72.95				610
	0020	Total project costs	C.F.	3.16	4.45	5.95				
	0100	Sitework	S.F.	3.17	5.35	8.15	5.20%	9.20%	13.70%	
	0500	Masonry		1.67	3.39	6.45	2.90	5.90%	8.60%	
	1800	Equipment		.60	1.01	2.69	1.30%	1.70%	4%	
	2720	Plumbing		1.65	2.48	3.54	3.60%	4.50%	6%	
	2770	Heating, ventilating, air conditioning		3.55	4.90	7.20	7.20%	10.40%	11.90%	
	2900	Electrical		3.61	5	6.90	7.40%	9.50%	11%	
	3100	Total: Mechanical & Electrical		7.50	11.15	16.10	14.70%	20.50%	26.80%	
620	0010	OFFICES Mid-Rise (5 to 10 story)	S.F.	47.90	59.05	80.60				620
	0020	Total project costs	C.F.	3.33	4.25	6.05				
	2720	Plumbing	S.F.	1.47	2.23	3.21	2.80%	3.60%	4.50%	
	2770	Heating, ventilating, air conditioning		3.62	5.15	8.25	7.60%	9.30%	11%	
	2900	Electrical		3.08	4.41	7.45	6.50%	8.20%	10%	
	3100	Total: Mechanical & Electrical		8.20	11	18.65	16.70%	21.10%	25.70%	
630	0010	OFFICES High-Rise (11 to 20 story)		57.50	73.70	92.80				630
	0020	Total project costs	C.F.	3.68	5.20	7.50				
	2900	Electrical	S.F.	2.87	4.18	6.45	5.80%	7%	10.50%	
	3100	Total: Mechanical & Electrical		9.75	13.35	24.35	16.50%	21.40%	29.40%	
640	0010	POLICE STATIONS		71.90	92.30	117				640
	0020	Total project costs	C.F.	5.25	6.85	9.05				
	0500	Masonry	S.F.	7.55	11.70	14.45	6.80%	10.60%	13.20%	
	1140	Roofing		2.27	2.43	5.45	2%	2.20%	4.20%	
	1350	Glass & glazing		.65	.87	.93	.60%	.80%	.80%	
	1570	Floor covering		.32	.65	.72	.30%	.70%	.70%	
	1580	Painting		1.17	1.40	1.94	1.10%	1.50%	1.90%	
	1800	Equipment		1.13	5.90	9.90	2%	6%	13.30%	
	2720	Plumbing		4.05	5.75	9.45	5.70%	6.80%	10.60%	
	2770	Heating, ventilating, air conditioning		5.75	7.80	11.10	7%	10.50%	11.90%	
	2900	Electrical		7.20	11.20	14.45	9.40%	11.60%	14.50%	
	3100	Total: Mechanical & Electrical		19.50	24.35	31.30	22.60%	27.50%	33%	
650	0010	POST OFFICES		57.90	69.75	88.45				650
	0020	Total project costs	C.F.	3.20	4	4.80				
	2720	Plumbing	S.F.	2.49	3.15	4	4.20%	5.30%	5.60%	
	2770	Heating, ventilating, air conditioning		3.61	4.81	7.34	6.60%	8%	9.80%	
	2900	Electrical		4.57	6.45	7.65	7.40%	9.50%	11%	
	3100	Total: Mechanical & Electrical		10.05	13.75	19.25	16.50%	22.20%	26.30%	
660	0010	POWER PLANTS		275	465	655				660
	0020	Total project costs	C.F.	9.35	17.25	47.85				
	2900	Electrical	S.F.	20.20	52.90	86.05	9.20%	12.70%	18.40%	
	8100	Total: Mechanical & Electrical		46.75	132	295	25.50%	32.50%	52.60%	
670	0010	RELIGIOUS EDUCATION		44.20	51.80	64.15				670
	0020	Total project costs	C.F.	2.55	3.65	4.80				
	2720	Plumbing	S.F.	1.86	2.75	3.95	4.10%	5.10%	7.10%	
	2770	Heating, ventilating, air conditioning		4.25	5.05	6.65	8.10%	9.90%	11.20%	
	2900	Electrical		3.32	4.41	5.80	6.90%	8.60%	10.20%	
	3100	Total: Mechanical & Electrical		7.35	10.40	15.05	14.90%	20.50%	24.60%	
690	0010	RESEARCH Laboratories and facilities		61	90.75	133				690
	0020	Total project costs	C.F.	3.55	7.10	10.65				
	1800	Equipment	S.F.	1.49	4.99	10.05	.90%	4.70%	9%	
	2720	Plumbing		6.35	8.75	12.25	5.20%	8.30%	10.80%	
	2770	Heating, ventilating, air conditioning		6.20	20.20	23.85	7.20%	16.40%	17.70%	
	2900	Electrical		7.40	10.35	23.05	9.20%	12.40%	16.20%	

362

For expanded coverage of these items see *Means Square Foot Cost Data 1988*

Figure 4.1

S.F & C.F. COSTS | C14.1-000 | Project Size Modifier

(92) Square Foot Project Size Modifier

One factor that affects the S.F. cost of a particular building is the size. In general, for buildings built to the same specifications in the same locality, the larger building will have the lower S.F. Cost. This is due mainly to the decreasing contribution of the exterior walls plus the economy of scale usually achievable in larger buildings. The Area Conversion Scale shown below will give a factor to convert costs for the typical size building to an adjusted cost for the particular project.

The Square Foot Base Size lists the median costs, most typical project size in our accumulated data and the range in size of the projects.

The Size Factor for your project is determined by dividing your project area in S.F. by the typical project size for the particular Building Type. With this factor, enter the Area Conversion Scale at the appropriate Size Factor and determine the appropriate cost multiplier for your building size.

Example: Determine the cost per S.F. for a 100,000 S.F. Mid-rise apartment building.

$$\frac{\text{Proposed building area} = 100{,}000 \text{ S.F.}}{\text{Typical size from below} = 50{,}000 \text{ S.F.}} = 2.00$$

Enter Area Conversion scale at 2.0, intersect curve, read horizontally the appropriate cost multiplier of 0.94. Size adjusted cost becomes 0.94x $51.15=$48.08 based on national average costs.

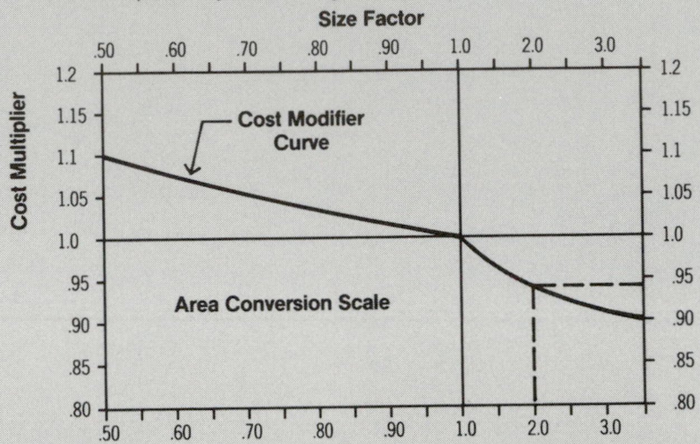

Building Type	Median Cost per S.F.	Typical Size Gross S.F.	Typical Range Gross S.F.	Building Type	Median Cost per S.F.	Typical Size Gross S.F.	Typical Range Gross S.F.
	Square Foot Base Size						
Apartments, Low Rise	$ 40.45	21,000	9,700 - 37,200	Jails	$119.00	13,700	7,500 - 28,000
Apartments, Mid Rise	51.15	50,000	32,000 - 100,000	Libraries	72.60	12,000	7,000 - 31,000
Apartments, High Rise	56.75	310,000	100,000 - 650,000	Medical Clinics	69.90	7,200	4,200 - 15,700
Auditoriums	68.25	25,000	7,600 - 39,000	Medical Offices	66.05	6,000	4,000 - 15,000
Auto Sales	43.15	20,000	10,800 - 28,600	Motels	51.55	27,000	15,800 - 51,000
Banks	94.20	4,200	2,500 - 7,500	Nursing Homes	69.75	23,000	15,000 - 37,000
Churches	62.60	9,000	5,300 - 13,200	Offices, Low Rise	55.40	8,600	4,700 - 19,000
Clubs, Country	60.55	6,500	4,500 - 15,000	Offices, Mid Rise	59.05	52,000	31,300 - 83,100
Clubs, Social	59.60	10,000	6,000 - 13,500	Offices, High Rise	73.70	260,000	151,000 - 468,000
Clubs, YMCA	64.00	28,300	12,800 - 39,400	Police Stations	92.30	10,500	4,000 - 19,000
Colleges (Class)	82.70	50,000	23,500 - 98,500	Post Offices	69.75	12,400	6,800 - 30,000
Colleges (Science Lab)	96.55	45,600	16,600 - 80,000	Power Plants	465.00	7,500	1,000 - 20,000
College (Student Union)	88.95	33,400	16,000 - 85,000	Religious Education	51.80	9,000	6,000 - 12,000
Community Center	64.55	9,400	5,300 - 16,700	Research	90.75	19,000	6,300 - 45,000
Court Houses	86.25	32,400	17,800 - 106,000	Restaurants	82.30	4,400	2,800 - 6,000
Dept. Stores	38.10	90,000	44,000 - 122,000	Retail Stores	40.00	7,200	4,000 - 17,600
Dormitories, Low Rise	61.45	24,500	13,400 - 40,000	Schools, Elementary	60.70	41,000	24,500 - 55,000
Dormitories, Mid Rise	79.20	55,600	36,100 - 90,000	Schools, Jr. High	60.50	92,000	52,000 - 119,000
Factories	33.35	26,400	12,900 - 50,000	Schools, Sr. High	60.05	101,000	50,500 - 175,000
Fire Stations	66.30	5,800	4,000 - 8,700	Schools, Vocational	57.60	37,000	20,500 - 82,000
Fraternity Houses	58.80	12,500	8,200 - 14,800	Sports Arenas	47.15	15,000	5,000 - 40,000
Funeral Homes	58.95	7,800	4,500 - 11,000	Supermarkets	39.40	20,000	12,000 - 30,000
Garages, Commercial	44.65	9,300	5,000 - 13,600	Swimming Pools	68.00	13,000	7,800 - 22,000
Garages, Municipal	47.25	8,300	4,500 - 12,600	Telephone Exchange	104.00	4,500	1,200 - 10,600
Garages, Parking	20.40	163,000	76,400 - 225,300	Terminals, Bus	52.50	11,400	6,300 - 16,500
Gymnasiums	56.35	19,200	11,600 - 41,000	Theaters	55.55	10,500	8,800 - 17,500
Hospitals	114.00	55,000	27,200 - 125,000	Town Halls	66.50	10,800	4,800 - 23,400
House (Elderly)	57.45	37,000	21,000 - 66,000	Warehouses	26.40	25,000	8,000 - 72,000
Housing (Public)	48.20	36,000	14,400 - 74,400	Warehouse & Office	30.10	25,000	8,000 - 72,000
Ice Rinks	45.80	29,000	27,200 - 33,600				

486

Figure 4.2

The size factor, 6.59, is off the horizontal scale on the Area Conversion Scale. Note that the curve has become horizontal at the high end of the size factor scale. This shows that any increase in size beyond a size factor of 3.5 does not result in any additional savings in square foot costs. Therefore, the cost multiplier for a size factor of 3.5, approximately .9, will be used to adjust the building costs for the Chambers project.

Building Cost	$2,604,231
Cost Multiplier	× .9
Adjusted Building Cost	**$2,343,808**

Next, the design fee, furniture, equipment, financing, and planning costs are estimated. Design fee information is given in Figure 4.3. This table results in a fee range of 6.7% to 6.4% of the construction costs for office buildings in the $2,500,000 to $5,000,000 cost range. Since the estimated construction costs for the Chambers project are just over $2,500,000, 6.7% will be used. Because only a shell is to be considered, furniture and equipment costs do not need to be included.

Financing costs for this estimate include only the costs for interim or construction financing. The permanent financing costs are part of the building's operating budget. The estimate of these costs, $425,000, is obtained by consulting a local savings and loan.

In Chapter 2, a range of 1% to 3% of the construction costs is allotted for planning costs. Three percent is used for projects costing less than $1,000,000, 2% for projects costing between $1,000,000 and $3,000,000, and 1% for projects costing more than $3,000,000. The construction cost of the Chambers project, $2,604,231, falls in the mid-range category; therefore, 2% is used.

Design, financing, and planning costs are added to the estimate as follows.

Construction Contract	$2,343,808
Design Fees at 6.7%	157,035
Total Construction Cost	**$2,500,843**
Financing Costs	425,000
Planning Costs at 2%	50,017
Total Project Cost	**$2,975,860**

This price was prepared using national average cost data. It must be adjusted to the Chambers location. Since Chambers is a suburb of Atlanta, it is assumed that construction prices there are the same as those in Atlanta. The Atlanta total cost index from Figure 4.4 is 89.2.

The Means cost data is based on January 1, 1988 and must be adjusted to reflect the price and bid conditions of the anticipated project bid date. Assuming that the date of the study is April, 1988, that the project will begin design the following June, and that conceptual design and detailed design will require approximately one year, the project will be bid in June, 1989. It is assumed that a plot of the construction inflation data for Atlanta shows construction inflation on a gentle upward trend exceeding 4% at the beginning of 1988. The trend indicates that construction inflation will reach 5% early in 1989. Consequently, an average inflation rate of 4.5% per year is used for adjustment. 4.5% will be added for 1988, and 2.25%, one half of 4.5%, will be added for the first half of 1989.

GENERAL CONDITIONS | C10.1-100 | Design & Engineering

Table 10.1-101 Architectural Fees

Tabulated below are typical percentage fees, below which adequate service cannot be expected. Fees may vary from those listed due to economic conditions.

Rates can be interpolated horizontally and vertically. Various portions of the same project requiring different rates should be adjusted proportionately. For alterations, add 50% to the fee for the first $500,000 of project cost and add 25% to the fee for project cost over $500,000.

Architectural fees tabulated below include Engineering Fees.

Building Type	Total Project Size in Thousands of Dollars						
	100	250	500	1,000	2,500	5,000	10,000
Factories, garages, warehouses repetitive housing	9.0%	8.0%	7.0%	6.2%	5.6%	5.3%	4.9%
Apartments, banks, schools, libraries, offices, municipal buildings	11.7	10.8	8.5	7.3	6.7	6.4	6.0
Churches, hospitals, homes, laboratories, museums, research	14.0	12.8	11.9	10.9	9.5	8.5	7.8
Memorials, monumental work, decorative furnishings	—	16.0	14.5	13.1	11.3	10.0	9.0

75

Table 10.1-102 Engineering Fees

Typical **Structural Engineering Fees** based on type of construction and total project size. These fees are included in Architectural Fees.

Type of Construction	Total Project Size			
	To $250,000	$250,000-$500,000	$1,000,000	$5,000,000 & Over
Industrial buildings, factories & warehouses	Technical payroll times 2.0 to 2.5	1.60%	1.25%	1.00%
Hotels, apartments, offices, dormitories, hospitals, public buildings, food stores		2.00%	1.70%	1.20%
Museums, banks, churches and cathedrals		2.00%	1.75%	1.25%
Thin shells, prestressed concrete, earthquake resistive		2.00%	1.75%	1.50%
Parking ramps, auditoriums, stadiums, convention halls, hangars & boiler houses		2.50%	2.00%	1.75%
Special buildings, major alterations, underpinning & future expansion	↓	Add to above 0.5%	Add to above 0.5%	Add to above 0.5%

For complex reinforced concrete or unusually complicated structures, add 20% to 50%.

Table 10.1-103 Mechanical and Electrical Fees

Typical **Mechanical and Electrical Engineering Fees** based on the size of the subcontract. These fees are included in Architectural Fees.

Type of Construction	Subcontract Size							
	$25,000	$50,000	$100,000	$225,000	$350,000	$500,000	$750,000	$1,000,000
Simple structures	6.4%	5.7%	4.8%	4.5%	4.4%	4.3%	4.2%	4.1%
Intermediate structures	8.0	7.3	6.5	5.6	5.1	5.0	4.9	4.8
Complex structures	12.0	9.0	9.0	8.0	7.5	7.5	7.0	7.0

For renovations, add 15% to 25% to applicable fee.

465

Figure 4.3

CITY COST INDEXES | C13.3-000 | Building Systems

DIV. NO.	BUILDING SYSTEMS	NEW HAVEN, CT			NORWALK, CT			STAMFORD, CT			WATERBURY, CT			WILMINGTON, DE		
		MAT.	INST.	TOTAL	MAT.	INST.	TOTAL	MAT.	INST.	TOTAL	MAT.	INST.	TOTAL	MAT.	INST.	TOTAL
1-2	FOUND/SUBSTRUCTURES	101.8	103.3	102.7	106.4	103.3	104.5	118.9	103.9	109.6	112.4	101.5	105.6	102.5	109.0	106.6
3	SUPERSTRUCTURES	95.1	103.7	99.6	96.3	103.6	100.1	101.5	104.3	103.0	99.8	101.6	100.8	94.8	108.4	101.9
4	EXTERIOR CLOSURE	106.9	106.0	106.3	99.2	104.5	102.6	103.1	104.4	103.9	97.3	103.4	101.2	91.4	96.8	94.9
5	ROOFING	88.8	105.4	94.5	89.9	105.9	95.4	89.0	106.9	95.1	90.0	102.4	94.3	91.2	103.1	95.3
6	INTERIOR CONSTRUCTION	106.4	105.3	105.8	109.5	105.2	107.2	109.6	106.1	107.7	105.4	102.4	103.8	97.4	101.0	99.4
7	CONVEYING	100.0	105.9	101.9	100.0	105.9	101.9	100.0	105.9	101.9	100.0	105.9	101.9	100.0	105.1	101.6
8	MECHANICAL	101.9	103.2	102.6	102.4	103.2	102.8	101.3	107.8	104.6	100.6	97.9	99.2	100.0	105.3	102.7
9	ELECTRICAL	93.7	105.4	101.9	97.7	104.9	102.7	95.5	106.2	103.0	92.5	102.5	99.5	106.9	105.2	105.7
11	EQUIPMENT	100.0	105.3	101.4	100.0	104.8	101.3	100.0	106.1	101.6	100.0	102.4	100.6	100.0	105.1	101.3
12	SITEWORK	120.1	99.5	110.7	126.0	101.6	114.9	126.9	102.2	115.6	108.8	100.5	105.1	118.7	101.7	111.0
1-12	WEIGHTED AVERAGE	101.4	104.4	103.0	102.1	104.1	103.2	104.2	105.4	104.9	100.9	101.6	101.3	98.9	103.9	101.6

DIV. NO.	BUILDING SYSTEMS	WASHINGTON, D.C.			FT LAUDERDALE, FL			JACKSONVILLE, FL			MIAMI, FL			ORLANDO, FL		
		MAT.	INST.	TOTAL	MAT.	INST.	TOTAL	MAT.	INST.	TOTAL	MAT.	INST.	TOTAL	MAT.	INST.	TOTAL
1-2	FOUND/SUBSTRUCTURES	105.3	87.9	94.6	96.8	92.7	94.3	96.6	82.0	87.5	94.7	94.1	94.3	97.4	83.2	88.6
3	SUPERSTRUCTURES	104.1	87.8	95.6	92.7	91.3	92.0	95.8	81.0	88.1	92.2	92.4	92.3	92.4	81.6	86.8
4	EXTERIOR CLOSURE	99.3	94.5	96.2	90.7	88.8	89.5	88.0	73.5	78.7	89.6	83.2	85.5	88.8	64.2	73.0
5	ROOFING	105.9	90.3	100.6	89.2	88.2	88.9	88.3	71.4	82.5	87.9	83.1	86.3	88.1	67.7	81.1
6	INTERIOR CONSTRUCTION	108.9	91.7	99.5	101.2	82.7	91.1	103.0	74.3	87.3	103.2	80.5	90.8	100.8	71.9	85.0
7	CONVEYING	100.0	91.9	97.4	100.0	85.7	95.4	100.0	74.5	91.8	100.0	84.6	95.0	100.0	73.6	91.5
8	MECHANICAL	101.6	84.7	92.9	100.7	85.8	93.0	100.0	78.8	89.1	97.6	89.4	93.4	97.0	78.3	87.4
9	ELECTRICAL	97.4	87.8	90.6	99.3	85.3	89.5	100.1	73.6	81.5	100.5	95.4	96.9	93.4	73.7	79.5
11	EQUIPMENT	100.0	90.5	97.4	100.0	85.7	96.1	100.0	74.5	93.1	100.0	84.6	95.8	100.0	73.6	92.9
12	SITEWORK	88.5	93.0	90.6	110.4	91.7	101.9	121.7	81.7	103.5	98.8	86.3	93.1	97.7	89.6	94.0
1-12	WEIGHTED AVERAGE	102.0	89.4	95.2	97.5	87.9	92.3	98.4	77.1	86.8	96.1	88.4	92.0	95.4	75.3	84.5

DIV. NO.	BUILDING SYSTEMS	TALLAHASSEE, FL			TAMPA, FL			ALBANY, GA			ATLANTA, GA			COLUMBUS, GA		
		MAT.	INST.	TOTAL	MAT.	INST.	TOTAL	MAT.	INST.	TOTAL	MAT.	INST.	TOTAL	MAT.	INST.	TOTAL
1-2	FOUND/SUBSTRUCTURES	106.2	75.0	86.9	99.4	96.3	97.5	107.0	81.6	91.3	88.8	84.5	86.1	105.2	79.3	89.2
3	SUPERSTRUCTURES	97.4	73.2	84.8	99.2	93.3	96.1	104.2	79.6	91.3	99.6	82.3	90.6	99.9	77.8	88.4
4	EXTERIOR CLOSURE	88.2	66.3	74.1	97.7	78.6	85.5	85.8	70.0	75.6	91.8	77.6	82.7	89.6	54.4	67.0
5	ROOFING	87.7	59.9	78.2	104.3	69.2	92.3	92.8	68.9	84.6	97.1	71.1	88.2	93.3	66.6	84.1
6	INTERIOR CONSTRUCTION	99.4	60.0	77.9	98.5	79.4	88.1	92.9	70.1	80.5	106.1	82.3	93.1	93.9	62.9	77.0
7	CONVEYING	100.0	67.9	89.7	100.0	79.3	93.3	100.0	78.9	93.2	100.0	78.9	93.2	100.0	78.9	93.2
8	MECHANICAL	99.5	69.2	83.9	97.2	80.1	88.4	100.7	72.4	86.1	102.6	78.7	90.3	98.3	68.5	83.0
9	ELECTRICAL	92.1	69.2	76.0	93.4	80.9	84.6	102.4	70.8	80.2	95.7	80.3	84.9	97.7	61.6	72.4
11	EQUIPMENT	100.0	67.9	91.4	100.0	79.3	94.4	100.0	70.0	91.9	100.0	78.1	94.1	100.0	64.9	90.6
12	SITEWORK	121.0	85.7	104.9	112.3	91.9	103.0	120.1	88.9	105.9	107.0	92.2	100.3	126.2	88.3	108.9
1-12	WEIGHTED AVERAGE	98.4	69.2	82.6	99.2	84.0	91.0	99.7	74.4	86.0	99.2	80.7	89.2	98.9	68.0	82.2

DIV. NO.	BUILDING SYSTEMS	MACON, GA			SAVANNAH, GA			HONOLULU, HI			CEDAR RAPIDS, IA			DES MOINES, IA		
		MAT.	INST.	TOTAL	MAT.	INST.	TOTAL	MAT.	INST.	TOTAL	MAT.	INST.	TOTAL	MAT.	INST.	TOTAL
1-2	FOUND/SUBSTRUCTURES	98.3	85.8	90.6	98.6	88.0	92.0	118.3	107.4	111.6	106.7	87.7	94.9	111.8	88.2	97.2
3	SUPERSTRUCTURES	93.7	83.5	88.4	99.7	85.1	92.1	114.8	107.2	110.9	98.9	85.8	92.1	99.6	87.1	93.1
4	EXTERIOR CLOSURE	89.8	69.3	76.6	89.3	70.7	77.4	117.7	115.5	116.3	95.8	79.4	85.2	93.5	82.0	86.1
5	ROOFING	93.0	68.1	84.4	90.7	71.3	84.0	109.5	114.5	111.2	89.9	80.4	86.6	89.1	82.0	86.6
6	INTERIOR CONSTRUCTION	93.5	69.4	80.4	106.9	70.7	87.1	134.9	116.8	125.1	106.1	80.3	92.1	105.4	82.0	92.6
7	CONVEYING	100.0	78.9	93.2	100.0	76.2	92.3	100.0	116.9	105.4	100.0	78.3	93.0	100.0	84.2	94.9
8	MECHANICAL	98.6	71.8	84.8	98.4	72.2	84.9	111.3	114.7	113.0	99.6	80.4	89.7	96.6	82.1	89.1
9	ELECTRICAL	108.5	70.8	82.0	101.8	76.7	84.2	106.2	105.5	105.7	97.6	80.4	85.5	99.4	82.0	87.2
11	EQUIPMENT	100.0	69.3	91.7	100.0	70.6	92.1	100.0	115.8	104.2	100.0	80.3	94.7	100.0	81.9	95.1
12	SITEWORK	119.2	88.1	105.1	121.9	87.2	106.1	135.6	105.9	122.1	103.3	92.6	98.4	102.5	92.4	97.9
1-12	WEIGHTED AVERAGE	97.6	75.1	85.4	99.8	76.8	87.4	115.6	111.7	113.5	100.0	82.3	90.4	99.6	84.0	91.1

DIV. NO.	BUILDING SYSTEMS	DAVENPORT, IA			SIOUX CITY, IA			WATERLOO, IA			BOISE, ID			POCATELLO, ID		
		MAT.	INST.	TOTAL	MAT.	INST.	TOTAL	MAT.	INST.	TOTAL	MAT.	INST.	TOTAL	MAT.	INST.	TOTAL
1-2	FOUND/SUBSTRUCTURES	95.5	94.4	94.8	100.3	82.2	89.1	103.8	84.1	91.6	102.9	92.8	96.7	109.7	92.7	99.2
3	SUPERSTRUCTURES	93.1	93.9	93.5	95.5	80.2	87.5	106.3	82.6	94.0	98.0	92.0	94.8	100.6	91.9	96.1
4	EXTERIOR CLOSURE	101.2	90.1	94.0	98.2	70.8	80.6	102.5	73.3	83.8	104.6	86.8	93.2	108.3	88.1	95.4
5	ROOFING	90.4	90.1	90.3	87.4	70.8	81.7	87.7	73.2	82.7	105.5	88.3	99.6	103.6	88.1	98.3
6	INTERIOR CONSTRUCTION	102.5	90.0	95.6	98.3	70.8	83.3	110.5	73.2	90.1	98.8	88.3	93.0	103.5	88.2	95.2
7	CONVEYING	100.0	78.3	93.0	100.0	73.3	91.4	100.0	78.3	93.0	100.0	87.5	96.0	100.0	87.5	96.0
8	MECHANICAL	99.8	90.1	94.8	97.9	71.0	84.0	101.1	73.3	86.7	96.6	88.4	92.4	101.2	88.3	94.6
9	ELECTRICAL	98.1	90.1	92.4	99.7	81.6	87.0	93.0	73.2	79.1	96.7	88.3	90.8	93.7	88.2	89.8
11	EQUIPMENT	100.0	90.0	97.3	100.0	70.7	92.1	100.0	73.1	92.8	100.0	88.2	96.8	100.0	88.1	96.8
12	SITEWORK	110.4	88.4	100.4	115.9	89.9	104.1	113.1	90.2	102.7	92.8	95.8	94.2	102.8	95.0	99.3
1-12	WEIGHTED AVERAGE	98.8	90.8	94.4	98.5	75.8	86.2	102.9	76.8	88.8	99.3	89.4	93.9	102.5	89.6	95.5

DIV. NO.	BUILDING SYSTEMS	CHICAGO, IL			DECATUR, IL			JOLIET, IL			PEORIA, IL			ROCKFORD, IL		
		MAT.	INST.	TOTAL	MAT.	INST.	TOTAL	MAT.	INST.	TOTAL	MAT.	INST.	TOTAL	MAT.	INST.	TOTAL
1-2	FOUND/SUBSTRUCTURES	98.8	102.4	101.0	103.4	95.2	98.3	100.8	102.1	101.6	97.5	88.1	91.7	105.2	95.5	99.2
3	SUPERSTRUCTURES	93.8	103.7	99.0	101.0	94.7	97.7	100.5	102.6	101.6	94.6	88.7	91.5	106.8	95.1	100.7
4	EXTERIOR CLOSURE	101.3	104.3	103.2	101.2	92.9	95.9	105.5	100.8	102.5	94.5	94.5	94.5	98.3	98.5	98.4
5	ROOFING	97.5	109.1	101.5	100.8	94.5	98.6	102.0	106.2	103.5	95.6	100.8	97.4	107.3	96.4	103.6
6	INTERIOR CONSTRUCTION	99.3	100.5	99.9	107.3	92.0	99.0	104.2	102.7	103.4	105.9	94.3	99.6	99.8	91.2	95.1
7	CONVEYING	100.0	103.9	101.2	100.0	88.7	96.3	100.0	103.9	101.2	100.0	86.3	95.6	100.0	100.2	100.0
8	MECHANICAL	96.4	96.5	96.4	95.6	93.1	94.3	97.2	95.0	96.1	96.1	92.4	94.2	101.7	99.7	100.7
9	ELECTRICAL	93.9	101.3	99.1	94.2	92.5	93.0	102.5	101.2	101.6	94.6	94.3	94.4	93.4	100.3	98.2
11	EQUIPMENT	100.0	102.6	100.7	100.0	92.4	97.9	100.0	101.1	100.3	100.0	94.2	98.4	100.0	100.2	100.0
12	SITEWORK	106.6	104.8	105.7	112.4	95.3	104.6	113.4	102.2	108.3	119.3	98.9	110.0	115.5	101.8	109.3
1-12	WEIGHTED AVERAGE	98.0	101.8	100.0	100.9	93.3	96.8	101.8	100.7	101.2	98.5	92.4	95.2	102.5	97.3	99.7

479

Figure 4.4

The amount of overhead and profit included in the Means prices is based on a moderate level of competition. To adjust for a relatively low level of competition, 3% to 5% should be added to project costs; for a high level of competition, 3% to 5% should be deducted. The leading indicator of construction volume, the level of business activity, is high at the time of the estimate, and shows every indication of remaining so. This suggests that construction volume will remain high, causing bid competition to be low. It is judged that an additional 3% will be a sufficient adjustment.

The adjustments for location, inflation, and bid conditions are as follows.

Total Project Cost	$2,975,860
Location Adjustment	× .892
Total Project Costs in Chambers, Ga.	**$2,654,467**
Inflation Adjustment 1/88 to 1/89	× 1.045
and 1/89 to 6/89	× 1.0225
Bid Conditions Adjustment	× 1.03
Adjusted Total Project Cost	**$2,921,421**

A square foot estimate has an accuracy range of around ±15%. The actual costs of the project could run between $2,921,421 +15% to $2,921,421 − 15%. This cost range should be calculated and used in the feasibility studies. The possible project cost range to be used for planning is determined as follows.

$2,921,421 + 15%	to	$2,921,421 − 15%
$3,359,634	to	$2,483,207
or		
$3,360,000	to	$2,483,000

Per Unit Cost

Project size is often not known in the early planning stages. However, a maximum or optimum amount that will provide an adequate return on investment may be known. Or, a maximum amount of available funding for a construction project may be known. In this case, the need is for a *per-unit total project cost* to determine an approximate size (facility) that can be built for the money. To give an example of this process, we will assume that the Chambers group has determined that $2,500,000 is available for the project. The same square foot cost data that was used to prepare the total project cost estimate is used for the per-unit cost estimate. Just as with the total project cost estimate, the cost of the interior finishes must be removed; and design, furniture, equipment, financing, and planning costs must be added. The project costs must be adjusted for location, inflation, and bid conditions. A possible project size range can then be calculated. Because project size is not known, a size adjustment cannot be made.

The same square foot cost as used in the total project cost estimate (the mid-range cost from line 171-610-0010 shown in Figure 4.1) is used for the *per-unit cost* estimate. Once again, the cost for interior finishes is removed in the same way. The square foot costs for site work, equipment, HVAC, plumbing, and electrical are deducted from the total project square foot costs. One fourth of the remainder is then subtracted from the total project square foot costs to determine the adjusted construction costs. The computations are as follows:

Total Building Costs	$55.40/S.F.
Less total Mechanical, Electrical, and Plumbing	11.15
Less equipment	1.01
Less site work	5.35
General Construction Cost	**$37.89/S.F.**

Of the $37.89, approximately one fourth is for interior finishes. Therefore:

Total Building Costs	$55.40/S.F.
Less interior finishes ($37.89 × .25)	9.47
Adjusted Building Costs	**$45.93/S.F.**

Next, the design, financing, and planning costs are estimated and added. Just as for the total project cost estimate, 6.7% is again used for design fees, 2% is used for planning costs, and a financing cost is obtained from a local loan official. This time, however, the percentages are applied to the adjusted square foot cost rather than to a total building cost. The estimate is as follows.

Adjusted Building Cost	$45.93/S.F.
Design Fees at 6.7%	3.08
Total Construction Cost	**$49.01/S.F.**
Approximate Finance Costs	8.33
Planning Costs at 2%	.98
Subtotal	**$58.32/S.F.**

Next, adjustments are made for location, inflation, and bid conditions. Again, the same percentages (as for the total project cost estimate) are used, but are now applied to the square foot price from above. These adjustments are as follows:

Subtotal	$58.32/S.F.
Locality Adjustment	× .892
Total Project Costs	**$52.02/S.F.**
Inflation Adjustment	
for 1/88 to 1/89	× 1.045
and 1/89 to 6/89	× 1.0225
Bid Conditions Adjustment	× 1.03
Adjusted Total Project Cost	**$57.25/S.F.**

The possible cost range is determined as follows:

$57.25/S.F. + 15% to $57.25/S.F. − 15%
or,
$65.84/S.F. to $48.66/S.F.

Finally, using this range of prices, a possible project size range may be calculated.

$2,500,000/$65.84/S.F. to $2,500,000/$48.66/S.F.
or,
37,971 S.F. to 51,377 S.F.

Summary

An effective cost control system must begin in the Need Identification Phase with realistic project cost estimates to be used in feasibility studies. The challenge is that at this phase, only the most basic information is available about the project. Further, the estimator performing this early estimate often has little training or experience in construction estimating. The Order of Magnitude and Square Foot/Cubic Foot methods are appropriate for these very early estimates. These methods are well suited to preliminary feasibility studies, as they provide a *total project cost range* or a *per-unit total project cost range* to determine the project range possible within a set budget. Need Identification estimates must include building costs, site development costs, design fees, and any furniture and equipment costs, as well as financing and planning costs. In addition, the estimate of total project costs must be adjusted for project size, project location, inflation, and bid conditions.

5

PROJECT PROGRAMMING

5

PROJECT PROGRAMMING

In the last chapter examples are given of the kinds of estimates required for the Need Identification Phase. These examples are based on a hypothetical project located in Chambers, Georgia. A total project cost estimate determined that a low-rise office building containing approximately 56,700 square feet could be built in Chambers for a cost in the $2,483,000 to $3,360,000 range. Preliminary feasibility studies determined that this project would provide an adequate return on investment. Accordingly, the investors in Chambers have decided to continue with the next step in the construction process, the *Need Analysis Phase*.

In the Need Analysis Phase, thorough studies are made to determine project and site requirements and limitations. The products of these studies include:

- general background information about the owner
- general building and zoning code requirements
- general design requirements of the owner
- a detailed listing of the spaces required in the facility
- a list of the furniture and equipment required for each space
- project schedule requirements
- a refined cost estimate, and
- detailed site drawings showing property lines, easements, topography, and existing improvements.

This information is incorporated into a written document, the *project program*. The project program is used for a number of purposes. The program budget, or refined cost estimate, is used in the final feasibility study to determine whether or not the project is indeed economically feasible. The program is also part of the information that is used to secure project funding. In addition, the program is made part of a design contract to define project design requirements and the project budget. It is particularly important that the project design requirements and budget limitations are clearly stated in the design contract. This statement of project scope and budget is the agreed upon basis for the eventual project design and provides the limits used in design cost control. The project program, essential to the cost control process, is further explained in the following paragraphs and an example is shown.

Traditionally, the program was developed informally by the architect as part of the design service. An architect or other design professional is well suited to the task of writing the program because some knowledge of design procedures and construction is necessary to do a complete and accurate job. Recently, however, the trend has been for someone other than the project architect to write the program. Large organizations with substantial real estate holdings often have their own facilities management departments which are capable of providing this service. Smaller firms may have construction management firms do the programming as part of a project management service.

Project programming may be seen as one of the more challenging tasks in the construction process. Effective communication is a vital, but potentially difficult part of this task. This challenge rests primarily with the design professional, who must be able to understand the user's needs—as expressed in layman's terms—and convert those needs into language that is meaningful to the project architect. The professional who creates the program must also educate the user on the following points: what will occur in the process, why it costs so much, and what to expect in the end product.

As stated in Chapter 1, a program includes a brief company and project background statement, a brief description of the project, the basic design standards to be followed, a detailed listing and description of the required spaces, a milestone schedule for the project, and the project budget. The program for the Chambers project is as follows.

Introductory Statement

The introductory statement should include a brief outline of the history, development, and goals of the company, and the reason for the project. This background information gives the designer a better understanding of the company and the project. An example for the Chambers project is as follows.

Introduction

Chambers Investment, Inc. was formed to answer the perceived need for quality professional office space in Chambers, Georgia. The corporation is comprised of professionals, residents of Chambers, who recognize the potential of the city. Chambers is located on a major interstate highway within minutes of the business center of Atlanta. It is near the regional airport, a high tech industrial development, and regional colleges, universities, and recreational facilities. Because Chambers is already a growing community populated by young professionals who commute into Atlanta, it was reasoned that an office development in Chambers would be a very good prospect.

Project Description

This section provides a brief description of the project, the location, and any initial concepts about materials and layout. For the Chambers project:

Project Description

The project is to be a building containing approximately 56,000 gross square feet, partially finished, with HVAC, plumbing, and electrical systems in place. The final design will convey an image attractive to growing professional firms. The project is to be located on a 2.6 acre tract on A Street near the intersection of I85 and A Street. See Exhibit A for the site plan. Some covered parking is desirable.

Note that the total required building area is given in *gross square feet*, calculated using the outside dimensions of the building. It includes the space occupied by walls, rest rooms, corridors, janitor's closets, chases, and equipment rooms, as well as the floor area usable for the building's intended function.

General Requirements

The General Requirements include a complete listing of design standards. Among these standards are references to applicable building codes and insurance standards. Also included are the standards and general requirements for site development, interior and exterior finishes, and structural, mechanical, plumbing, electrical, and communication systems. Attention must also be clearly drawn to the need for life cycle cost studies appropriate for the type of construction. Different basic assumptions are used for the maintenance of commercial, as compared to government and institutional, facilities. Maintenance is a deductible business expense for commercial facilities, and a budget appropriation line item for government facilities. In the latter case, it may be to the user's advantage to choose more expensive finishes and systems that require less maintenance than those normally used in commercial construction. This difference must be clearly stated. General Requirements for the Chambers project are as follows.

General Requirements

CODES: The facility will be designed to comply with the Standard Building Code as modified by the building code of the City of Chambers, Georgia, and by the Life Safety Code of the National Fire Underwriters Association.

SITE: The site shall contain hard surface parking to meet city building code requirements. Approximately 10% of the parking shall be covered and have immediate access to the building. Access to A Street shall be coordinated with and meet the standards of the city street department. A screened, concrete paved area will be provided for trash container with a heavy duty, paved access route from A Street. Sidewalks will be provided as necessary for circulation from the street and parking to the building. The designer will coordinate his design with the owner's landscape consultant and provide utility hookups for the lawn sprinkler system.

STRUCTURE: The designer will provide a structure adequate to meet code live load, dead load, and wind load requirements as economically as possible. The owner will pay for the necessary soil testing.

FINISHES:	The choices for both interior and exterior finishes shall be the most economical and durable available in keeping with the nature and image of the building. Life cycle cost studies will be performed to aid in the selection. The interior lease spaces will be finished by the tenants. Rest room facilities and lobbies will be finished as part of this project.
SPECIAL REQUIREMENTS:	The guidelines established by the state of Georgia for handicapped access will be followed.
MECHANICAL:	Complete HVAC and plumbing systems will be provided as part of this project. Based on life cycle cost studies, the most practical and economical design possible will be provided in accordance with the National Plumbing Code, ASHRAE design standards, and local building codes. The design will incorporate the latest energy conservation technology economically feasible.
ELECTRICAL:	A complete electrical system shall be furnished including service, transformation, power distribution, wiring devices, lighting, "clean" circuits for computers, and conduit and outlets for communication systems. The design shall incorporate the latest engineering technology economically feasible in accordance with the National Electric Code, local building codes, and local power company standards.

Requested Facilities

The Requested Facilities section gives a complete listing of all the required spaces and the requirements for each space. The information should include the name, function, size, occupancy, finish requirements, utility requirements, and furniture and equipment needs for each space. For the Chambers office building, little need be said except for lobbies, rest rooms, and lease space. The individual listings are as follows:

Requested Facilities

SPACE:	Entrance and elevator lobbies.
SIZE:	To meet code requirements.
PURPOSE:	To provide space for circulation, display, and waiting.
ACCESS:	From building exterior and covered parking to elevators, tenant spaces, and rest room facilities.
FINISHES:	All finishes to be durable, easily maintained, and in keeping with character of building. Ceiling: Modular metal panels and/or high quality lay-in where appropriate. Walls: Durable materials such as stone veneer and vinyl covering shall be investigated. Flooring: Durable materials such as quarry tile and brick will be investigated.

UTILITIES:	Water fountains, general lighting, emergency lighting, convenience outlets, pay telephones, comfort conditioned, fire sprinklers per code requirements.
FURNITURE:	Settees, cigarette urns.
EQUIPMENT:	None.

Requested Facilities

SPACE:	Tenant Space.
SIZE:	Design to achieve highest efficiency possible for maximum amount of lease space.
PURPOSE:	Lease space.
ACCESS:	Direct access to lobbies, indirect access to rest rooms.
FINISHES:	Unfinished. Ceiling: Acoustical lay-in to support lighting. Flooring: Unfinished.
UTILITIES:	General lighting, emergency lighting per code, convenience outlets, comfort conditioned, fire sprinklers per code.
FURNITURE:	None.
EQUIPMENT:	None.

Requested Facilities

SPACE:	Rest rooms.
SIZE:	To meet code requirements.
PURPOSE:	To provide rest room facilities for building.
ACCESS:	Direct access to lobbies, indirect access to tenant spaces.
FINISHES:	All finishes to be durable, easily maintained, and in keeping with character of building. Ceiling: Moisture resistant. Walls: Durable materials such as ceramic tile and vinyl wall covering shall be investigated. Flooring: Durable materials such as ceramic tile and vinyl composition tile shall be investigated.
UTILITIES:	Plumbing, ventilation, and lighting to per code requirements, comfort conditioned.
FURNITURE:	None.
EQUIPMENT:	None.

Note that some finishes are included in the Requested Facilities section. Further research by the Chambers group determined that certain spaces, such as lobbies and rest rooms, would have to be finished to provide a usable facility for the tenants. These kinds of discoveries— and resulting changes—are to be expected in the planning process. Such requirements are included in the Requested Facilities and their impact on project cost evaluated when preparing the Project Budget.

Each space description usually includes a size requirement in *net square feet*. It should be remembered that *net square feet* includes only the usable floor area required for the intended function of the space, and excludes the floor area occupied by walls, corridors, rest rooms, et cetera. Generally, a program contains a summary of the required spaces with a total of the net square feet required for the building. This total is multiplied by a floor area ratio to determine the gross square feet required for a building of that category to house the net square feet listed in the program. Figure 5.1 from the 1988 edition of *Means Assemblies Cost Data* provides the floor area ratios for a number of building types. The percentages given in the *gross to net ratio* are used to convert net area to gross area. Assuming that our office building had a total of 42,000 net square feet listed in the program, the gross building area would be calculated as follows:

$$42{,}000 \text{ N.S.F.} \times 1.35 \text{ G.S.F./N.S.F.} = 56{,}700 \text{ G.S.F.}$$

The floor area requirements are not given for the Chambers project because all of the areas other than the tenant space are determined by building codes. A note in the tenant space requirements specifies that the designer furnish the highest efficiency, or maximum net to gross ratio, possible in the building design.

REFERENCE AIDS | **C15.2-100** | **Area Requirements**

(100)

Table 15.2-101 Floor Area Ratios: commonly used gross to net area and net to gross area ratios expressed in % for various building types.

Building Type	Gross to Net Ratio	Net to Gross Ratio	Building Type	Gross to Net Ratio	Net to Gross Ratio
Apartment	156	64	School Buildings (campus type)		
Bank	140	72	Administrative	150	67
Church	142	70	Auditorium	142	70
Courthouse	162	61	Biology	161	62
Department Store	123	81	Chemistry	170	59
Garage	118	85	Classroom	152	66
Hospital	183	55	Dining Hall	138	72
Hotel	158	63	Dormitory	154	65
Laboratory	171	58	Engineering	164	61
Library	132	76	Fraternity	160	63
Office	135	75	Gymnasium	142	70
Restaurant	141	70	Science	167	60
Warehouse	108	93	Service	120	83
			Student Union	172	59

The gross area of a building is the total floor area based on outside dimensions.
The net area of a building is the usable floor area for the function intended and excludes such items as stairways, corridors and mechanical rooms. In the case of a commercial building, it might be considered as the "leasable area."

Figure 5.1

Project Schedule

A brief milestone schedule should also be included to give the designer some idea as to the time frame of the project. The items covered should include:

- a final presentation of the schematic design
- several milestone points in the design development phase when a brief review and a check estimate are due
- a date when the complete plans and specifications are due
- a bid date
- a construction completion date

It should be stated in the design contract that time is of the essence and that the schedule must be met—with penalties accrued for delays. Assuming a mid-1988 date for the program, a schedule for the Chambers office building is as follows.

Project Schedule	
Final Conceptual Design Presentation	October 1, 1988
50% Design Development Review	January 5, 1989
80% Design Development Review	March 1, 1989
Final Design Development Review	April 15, 1989
Bid	June 1, 1989
Construction Completion	March 1, 1990

Project Budget

The purpose of the budget is to establish for the owner the total cost of the project, and to establish for the A/E the total amount that he has to spend on the project. It is not necessary or even desirable show a breakdown of the building cost and site development cost into component elements. Although much more detail may be used in preparing the estimate, it is not needed in the program. The estimate should include, as necessary, the items listed in Chapter 1, beginning with construction contract cost. This estimate will be prepared in the following chapters.

Exhibits

Exhibits provide the designer with additional detailed information about existing conditions or project requirements. The exhibits section should include a schematic location and site plan. It may also include: drawings for specific space layout requirements, such as the desired arrangement of furniture in an office, or the layout of a laboratory showing the desired location of required equipment; or, a schematic showing the flow of work in an assembly line, or the desired arrangement of equipment in an athletic facility. The Chambers program includes a rough site plan as shown in Figure 5.2.

Summary

In the Need Analysis Phase, studies are conducted to determine the requirements and limitations of the construction project. The results of these studies are compiled into a written document, the Project Program. The program contains background information about the owner and the project, a listing of general and specific project requirements, a project schedule, a project budget, and schematic drawings providing additional information about special requirements. The program is used in the final feasibility study, as part of the information used to secure financing, and as the basis for a design contract. The Project Program is a clear statement of design and budget limitations, an essential part of a successful design cost control system.

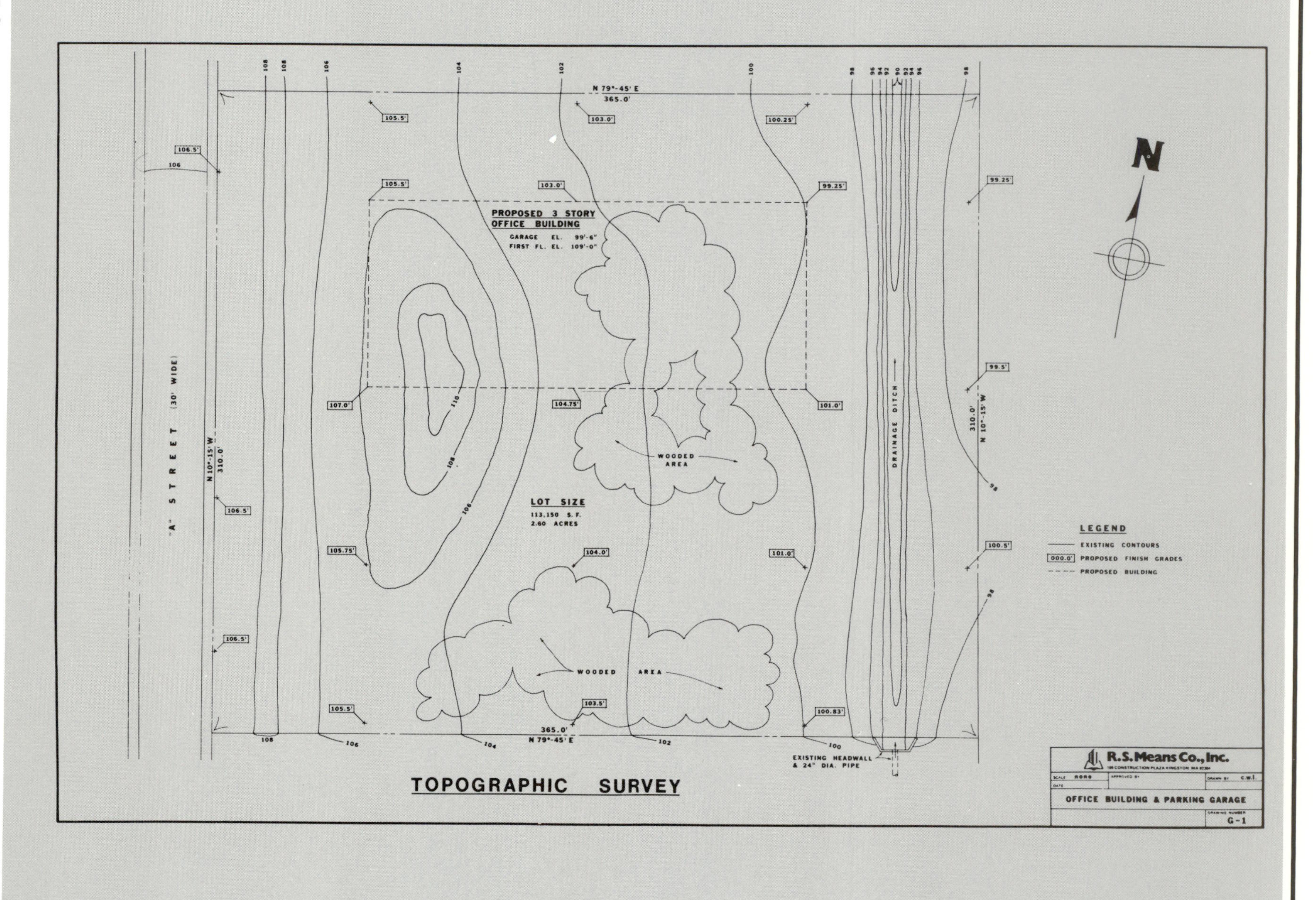
TOPOGRAPHIC SURVEY
PROPOSED 3 STORY
OFFICE BUILDING
GARAGE EL. 99'-6"
FIRST FL. EL. 109'-0"
LOT SIZE
113,150 S.F.
2.60 ACRES
WOODED AREA
WOODED AREA
DRAINAGE DITCH
EXISTING HEADWALL
& 24" DIA. PIPE
"A" STREET (30' WIDE)
N 79°-45' E
365.0'
N 10°-15' W
310.0'
LEGEND
EXISTING CONTOURS
000.0' PROPOSED FINISH GRADES
PROPOSED BUILDING
N
R.S. Means Co., Inc.
OFFICE BUILDING & PARKING GARAGE
G-1

Figure 5.2

6

THE NEED ANALYSIS ESTIMATE

6

THE NEED ANALYSIS ESTIMATE

In the Need Analysis Phase, the results of thorough studies to determine project requirements and limitations were synthesized into a Project Program. The next step in the Need Analysis Phase is to prepare a refined cost estimate, using information from the Project Program. This estimate is used to set a project budget that is adequate to build and furnish the facility, while still providing something of a challenge for the designer.

At this stage, the detailed requirements for the facility—including size, finishes, functions, utility requirements, and site requirements—are known. While there is sufficient information to use the Assemblies/Systems method, the choice of estimating method depends more on the estimator's level of expertise. If the estimator's knowledge of building systems is limited, then the Square Foot method outlined in Chapter 4 should be used. If the estimator is familiar with building systems, a modeling exercise using a combination of the Assemblies and Square Foot methods can be used to provide a more accurate estimate of project costs. Although the Square Foot method often provides an estimate accurate enough to set a realistic budget, there are some situations in which it will not provide that kind of accuracy. The Chambers project represents such a situation. The modeling method combining Assemblies and Square Foot estimating provides the means to identify the specific project conditions, determine the effect that those conditions have on project cost, and provide a more refined estimate, used to set the project budget. This modeling exercise is demonstrated in the following pages.

Modeling

In the modeling exercise, an estimate of project cost is prepared based on a "broad stroke" conceptual design that satisfies program requirements. There are three steps involved in preparing an estimate by modeling. First, the model is designed. Second, an Assemblies/Square Foot estimate is made using the model as a basis. Third, the estimate is adjusted for sales tax, general conditions, overhead and profit, inflation, bid conditions, and location.

Model Design

To create a model for a project one must:

- Determine the area needed for parking, sidewalks, and landscaping, based on code and zoning requirements.
- Subtract the total area needed for parking, sidewalks, and landscaping from the total area of the site in order to determine the maximum building "footprint" area.
- Divide the area of the building by the "footprint" area to determine the number of floors required.
- Set the bay spacing and floor-to-floor heights.
- Choose the structural systems based on local construction practice.
- Perform a rough structural analysis to size members.
- Draw rough typical floor plans.
- Choose exterior closure systems.
- Choose interior finishes.
- Choose equipment and specialty items.
- Choose conveying systems.
- Establish mechanical, electrical, and plumbing pricing criteria.
- Establish site utility needs.

Parking and Landscaping

The area required for parking and landscaping is prescribed by local building code and zoning requirements. The building code usually specifies the occupancy rate (the number of square feet of building per person) and number of parking spaces required per occupant for different types of buildings. Zoning regulations often require that a certain minimum percentage of the site must be landscaped; or, to put it another way, that the building may occupy only a limited portion of the site. For the Chambers project, we will assume that local zoning and building code regulations set the occupancy of low-rise office buildings at 135 gross square feet per person, require .66 parking spaces per occupant, and require that 10% of the site must be landscaped. We will assume that, based on the designer's experience, 2% of the area available for the building and parking will be required for sidewalks. The procedure for determining these requirements is as follows:

1. Determine the total square foot area of the site.
2. Deduct the percentage required for landscaping from the total area to determine the area available for building, parking, and sidewalks.
3. Calculate the area available for parking and building by subtracting the percentage needed for sidewalks from the area available for building, parking, and sidewalks.
4. Determine building occupancy by dividing the building area by the occupancy rate.
5. Calculate the number of parking spaces required by multiplying the building occupancy by the *parking space per occupant* requirement.
6. Calculate the parking area requirements by multiplying the number of parking spaces by a square foot/parking space conversion factor.

The proposed site for the Chambers project shown in Figure 6.1 has a frontage of 310′ and a depth of 365′. The space required for parking and maneuvering is between 300 and 400 square feet per car, depending on design. An average of 350 square feet will be used for this planning exercise. The calculations for parking and landscaping are as follows:

Site Area: 310′ × 365′ =	113,150 S.F.
Landscape: 113,150 S.F. × 10% =	11,315
Area Available for Parking, Sidewalks & Building:	101,835
Sidewalks: 101,835 S.F. × 2% =	2,036
Area Available for Parking & Building	99,799 S.F.
Building Occupancy: 56,700 G.S.F./135 =	420 people
Parking Area Requirements: 420 × .66 =	277.2 cars

278 Cars × 350 S.F./Car = 97,300 S.F. Required for Parking

Building Footprint/Number of Floors

To determine the area remaining for the building footprint, the required parking area is subtracted from the site area. The gross building area is then divided by the footprint area to determine the number of floors required.

99,799 S.F. − 97,300 = 2,499 S.F.

$$\frac{56{,}700 \text{ S.F.}}{2{,}499 \text{ S.F./Floor}} = 22 \text{ Floors}$$

A building 22 stories high for this size project is not feasible. A more usable configuration would be to have some of the required parking under the building. This satisfies the requirement for covered parking, but has a significant impact on project cost. Rather than having a relatively inexpensive slab on grade for a first floor, a more expensive supported system is now required. Also, the slope of the site from the street to the ditch along the east property line suggests a partial basement will be required for parking, thereby adding the expense of basement walls. The configuration does allow much more flexibility in determining footprint size and number of floors. Three floors will be assumed. Footprint size is determined as follows:

$$\frac{56{,}700 \text{ S.F.}}{3} = 18{,}900 \text{ S.F.}$$

One bay will be carried up an additional floor to allow for roof access and an elevator equipment room.

Bay Spacing and Floor-to-Floor Height

To determine bay spacing, various trials are made to find a bay size and a building configuration that will meet project requirements. A building that is nine 25′ bays long and three 28′ bays wide (225′ × 84′) provides the required footprint area for the Chambers project. Therefore, a 25′ × 28′ bay will be used. The choice of floor-to-floor height is a matter of experience. A height of 14′ allows for 10′ ceiling heights with space remaining for floor structure and air conditioning duct.

Structural Systems

The choice of structural system is based on a knowledge of local conditions and construction practice. The systems most commonly used in the local area are generally the most economical for local conditions. In the Chambers area, the average soil bearing capacity is 6 kips/S.F., and spread footings and strip footings are commonly used to support low-rise construction. Basement walls are normally cast-in-place concrete. Concrete floor structures could be cast-in-place concrete multispan joist slabs. A steel joists-on-beams-and-columns system will be used for floors above the first and for roofs.

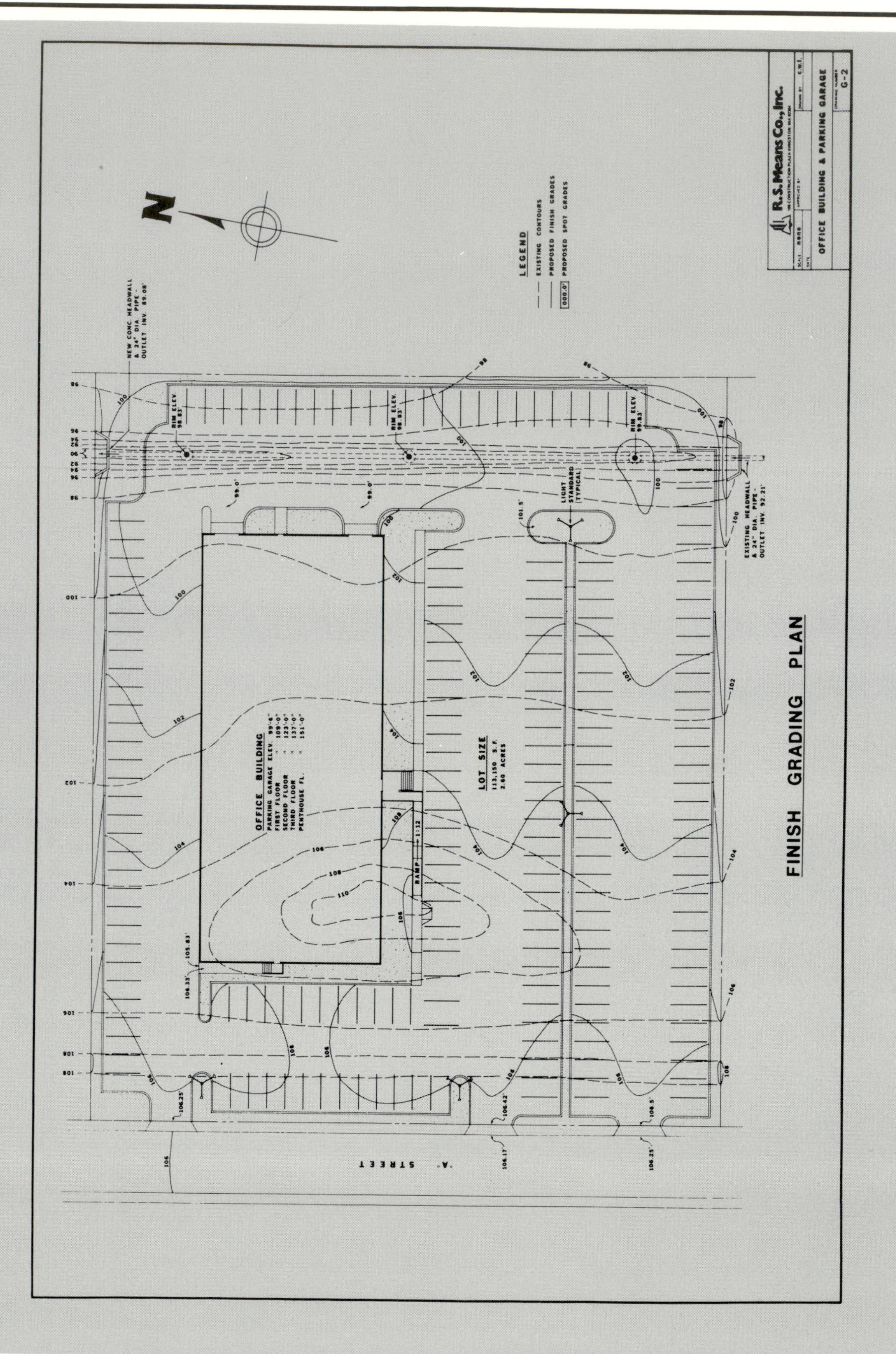
FINISH GRADING PLAN
OFFICE BUILDING
LOT SIZE
113,150 S.F.
2.60 ACRES
LEGEND
EXISTING CONTOURS
PROPOSED FINISH GRADES
PROPOSED SPOT GRADES
R.S. Means Co., Inc.
OFFICE BUILDING & PARKING GARAGE
G-2
N
RAMP
'A' STREET

Figure 6.1

Structural Analysis

A quick and rough structural analysis is needed to size the structural members for estimating. This analysis is for estimating purposes only and is not intended for use in the actual design of the structure. The steps in the structural analysis process are:

1. Determine the required roof and floor live load requirements.
2. Determine the roof and floor structural system weights to meet the live load requirements for the chosen systems and bay spacings from the tables in *Means Assemblies Cost Data*.
3. Perform load calculations to determine column sizes and total loads to foundations.
4. Size the foundation system based on the total loads from the load calculations.

The live load requirements are established by the local building code. They are specified in pounds per square foot for different types of occupancies and different occupancy applications. Examples of these requirements are shown in the Live Load table from *Means Assemblies Cost Data* in Figure 6.2. This table provides the floor and roof load requirements from several of the major standard building codes for a number of building occupancies. We will use the Uniform Building Code (UBC) requirements from the table to determine the load requirements for the Chambers office building. The UBC live load requirement for office buildings is 50 P.S.F., and for flat roofs, 20 P.S.F. The weights of partitions, ceilings, and mechanical equipment must be added to these loads to determine the total superimposed floor load. The weights of ceilings and mechanical equipment must be added to determine the total superimposed roof load.

When using the cost tables in *Means Assemblies Cost Data*, these loads can be approximated by using a superimposed load higher than the required live load. For example, the tables for a steel joists and beams on columns roof structure are shown in Figure 6.3a and b. The superimposed load categories are 20 P.S.F., 30 P.S.F., and 40 P.S.F. The codes require a live load of 20 P.S.F. for the roof. The next highest superimposed load category is 30 P.S.F. Therefore, a 30 P.S.F. superimposed load will be used for the roof. Figure 6.4 gives the tables for a steel joists and beams on columns floor structure. The superimposed load categories given are 40 P.S.F., 65 P.S.F., 75 P.S.F., 100 P.S.F, and 125 P.S.F. The codes require a floor loading of 50 P.S.F. A higher superimposed load category is 75 P.S.F. Therefore, a 75 P.S.F. superimposed load will be used for floors.

Next, the weights for the roof and floor structural systems are taken from tables in *Means Assemblies Cost Data*. The superimposed load requirements for the Chambers office building are 75 pounds per square foot for the building floors and 30 pounds per square foot for the roof. The bay spacing for both is 25′ by 28′. The tables shown in Figures 6.3a and b are typical of Means floor and roof structure tables in that they provide the depth of the structural system, total load (total weight of the floor structure and superimposed load), system price per square foot, and column price per square foot for a number of bay sizes and superimposed loads. The table does not specify values for a 25′ × 28′ bay. For the modeling exercise, however, using the values for the next largest bay size provides an acceptable degree of accuracy.

REFERENCE AIDS | **C15.1-100** | **Live Loads**

Table 15.1-101 Minimum Design Live Loads in Pounds Per S.F. for Various Building Codes

Occupancy	Description	Minimum Live Loads, Pounds Per S.F.				
		BOCA	ANSI	UBC	Chicago	New York
Armories		150	150			150
Assembly	Fixed seats	60	60	50	60	60
	Movable seats	100	100	100	100	100
	Platforms or stage floors	100	100	125	150	
Commercial & Industrial	Light manufacturing	125	125	75	100	100
	Heavy manufacturing	250	250	125	100	100
	Light storage	125	125	75	100	100
	Heavy storage	250	250	100	100	150
	Stores, retail, first floor	100	100	75	100	100
	Stores, retail, upper floors	75	75			75
	Stores, wholesale	125	125	100	100	100
Court rooms		100				
Dance halls	Ballrooms	100	100			100
Dining rooms	Restaurants	100	100			100
Fire escapes	Other than below	100	100			100
	Multi or single family residential	40				40
Garages	Passenger cars only	50	50	50	50	50
Gymnasiums	Main floors and balconies	100	100			100
Hospitals	Operating rooms, laboratories	60	60			60
	Private room	40	40			40
	Wards	40	40			40
	Corridors, above first floor	80	80			
Libraries	Reading rooms	60	60			60
	Stack rooms	150	150	125		
	Corridors, above first floor		80			
Marquees		75	75			
Office Buildings	Offices	50	50	50	50	50
	Lobbies	100	100			100
	Corridors, above first floor	80	*	100	100	75
Residential	Multi family private apartments	40	40	40	40	40
	Multi family, public rooms	100	100			
	Multi family, corridors	80	*	100	100	100
	Dwellings, first floor	40	40	40	40	40
	Dwellings, second floor & habitable attics	30	30			30
	Dwellings, uninhabitable attics	20	20			20
	Hotels, guest rooms	40	40	40	40	40
	Hotels, public rooms	100	100			
	Hotels, corridors serving public rooms	100	100			
	Hotels, corridors	80	80			
Roofs	Flat	12-20		20	25	40
	Pitched	12-16				
Schools	Classrooms	40	40	40	40	40
	Corridors	80	80	100	100	100
Sidewalks	Driveways, etc. subject to trucking	250	250	250		600
Stairs	Exits	100		100	100	100
Theaters	Aisles, corridors and lobbies	100	100			
	Orchestra floors	60	100			
	Balconies	60	100			
	Stage floors	150	150		150	
Yards	Terrace, Pedestrian	100	100			

BOCA = Building Officials & Code Administration International, Inc. National Building Code 1985
ANSI = Standard A58.1-1982
UBC = Uniform Building Code, International Conference of Building Officials, 1967
Chicago = Chicago, Ill., amended to July 1, 1967
New York = New York City, amended to 1970
* Corridor loading equal to occupancy loading.

496

Figure 6.2

ROOFS	B3.7-420	Steel Joists & Beams On Cols

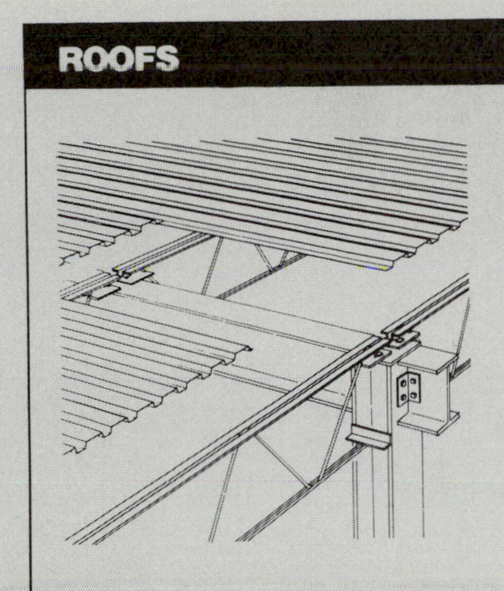

Description: Table 3.7-420 lists the cost per S.F. for a roof system with steel columns, beams, and deck, using open web steel joists and 1-1/2" galvanized metal deck.

Design and Pricing Assumptions:
Columns are 18' high.
Building is 4 bays long by 4 bays wide.
Joists are 5'-0" O.C. and span the long direction of the bay.
Joists at columns have bottom chords extended and are connected to columns.
Column costs are not included but are listed separately per S.F. of floor.

Roof deck is 1-1/2", 22 gauge galvanized steel. Joist cost includes appropriate bridging. Deflection is limited to 1/240 of the span. Fireproofing is not included.

Design Loads	Min.	Max.
Joists & Beams	3 PSF	5 PSF
Deck	2	2
Insulation	3	3
Roofing	6	6
Misc.	6	6
Total Dead Load	20 PSF	22 PSF

System Components	QUANTITY	UNIT	COST PER S.F. MAT.	INST.	TOTAL
SYSTEM 03.4-420-1100					
METAL DECK AND JOISTS,15'X20' BAY,20 PSF S. LOAD					
Structural steel	.954	Lb.	.45	.15	.60
Open web joists	1.260	Lb.	.49	.21	.70
Metal decking, open, galvanized, 1-½" deep, 22 gauge	1.050	S.F.	.70	.27	.97
TOTAL			1.64	.63	2.27

3.7-420 Steel Joists, Beams, & Deck On Columns

	BAY SIZE (FT.)	SUPERIMPOSED LOAD (P.S.F.)	DEPTH (IN.)	TOTAL LOAD (P.S.F.)	COLUMN ADD	COST PER S.F. MAT.	INST.	TOTAL
1100	15x20	20	16	40		1.64	.63	2.27
1200					columns	.73	.25	.98
1300		30	16	50		1.78	.69	2.47
1400					columns	.73	.25	.98
1500		40	18	60		1.82	.71	2.53
1600					columns	.73	.25	.98
1700	20x20	20	16	40		1.74	.67	2.41
1800					columns	.55	.19	.74
1900		30	18	50		1.90	.72	2.62
2000					columns	.55	.19	.74
2100		40	18	60		2.12	.80	2.92
2200					columns	.55	.19	.74
2300	20x25	20	18	40		1.85	.71	2.56
2400					columns	.44	.15	.59
2500		30	18	50		1.97	.84	2.81
2600					columns	.44	.15	.59
2700		40	20	60		2.07	.81	2.88
2800					columns	.58	.20	.78
2900	25x25	20	18	40		2.12	.81	2.93
3000					columns	.35	.12	.47
3100		30	22	50		2.20	.92	3.12
3200					columns	.47	.10	.03
3300		40	20	60		2.43	.92	3.35
3400					columns	.47	.16	.63

For expanded coverage of these items see *Means Concrete Cost Data 1988*

103

Figure 6.3a

ROOFS	B3.7-420	Steel Joists & Beams On Cols

3.7-420 Steel Joists, Beams, & Deck On Columns

	BAY SIZE (FT.)	SUPERIMPOSED LOAD (P.S.F.)	DEPTH (IN.)	TOTAL LOAD (P.S.F.)	COLUMN ADD	COST PER S.F. MAT.	INST.	TOTAL
3500	25x30	20	22	40		2.01	.73	2.74
3600					columns	.39	.13	.52
3700		30	20	50		2.20	.94	3.14
3800					columns	.39	.13	.52
3900		40	25	60		2.40	.87	3.27
4000					columns	.47	.16	.63
4100	30x30	20	25	42		2.26	.81	3.07
4200					columns	.32	.11	.43
4300		30	22	52		2.47	.89	3.36
4400					columns	.39	.13	.52
4500		40	28	62		2.58	.92	3.50
4600					columns	.39	.13	.52
4700	30x35	20	22	42		2.33	.84	3.17
4800					columns	.33	.11	.44
4900		30	28	52		2.46	.88	3.34
5000					columns	.33	.11	.44
5100		40	25	62		2.67	.96	3.63
5200					columns	.39	.13	.52
5300	35x35	20	28	42		2.45	.88	3.33
5400					columns	.28	.10	.38
5500		30	25	52		2.73	.98	3.71
5600					columns	.33	.11	.44
5700		40	28	62		2.93	1.05	3.98
5800					columns	.37	.13	.50

104

For expanded coverage of these items see *Means Concrete Cost Data 1988*

Figure 6.3b

The actual structural design for the Chambers project is not known, but the modeling exercise provides an approximation. In this approximation some leeway is acceptable, and even desirable, in order to provide an adequate budget. The cost difference is small between the calculated sizes and loads and the next size larger unit listed in the Means tables. This difference provides a little "cushion" to cover errors and oversights that may occur in the need analysis estimate.

The next largest bay size from the table in Figure 6.3b for a 30 P.S.F. superimposed load is the 25′ × 30′ bay in line 3700. The 50 P.S.F. total load from this line is used as the roof structure and superimposed load weight in the structural analysis. This procedure can be used to determine the total loads of the floors as well. The weights for a steel joists and beams on columns floor structure are listed in Figures 6.4a and b. The weight of 120 pounds per square foot, total load from line 6050, is used for the second and third floors and the penthouse. One hundred sixty pounds per square foot superimposed load (line 5600 in Figure 6.5b) is used for the weight of the cast-in-place multispan joist slab for the first floor.

Next, load calculations are made for each of the typical column conditions in the building to determine column sizes and total footing loads. There are three typical column conditions:

1) An interior column supporting an area equal to a whole bay
2) A column on the exterior wall supporting an area equal to a half bay
3) A column at an exterior corner supporting an area equal to a quarter bay

The analysis is performed for the interior column, and percentages of the total are used to approximate the total footing loads for the other conditions. The procedure for performing these calculations is shown in Figures 6.6a and b from *Means Assemblies Cost Data*. The total footing load obtained by applying this procedure is multiplied—first by 60% to approximate the exterior footing load, and then by 45% to approximate the corner footing load. These percentages allow for the contribution of the exterior wall loads to the footings.

The procedure listed above assumes that the entire roof is at one level. Since the Chambers building has one bay raised above the rest of the roof to provide roof access and space for elevator equipment, an additional analysis will have to be made. In addition to the analysis to size footings to support the load of a roof and three floors, an analysis must also be made to size the footings that will support a roof and four floors. Figure 6.7 shows the calculations for the four-floor condition, and Figure 6.8 shows the calculations for the three-floor condition. In these figures the weight and load capacities of the steel columns are taken from Figures 6.9a–d. The same information for concrete columns is taken from Figures 6.10a and b. These tables provide the load capacity, unsupported height, weight of the column in pounds per linear foot, and section size for seven types of steel column assemblies. They also list the load capacity, story height, column size, column weight in pounds per linear foot, and concrete strength for square tied concrete columns. In each case, the column chosen has a load capacity equal to or in the next category greater than the calculated load. For example, the column supporting the penthouse level in Figure 6.7 has a calculated load of 56.3K. The Steel Column tables in Figures 6.9a and b show column assemblies to support 50K and 75K. The calculated load exceeding 50K

FLOORS	B3.5-460	Steel Joists & Beams On Cols.

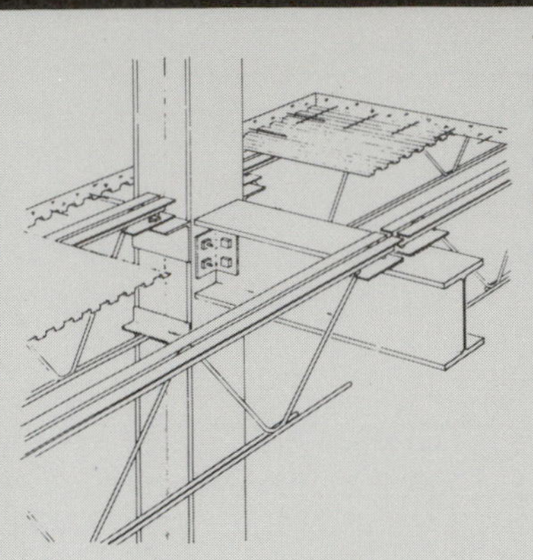

Table 3.5-460 lists costs for a floor system on steel columns and beams using open web steel joists, galvanized steel slab form, and 2-1/2" concrete slab reinforced with welded wire fabric.

Design and Pricing Assumptions:
Structural Steel is A36.
Concrete f'c = 3 KSI placed by pump.
WWF 6 x 6 #10/#10 (W1.4/W1.4).
Columns are 12' high.
Building is 4 bays long by 4 bays wide.
Joists are 2' O.C. ± and span the long direction of the bay.

Joists at columns have bottom chords extended and are connected to columns.

Slab form is 28 gauge galvanized. Columns costs in table are for columns to support 1 floor plus roof loading in a 2-story building; however, column costs are from ground floor to 2nd floor only. Joist costs include appropriate bridging. Deflection is limited to 1/360 of the span. Screeds and steel trowel finish.

Design Loads	Min.	Max.
S.S. & Joists	6.3 PSF	15.3 PSF
Slab Form	1.0	1.0
2-1/2" Concrete	27.0	27.0
Ceiling	3.0	3.0
Misc.	5.7	1.7
	43.0 PSF	48.0 PSF

System Components	QUANTITY	UNIT	COST PER S.F. MAT.	INST.	TOTAL
SYSTEM 03.5-460-2350					
15'X20'BAY 40 PSF S. LOAD, 17" DEPTH, 83 PSF TOTAL LOAD					
Structural steel	1.974	Lb.	.93	.39	1.32
Open web joists	3.140	Lb.	1.22	.53	1.75
Slab form, galvanized steel 9/16" deep, 28 gauge	1.020	S.F.	.43	.23	.66
Welded wire fabric rolls, 6 x 6, #10/10 (w1.4/w1.4) 21 lb/csf	1.000	S.F.	.09	.16	.25
Concrete ready mix, regular weight, 3000 PSI	.210	C.F.	.44		.44
Place and vibrate concrete, elevated slab less than 6", pumped	.210	C.F.		.16	.16
Finishing floor, monolithic steel trowel finish for finish floor	1.000	S.F.		.48	.48
Curing with sprayed membrane curing compound	.010	S.F.	.02	.04	.06
TOTAL			3.13	1.99	5.12

3.5-460 Steel Joists, Beams & Slab On Columns

	BAY SIZE (FT.)	SUPERIMPOSED LOAD (P.S.F.)	DEPTH (IN.)	TOTAL LOAD (P.S.F.)	COLUMN ADD	COST PER S.F. MAT.	INST.	TOTAL
2350	15x20 (18)	40	17	83		3.13	1.99	5.12
2400					column	.44	.19	.63
2450	15x20	65	19	108		3.43	2.12	5.55
2500					column	.44	.19	.63
2550	15x20	75	19	119		3.59	2.19	5.78
2600					column	.49	.21	.70
2650	15x20	100	19	144		3.81	2.29	6.10
2700					column	.49	.21	.70
2750	15x20	125	19	170		4.02	2.58	6.60
2800					column	.65	.28	.93
2850	20x20	40	19	83		3.35	2.09	5.44
2900					column	.36	.15	.51
2950	20x20	65	23	109		3.70	2.24	5.94
3000					column	.49	.21	.70
3100	20x20	75	26	119		3.89	2.32	6.21
3200					column	.49	.21	.70
3400	20x20	100	23	144		4.03	2.38	6.41
3450					column	.49	.21	.70
3500	20x20	125	23	170		4.49	2.58	7.07
3600					column	.58	.25	.83

For expanded coverage of these items see *Means Concrete Cost Data 1988*

Figure 6.4a

FLOORS | **B3.5-460** | **Steel Joists & Beams On Cols.**

3.5-460 | **Steel Joists, Beams & Slab On Columns**

	BAY SIZE (FT.)	SUPERIMPOSED LOAD (P.S.F.)	DEPTH (IN.)	TOTAL LOAD (P.S.F.)	COLUMN ADD	COST PER S.F.		
						MAT.	INST.	TOTAL
3700	20x25	40	44	83		3.64	2.41	6.05
3800					column	.39	.17	.56
3900	20x25	65	26	110		3.97	2.55	6.52
4000					column	.39	.17	.56
4100	20x25	75	26	120		4.13	2.42	6.55
4200					column	.47	.20	.67
4300	20x25	100	26	145		4.37	2.53	6.90
4400					column	.47	.20	.67
4500	20x25	125	29	170		4.90	2.75	7.65
4600					column	.54	.23	.77
4700	25x25	40	23	84		3.90	2.51	6.41
4800					column	.37	.16	.53
4900	25x25	65	29	110		4.12	2.62	6.74
5000					column	.37	.16	.53
5100	25x25	75	26	120		4.54	2.60	7.14
5200					column	.43	.18	.61
5300	25x25	100	29	145		5.10	2.83	7.93
5400					column	.43	.18	.61
5500	25x25	125	32	170		5.35	2.95	8.30
5600					column	.48	.20	.68
5700	25x30	40	29	84		4.01	2.59	6.60
5800					column	.36	.15	.51
5900	25x30	65	29	110		4.18	2.70	6.88
6000					column	.36	.15	.51
6050	25x30	75	29	120		4.53	2.48	7.01
6100					column	.40	.17	.57
6150	25x30	100	29	145		4.91	2.62	7.53
6200					column	.40	.17	.57
6250	25x30	125	32	170		5.20	3.20	8.40
6300					column	.46	.20	.66
6350	30x30	40	29	84		4.23	2.36	6.59
6400					column	.33	.14	.47
6500	30x30	65	29	110		4.80	2.59	7.39
6600					column	.33	.14	.47
6700	30x30	75	32	120		4.89	2.63	7.52
6800					column	.38	.16	.54
6900	30x30	100	35	145		5.45	2.84	8.29
7000					column	.45	.19	.64
7100	30x30	125	35	172		5.85	3.50	9.35
7200					column	.50	.21	.71
7300	30x35	40	29	85		4.77	2.57	7.34
7400					column	.29	.12	.41
7500	30x35	65	29	111		5.25	3.22	8.47
7600					column	.37	.16	.53
7700	30x35	75	32	121		5.25	3.22	8.47
7800					column	.38	.16	.54
7900	30x35	100	35	148		5.75	2.95	8.70
8000					column	.46	.20	.66
8100	30x35	125	38	173		6.35	3.19	9.54
8200					column	.47	.20	.67
8300	35x35	40	32	85		4.89	2.63	7.52
8400					column	.33	.14	.47
8500	35x35	65	35	111		5.50	3.35	8.85
8600					column	.40	.17	.57
9300	35x35	75	38	121		5.65	3.41	9.06
9400					column	.40	.17	.57

For expanded coverage of these items see *Means Concrete Cost Data 1988*

87

Figure 6.4b

FLOORS	B3.5-160	C.I.P. Multispan Joist Slab

General: Combination of thin concrete slab and monolithic ribs at uniform spacing to reduce dead weight and increase rigidity. The ribs (or joists) are arranged parallel in one direction between supports.

Square end joists simplify forming. Tapered ends can increase span or provide for heavy load.

Costs for multiple span joists are provided in this section. Single span joist costs are not provided here.

Design and Pricing Assumptions:
Concrete f'c = 4 KSI, normal weight placed by concrete pump.
Reinforcement, fy = 60 KSI.
Forms, four use.
4-1/2" slab.
30" pans, sq. ends (except for shear req.).
6" rib thickness.
Distribution ribs as required.
Finish, steel trowel.
Curing, spray on membrane.
Based on 4 bay x 4 bay structure.

System Components	QUANTITY	UNIT	COST PER S.F. MAT.	INST.	TOTAL
SYSTEM 03.5-160-2000					
15'X15' BAY, 40 PSF, S. LOAD 12" MIN. COLUMN					
Forms in place, floor slab with 30" metal pans, 4 use	.905	S.F.	.91	2.73	3.64
Forms in place, exterior spandrel, 12" wide, 4 uses	.170	SFCA	.13	.83	.96
Forms in place, interior beam. 12" wide, 4 uses	.095	SFCA	.07	.38	.45
Edge forms, 7"-12" high on elevated slab, 4 uses	.010	L.F.		.03	.03
Reinforcing in place, elevated slabs #4 to #7	.628	Lb.	.19	.13	.32
Concrete ready mix, regular weight, 4000 psi	.555	C.F.	1.19		1.19
Place and vibrate concrete, elevated slab, 6" to 10" pump	.555	C.F.		.35	.35
Finish floor, monolithic steel trowel finish for finish floor	1.000	S.F.		.48	.48
Cure with sprayed membrane curing compound	.010	S.F.	.02	.04	.06
TOTAL			2.51	4.97	7.48

3.5-160	Cast In Place Multispan Joist Slab							
	BAY SIZE (FT.)	SUPERIMPOSED LOAD (P.S.F.)	MINIMUM COLUMN SIZE (IN.)	RIB DEPTH (IN.)	TOTAL LOAD (P.S.F.)	COST PER S.F. MAT.	INST.	TOTAL
2000	15 x 15	40	12	8	115	2.51	4.97	7.48
2100	(17) (18)	75	12	8	150	2.53	4.98	7.51
2200		125	12	8	200	2.60	5.05	7.65
2300		200	14	8	275	2.69	5.20	7.89
2600	15 x 20	40	12	8	115	2.56	4.97	7.53
2800		75	12	8	150	2.63	5.05	7.68
3000		125	14	8	200	2.77	5.35	8.12
3300		200	16	8	275	2.92	5.40	8.32
3600	20 x 20	40	12	10	120	2.63	4.85	7.48
3900		75	14	10	155	2.79	5.15	7.94
4000		125	16	10	205	2.81	5.25	8.06
4100		200	18	10	280	2.99	5.45	8.44
4300	20 x 25	40	12	10	120	2.67	4.92	7.59
4400		75	14	10	155	2.80	5.15	7.95
4500		125	16	10	205	2.98	5.45	8.43
4600		200	18	12	280	3.16	5.70	8.86
4700	25 x 25	40	12	12	125	2.65	4.80	7.45
4800		75	16	12	160	2.86	5.05	7.91
4900		125	18	12	210	3.25	5.60	8.85
5000		200	20	14	291	3.49	5.70	9.19

64

For expanded coverage of these items see *Means Concrete Cost Data 1988*

Figure 6.5a

FLOORS	B3.5-160	C.I.P. Multispan Joist Slab

3.5-160 Cast In Place Multispan Joist Slab

	BAY SIZE (FT.)	SUPERIMPOSED LOAD (P.S.F.)	MINIMUM COLUMN SIZE (IN.)	RIB DEPTH (IN.)	TOTAL LOAD (P.S.F.)	COST PER S.F. MAT.	INST.	TOTAL
5400	25 x 30	40	14	12	125	2.82	5	7.82
5600		75	16	12	160	2.95	5.20	8.15
5800		125	18	12	210	3.21	5.55	8.76
6000		200	20	14	291	3.50	5.80	9.30
6200	30 x 30	40	14	14	131	2.99	5.10	8.09
6400		75	18	14	166	3.11	5.25	8.36
6600		125	20	14	216	3.34	5.50	8.84
6700		200	24	16	297	3.62	5.70	9.32
6900	30 x 35	40	16	14	131	3.09	5.30	8.39
7000		75	18	14	166	3.19	5.35	8.54
7100		125	22	14	216	3.32	5.60	8.92
7200		200	26	16	297	3.72	5.90	9.62
7400	35 x 35	40	16	16	137	3.19	5.25	8.44
7500		75	20	16	172	3.41	5.45	8.86
7600		125	24	16	222	3.45	5.45	8.90
7700		200	26	20	309	3.83	5.80	9.63
8000	35 x 40	40	18	16	137	3.33	5.45	8.78
8100		75	22	16	172	3.47	5.65	9.12
8300		125	26	16	222	3.55	5.60	9.15
8400		200	30	20	309	3.96	5.85	9.81
8750	40 x 40	40	18	20	149	3.59	5.40	8.99
8800		75	24	20	184	3.75	5.55	9.30
8900		125	26	20	234	3.94	5.75	9.69
9100	40 x 45	40	20	20	149	3.78	5.60	9.38
9500		75	24	20	184	3.83	5.65	9.48
9800		125	28	20	234	3.99	5.85	9.84

For expanded coverage of these items see *Means Concrete Cost Data 1988*

65

Figure 6.5b

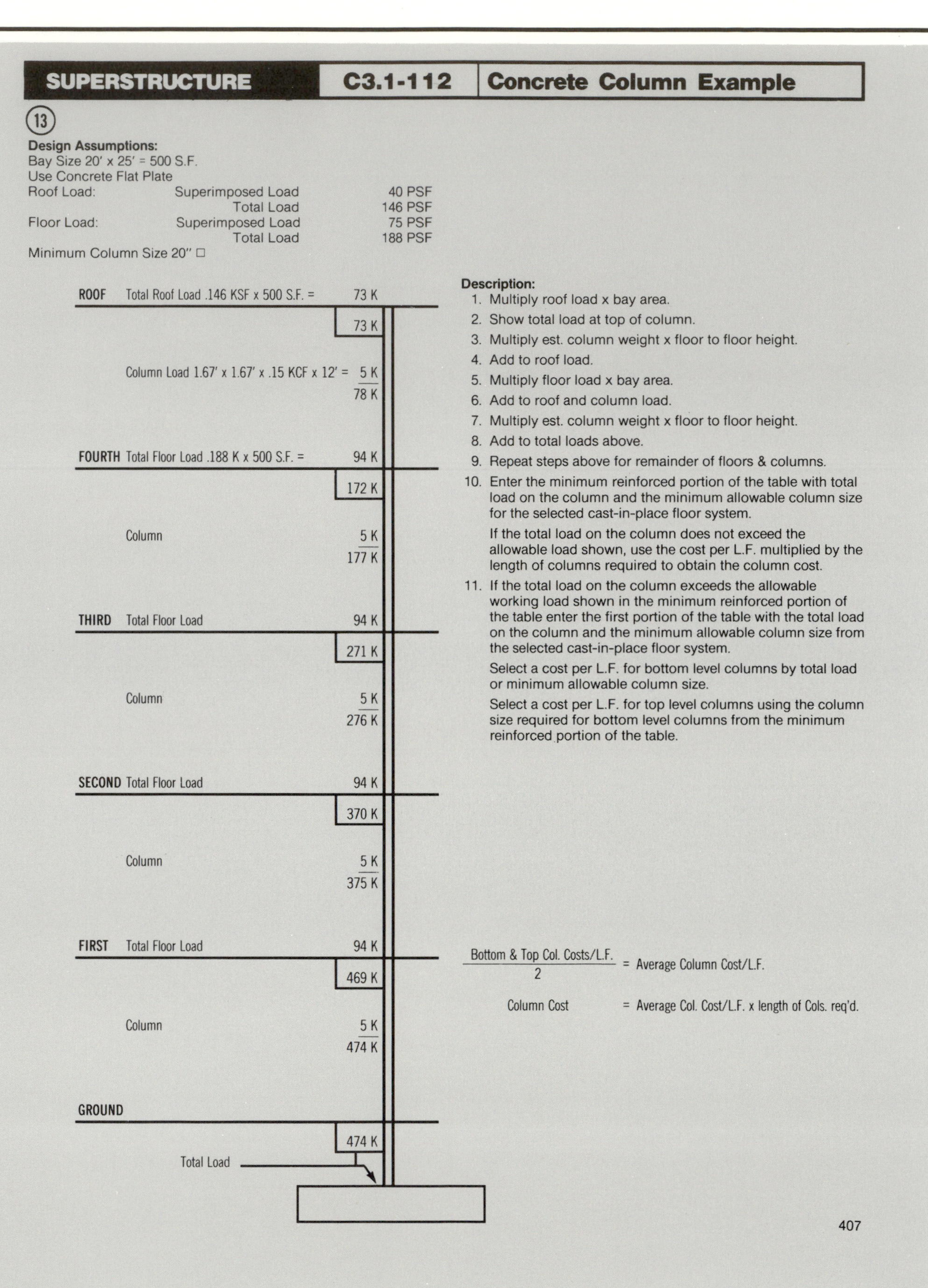

SUPERSTRUCTURE
C3.1-112
Concrete Column Example
13
Design Assumptions:
Bay Size 20' x 25' = 500 S.F.
Use Concrete Flat Plate
Roof Load: Superimposed Load 40 PSF
Total Load 146 PSF
Floor Load: Superimposed Load 75 PSF
Total Load 188 PSF
Minimum Column Size 20" □
ROOF Total Roof Load .146 KSF x 500 S.F. = 73 K
73 K
Column Load 1.67' x 1.67' x .15 KCF x 12' = 5 K
78 K
FOURTH Total Floor Load .188 K x 500 S.F. = 94 K
172 K
Column 5 K
177 K
THIRD Total Floor Load 94 K
271 K
Column 5 K
276 K
SECOND Total Floor Load 94 K
370 K
Column 5 K
375 K
FIRST Total Floor Load 94 K
469 K
Column 5 K
474 K
GROUND
474 K
Total Load
Description:
1. Multiply roof load x bay area.
2. Show total load at top of column.
3. Multiply est. column weight x floor to floor height.
4. Add to roof load.
5. Multiply floor load x bay area.
6. Add to roof and column load.
7. Multiply est. column weight x floor to floor height.
8. Add to total loads above.
9. Repeat steps above for remainder of floors & columns.
10. Enter the minimum reinforced portion of the table with total load on the column and the minimum allowable column size for the selected cast-in-place floor system.
If the total load on the column does not exceed the allowable load shown, use the cost per L.F. multiplied by the length of columns required to obtain the column cost.
11. If the total load on the column exceeds the allowable working load shown in the minimum reinforced portion of the table enter the first portion of the table with the total load on the column and the minimum allowable column size from the selected cast-in-place floor system.
Select a cost per L.F. for bottom level columns by total load or minimum allowable column size.
Select a cost per L.F. for top level columns using the column size required for bottom level columns from the minimum reinforced portion of the table.
Bottom & Top Col. Costs/L.F. / 2 = Average Column Cost/L.F.
Column Cost = Average Col. Cost/L.F. x length of Cols. req'd.
407

Figure 6.6a

SUPERSTRUCTURE	C3.1-130	Steel Columns

(14)

Description: Below is an example of steel column determination when roof and floor loads are known.

Design Assumptions:

Bay Size	35′ x 35′ =1,225 S.F.	
Roof Load:	Superimposed Load	40 PSF
	Dead Load	44 PSF
	Total Load	84 PSF
Floor Load:	Superimposed Load	125 PSF
	Dead Load	45 PSF
	Total Load	170 PSF

CALCULATION OF TOTAL LOADS

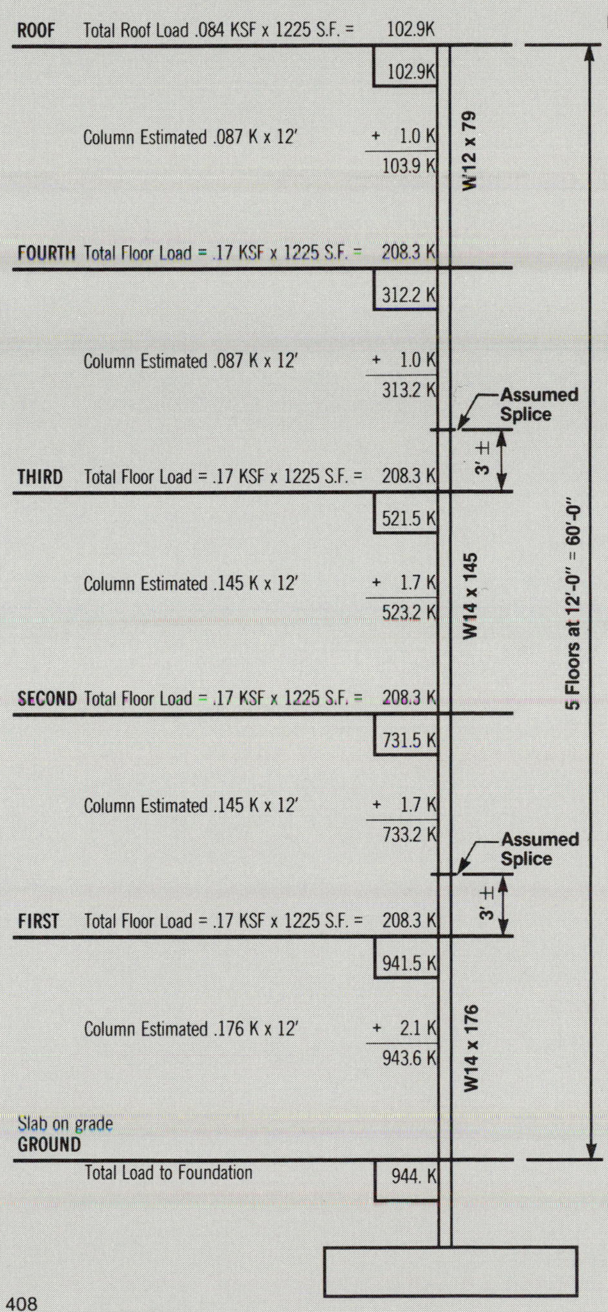

Description:

1. Multiply roof load x bay area.
2. Show total load at top of column.
3. Multiply estimated column weight* x floor to floor height.
4. Add to roof load.
5. Multiply floor load x bay area.
6. Add to roof and column load.
7. Multiply estimated column weight x floor to floor height.
8. Add to total loads above.
9. Choose column from **System 3.1-130** using unsupported height.
10. Interpolate or use higher loading to obtain cost/L.F.
11. Repeat steps above for remainder of floors and columns.
12. Multiply number of columns by the height of the column times the cost per foot to obtain the cost of each type of column.

***To Estimate Column Weight**

Roof Load	.084 KSF
Floor Load x No Floors above Splice 170 x 1	.170 KSF
Total	.254 KSF

Total Load (KSF) x Bay Area (S.F.) = Load to Col.
.254 KSF x 1,225 KSF = 311 K

From **System 3.1-130,** choose a column by:

load,	height,	weight
400 K	10′	79 lb.

408

Figure 6.6b

(the weight for the column assembly capable of supporting 75K) is used in the analysis. Also, since the unsupported length of column in the building exceeds 10′, the next category in the table (for 16′) is used.

To determine the total load to the foundation at the exterior and corner columns, the total load from the structural analysis is multiplied by 60% and 45% in turn. It is assumed that the penthouse will be over an interior bay. Therefore, the exterior footings support only three floors, and the calculations are made using the total load from the analysis in Figure 6.8.

320.6K × .6 = 192.4K @ Exterior
320.6K × .45 = 144.3K @ Corner

The total load calculations are used to size the spread footings that support the columns. The footing size and load bearing information is taken from the Spread Footing tables in *Means Assemblies Cost Data* shown in Figure 6.11. The footing that most closely meets the load and soil bearing conditions—without being exceeded by those conditions—is used. For example, the load to the interior footings is 341.9K at the penthouse and 320.6K at the other interior footings. Both exceed the 300K load limit for the 6KSF soil capacity of the footing in line 7810. Therefore, the next highest unit, the 400K capacity footing in line 7900, is used for both types of interior footings. Accordingly, the 200K capacity footing in line 7700 is used to support the 192.4K load from the exterior columns, and the 150K capacity footing in line 7610 is used to support the 144.3K load from the corner columns.

Basement walls and footings must also be selected, along with the basement floor slab. Experience indicates that 12″ cast-in-place concrete walls on 24″ × 12″ strip footings will be required for the basement walls. A cast-in-place slab on grade will be needed for the basement parking area. Figure 6.12 provides information for making the slab choice. Again, experience suggests that a type "B" slab for a 4K loading is required; thus, a 5″ slab with reinforcing is chosen.

Floor Plans

In many instances, rough floor plans are not needed. The total gross square feet of floor area to receive finishes can be used with the Partition/Door Guidelines chart (Figure 6.13 from *Means Assemblies Cost Data*) to determine an approximate square footage of partitions and number of doors for pricing. The Partition/Door Guidelines table has columns listing building types, number of stories, partition factors in square feet of floor area per linear foot of partition, door factors in square feet of floor area per door, and descriptions of the partition types. If the tenant space in the Chambers office building were to be finished, the factors specified for three-to-five story office buildings would be used. The partition and door requirements are as follows:

$$\frac{56{,}700 \text{ S.F.}}{20 \text{ S.F./L.F.}} = 2835 \text{ L.F. of Partitions} \qquad \frac{56{,}700 \text{ S.F.}}{350 \text{ S.F./Dr}} = 162 \text{ Doors}$$

The partition assemblies prices in *Means Assemblies Cost Data* are given per square foot of partition system. Linear feet of partitions must be converted to square feet. Assuming an average partition height of 11′, the total square footage of partitions is:

2835 L.F. × 11′ = 31,185 S.F.

This square footage is then prorated for the different partition types required in the Requested Facilities section of the program.

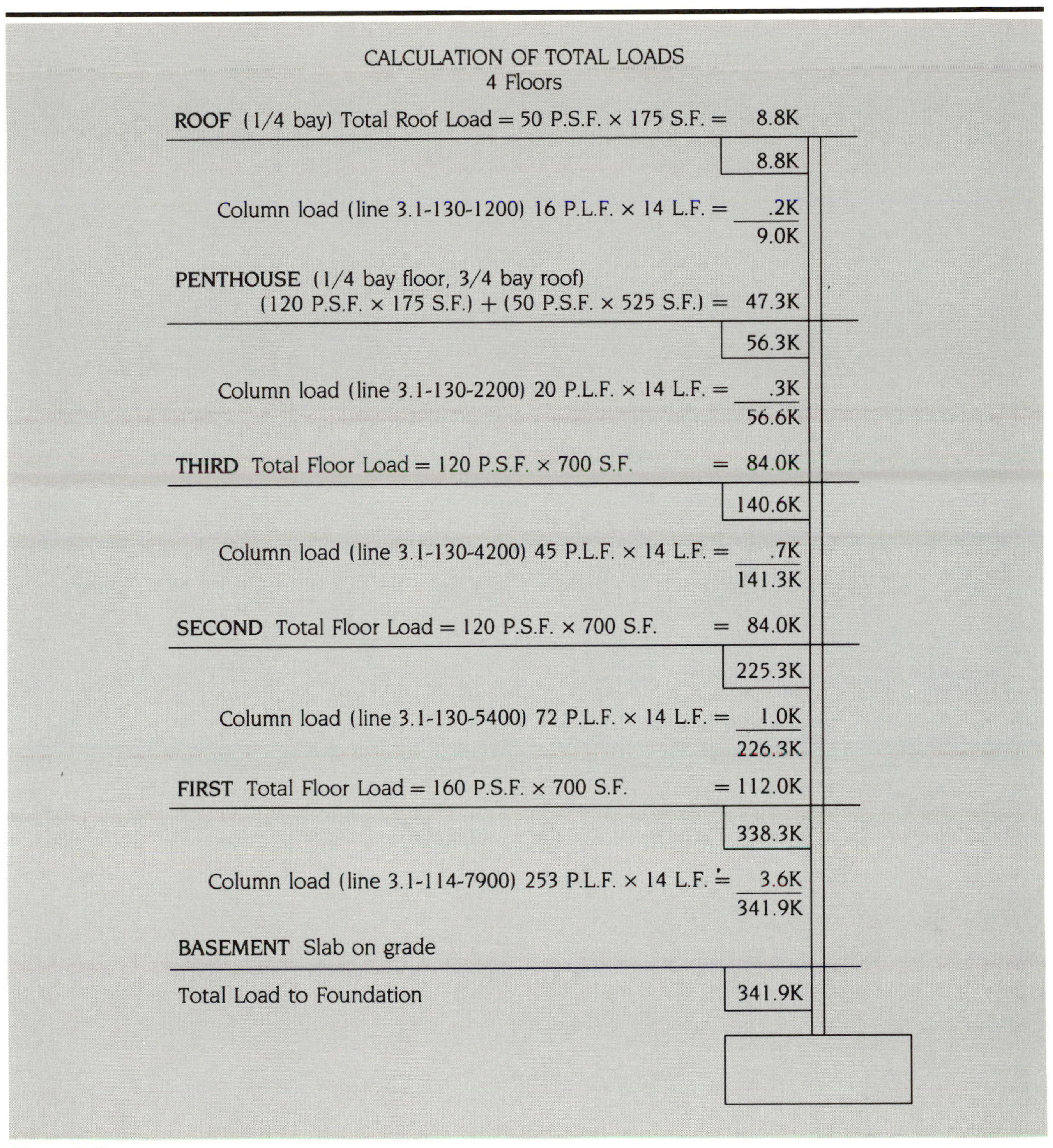

CALCULATION OF TOTAL LOADS
4 Floors
ROOF (1/4 bay) Total Roof Load = 50 P.S.F. × 175 S.F. = 8.8K
8.8K
Column load (line 3.1-130-1200) 16 P.L.F. × 14 L.F. = .2K
9.0K
PENTHOUSE (1/4 bay floor, 3/4 bay roof)
(120 P.S.F. × 175 S.F.) + (50 P.S.F. × 525 S.F.) = 47.3K
56.3K
Column load (line 3.1-130-2200) 20 P.L.F. × 14 L.F. = .3K
56.6K
THIRD Total Floor Load = 120 P.S.F. × 700 S.F. = 84.0K
140.6K
Column load (line 3.1-130-4200) 45 P.L.F. × 14 L.F. = .7K
141.3K
SECOND Total Floor Load = 120 P.S.F. × 700 S.F. = 84.0K
225.3K
Column load (line 3.1-130-5400) 72 P.L.F. × 14 L.F. = 1.0K
226.3K
FIRST Total Floor Load = 160 P.S.F. × 700 S.F. = 112.0K
338.3K
Column load (line 3.1-114-7900) 253 P.L.F. × 14 L.F. = 3.6K
341.9K
BASEMENT Slab on grade
Total Load to Foundation
341.9K

Figure 6.7

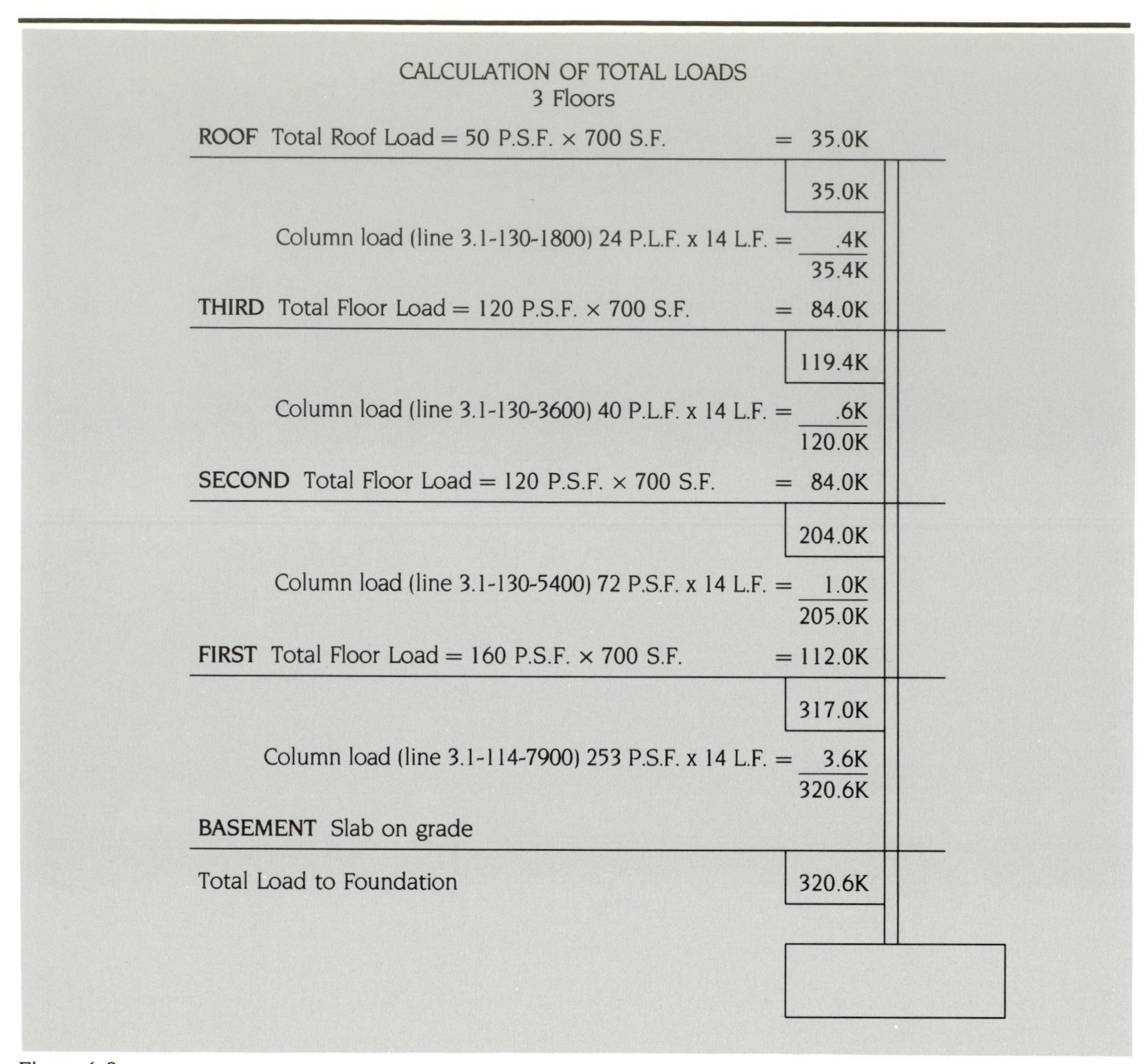
CALCULATION OF TOTAL LOADS
3 Floors
ROOF Total Roof Load = 50 P.S.F. × 700 S.F. = 35.0K
35.0K
Column load (line 3.1-130-1800) 24 P.L.F. x 14 L.F. = .4K
35.4K
THIRD Total Floor Load = 120 P.S.F. × 700 S.F. = 84.0K
119.4K
Column load (line 3.1-130-3600) 40 P.L.F. x 14 L.F. = .6K
120.0K
SECOND Total Floor Load = 120 P.S.F. × 700 S.F. = 84.0K
204.0K
Column load (line 3.1-130-5400) 72 P.S.F. x 14 L.F. = 1.0K
205.0K
FIRST Total Floor Load = 160 P.S.F. × 700 S.F. = 112.0K
317.0K
Column load (line 3.1-114-7900) 253 P.S.F. x 14 L.F. = 3.6K
320.6K
BASEMENT Slab on grade
Total Load to Foundation
320.6K

Figure 6.8

COLUMNS & BEAMS	B3.1-130	Steel Columns

(A) Wide Flange

(B) Pipe

(C) Pipe, Concrete Filled

(E) Square Tube Concrete Filled

(F) Rectangular Tube

(G) Rectangular Tube, Concrete Filled

General: The following pages provide data for seven types of steel columns: wide flange, round pipe, round pipe concrete filled, square tube, square tube concrete filled, rectangular tube and rectangular tube concrete filled.

Design Assumptions: Loads are concentric; wide flange and round pipe bearing capacity is for 36 KSI steel. Square and rectangular tubing bearing capacity is for 46 KSI steel.

The effective length factor K=1.1 is used for determining column values in the tables. K=1.1 is within a frequently used range for pinned connections with cross bracing.

How To Use Tables:

a. Steel columns usually extend through two or more stories to minimize splices. Determine floors with splices.

b. Enter Table 3.1-130 with load to column at the splice. Use the unsupported height.

c. Determine the column type desired by price or design.

Cost:

a. Multiply number of columns at the desired level by the total height of the column by the cost/VLF.

b. Repeat the above for all tiers.

Please see (14) in reference section for further design and cost information.

3.1-130	Steel Columns							
	LOAD (KIPS)	UNSUPPORTED HEIGHT (FT.)	WEIGHT (P.L.F.)	SIZE (IN.)	TYPE	COST PER V.L.F. MAT.	INST.	TOTAL
1000	25	10	13	4	A	7.40	5.05	12.45
1020	(14)		7.58	3	B	6.35	5.05	11.40
1040			15	3-½	C	7.70	5.05	12.75
1060			6.87	3	D	8.15	5.05	13.20
1080			15	3	E	10.55	5.05	15.60
1100			8.15	4x3	F	7.90	5.05	12.95
1120			20	4x3	G	10.30	5.05	15.35
1200		16	16	5	A	8.40	3.78	12.18
1220			10.79	4	B	8.35	3.78	12.13
1240			36	5 ½	C	11.60	3.78	15.38
1260			11.97	5	D	13.15	3.78	16.93
1280			36	5	E	17.45	3.78	21.23
1300			11.97	6x4	F	10.70	3.78	14.48
1320			64	8x6	G	21	3.78	24.78
1400		20	20	6	A	9.95	3.78	13.73
1420			14.62	5	B	10.70	3.78	14.48
1440			49	6-⅝	C	13.95	3.78	17.73
1460			11.97	5	D	12.45	3.78	16.23
1480			49	6	E	20	3.78	23.78
1500			14.53	7x5	F	12.30	3.78	16.08
1520			64	8x6	G	19.70	3.78	23.48
1600	50	10	16	5	A	9.10	5.05	14.15
1620			14.62	5	B	12.20	5.05	17.25
1640			24	4-½	C	9.20	5.05	14.25
1660			12.21	4	D	14.50	5.05	19.55
1680			25	4	E	14.70	5.05	19.75
1700			11.97	6x4	F	11.60	5.05	16.65
1720			28	6x3	G	13.70	5.05	18.75
1800		16	24	8	A	12.65	3.78	16.43
1820			18.97	6	B	14.25	3.78	18.03
1840			36	5-½	C	11.60	3.78	15.38
1860			14.63	6	D	16.05	3.78	19.83
1880			36	5	E	17.45	3.78	21.23
1900			14.53	7x5	F	13	3.78	16.78
1920			64	8x6	G	21	3.78	24.78
1940								

For expanded coverage of these items see *Means Concrete Cost Data 1988* 35

Figure 6.9a

COLUMNS & BEAMS	B3.1-130	Steel Columns

3.1-130 Steel Columns

	LOAD (KIPS)	UNSUPPORTED HEIGHT (FT.)	WEIGHT (P.L.F.)	SIZE (IN.)	TYPE	COST PER V.L.F.		
						MAT.	INST.	TOTAL
2000	50	20	28	8	A	13.95	3.78	17.73
2020			18.97	6	B	13.45	3.78	17.23
2040			49	6-⅝	C	13.95	3.78	17.73
2060			19.02	6	D	19.75	3.78	23.53
2080			49	6	E	20	3.78	23.78
2100			22.42	8x6	F	19	3.78	22.78
2120			64	8x6	G	19.70	3.78	23.48
2200	75	10	20	6	A	11.35	5.05	16.40
2220			18.97	6	B	15.40	5.05	20.45
2240			36	4-½	C	23	5.05	28.05
2260			14.53	6	D	17.25	5.05	22.30
2280			28	4	E	14.65	5.05	19.70
2300			14.33	7x5	F	13.85	5.05	18.90
2320			35	6x4	G	15.45	5.05	20.50
2400		16	31	8	A	16.30	3.78	20.08
2420			28.55	8	B	21	3.78	24.78
2440			49	6-⅝	C	14.70	3.78	18.48
2460			17.08	7	D	18.75	3.78	22.53
2480			36	5	E	17.45	3.78	21.23
2500			23.34	7x5	F	21	3.78	24.78
2520			64	8x6	G	21	3.78	24.78
2600		20	31	8	A	15.45	3.78	19.23
2620			28.55	8	B	20	3.78	23.78
2640			81	8-⅝	C	21	3.78	24.78
2660			22.42	7	D	23	3.78	26.78
2680			49	6	E	20	3.78	23.78
2700			22.42	8x6	F	19	3.78	22.78
2720			64	8x6	G	19.70	3.78	23.48
2800	100	10	24	8	A	13.65	5.05	18.70
2820			28.57	6	B	23	5.05	28.05
2840			35	4-½	C	23	5.05	28.05
2860			17.08	7	D	20	5.05	25.05
2880			36	5	E	18.85	5.05	23.90
2900			19.02	7x5	F	18.40	5.05	23.45
2920			46	8x4	G	18.85	5.05	23.90
3000		16	31	8	A	16.30	3.78	20.08
3020			28.55	8	B	21	3.78	24.78
3040			56	6-⅝	C	22	3.78	25.78
3060			22.42	7	D	25	3.78	28.78
3080			49	6	E	21	3.78	24.78
3100			22.42	8x6	F	20	3.78	23.78
3120			64	8x6	G	21	3.78	24.78
3200		20	40	8	A	19.95	3.78	23.73
3220			28.55	8	B	20	3.78	23.78
3240			81	8-⅝	C	21	3.78	24.78
3260			25.82	8	D	27	3.78	30.78
3280			66	7	E	24	3.78	27.78
3300			27.59	8x6	F	23	3.78	26.78
3320			70	8x6	G	28	3.78	31.78
3400	125	10	31	8	A	17.65	5.05	22.70
3420			28.57	6	B	23	5.05	28.05
3440			81	8	C	24	5.05	29.05
3460			22.42	7	D	27	5.05	32.05
3480			49	6	E	23	5.05	28.05
3500			22.42	8x6	F	22	5.05	27.05
3520			64	8x6	G	22	5.05	27.05

36

For expanded coverage of these items see *Means Concrete Cost Data 1988*

Figure 6.9b

COLUMNS & BEAMS	B3.1-130	Steel Columns

3.1-130 Steel Columns

	LOAD (KIPS)	UNSUPPORTED HEIGHT (FT.)	WEIGHT (P.L.F.)	SIZE (IN.)	TYPE	COST PER V.L.F.		
						MAT.	INST.	TOTAL
3600	125	16	40	8	A	21	3.78	24.78
3620			28.55	8	B	21	3.78	24.78
3640			81	8	C	22	3.78	25.78
3660			25.82	8	D	28	3.78	31.78
3680			66	7	E	25	3.78	28.78
3700			27.59	8x6	F	25	3.78	28.78
3720			64	8x6	G	21	3.78	24.78
3800		20	48	8	A	24	3.78	27.78
3820			40.48	10	B	29	3.78	32.78
3840			81	8	C	21	3.78	24.78
3860			25.82	8	D	27	3.78	30.78
3880			66	7	E	24	3.78	27.78
3900			37.59	10x6	F	32	3.78	35.78
3920			60	8x6	G	28	3.78	31.78
4000	150	10	35	8	A	19.90	5.05	24.95
4020			40.48	10	B	33	5.05	38.05
4040			81	8-⅝	C	24	5.05	29.05
4060			25.82	8	D	31	5.05	36.05
4080			66	7	E	27	5.05	32.05
4100			27.48	7x5	F	27	5.05	32.05
4120			64	8x6	G	22	5.05	27.05
4200		16	45	10	A	24	3.78	27.78
4220			40.48	10	B	30	3.78	33.78
4240			81	8-⅝	C	22	3.78	25.78
4260			31.84	8	D	35	3.78	38.78
4280			66	7	E	25	3.78	28.78
4300			37.69	10x6	F	34	3.78	37.78
4320			70	8x6	G	30	3.78	33.78
4400		20	49	10	A	24	3.78	27.78
4420			40.48	10	B	29	3.78	32.78
4440			123	10-¾	C	30	3.78	33.78
4460			31.84	8	D	33	3.78	36.78
4480			82	8	E	28	3.78	31.78
4500			37.69	10x6	F	32	3.78	35.78
4520			86	10x6	G	28	3.78	31.78
4600	200	10	45	10	A	26	5.05	31.05
4620			40.48	10	B	33	5.05	38.05
4640			81	8-⅝	C	24	5.05	29.05
4660			31.84	8	D	38	5.05	43.05
4680			82	8	E	32	5.05	37.05
4700			37.69	10x6	F	36	5.05	41.05
4720			70	8x6	G	32	5.05	37.05
4800		16	49	10	A	26	3.78	29.78
4820			49.56	12	B	37	3.78	40.78
4840			123	10-¾	C	32	3.78	35.78
4860			37.60	8	D	41	3.78	44.78
4880			90	8	E	42	3.78	45.78
4900			42.79	12x6	F	38	3.78	41.78
4920			85	10x6	G	35	3.78	38.78
5000		20	58	12	A	29	3.78	32.78
5020			49.56	12	B	35	3.78	38.78
5040			123	10-¾	C	30	3.78	33.78
5060			40.35	10	D	42	3.78	45.78
5080			90	8	E	40	3.78	43.78
5100			47.90	12x8	F	41	3.78	44.78
5120			93	10x6	G	42	3.78	45.78

For expanded coverage of these items see *Means Concrete Cost Data 1988*

37

Figure 6.9c

COLUMNS & BEAMS	B3.1-130	Steel Columns

3.1-130 Steel Columns

	LOAD (KIPS)	UNSUPPORTED HEIGHT (FT.)	WEIGHT (P.L.F.)	SIZE (IN.)	TYPE	COST PER V.L.F.		
						MAT.	INST.	TOTAL
5200	300	10	61	14	A	35	5.05	40.05
5220			65.42	12	B	53	5.05	58.05
5240			169	12-¾	C	42	5.05	47.05
5260			47.90	10	D	57	5.05	62.05
5280			90	8	E	45	5.05	50.05
5300			47.90	12x8	F	46	5.05	51.05
5320			86	10x6	G	48	5.05	53.05
5400		16	72	12	A	38	3.78	41.78
5420			65.42	12	B	49	3.78	52.78
5440			169	12-¾	C	39	3.78	42.78
5460			58.10	12	D	64	3.78	67.78
5480			135	10	E	54	3.78	57.78
5500			58.10	14x10	F	52	3.78	55.78
5600		20	79	12	A	39	3.78	42.78
5620			65.42	12	B	46	3.78	49.78
5640			169	12-¾	C	37	3.78	40.78
5660			58.10	12	D	60	3.78	63.78
5680			135	10	E	51	3.78	54.78
5700			58.10	14x10	F	49	3.78	52.78
5800	400	10	79	12	A	45	5.05	50.05
5840			178	12-¾	C	55	5.05	60.05
5860			68.31	14	D	81	5.05	86.05
5880			135	10	E	58	5.05	63.05
5900			62.46	14x10	F	60	5.05	65.05
6000		16	87	12	A	46	3.78	49.78
6040			178	12-¾	C	51	3.78	54.78
6060			68.31	14	D	75	3.78	78.78
6080			145	10	E	70	3.78	73.78
6100			76.07	14x10	F	68	3.78	71.78
6200		20	90	14	A	45	3.78	48.78
6240			178	12-¾	C	48	3.78	51.78
6260			68.31	14	D	71	3.78	74.78
6280			145	10	E	66	3.78	69.78
6300			76.07	14x10	F	65	3.78	68.78
6400	500	10	99	14	A	56	5.05	61.05
6460			76.07	12	D	90	5.05	95.05
6480			145	10	E	75	5.05	80.05
6500			76.07	14x10	F	74	5.05	79.05
6600		16	109	14	A	57	3.78	60.78
6660			89.68	14	D	98	3.78	101.78
6700			89.68	16x12	F	80	3.78	83.78
6800		20	120	12	A	60	3.78	63.78
6860			89.68	14	D	93	3.78	96.78
6900			89.68	16x12	F	76	3.78	79.78
7000	600	10	120	12	A	68	5.05	73.05
7060			89.68	14	D	105	5.05	110.05
7100			89.68	16x12	F	87	5.05	92.05
7200		16	132	14	A	69	3.78	72.78
7260			103.30	16	D	115	3.78	118.78
7400		20	132	14	A	66	3.78	69.78
7460			103.30	16	D	105	3.78	108.78
7600	700	10	136	12	A	77	5.05	82.05
7660			103.30	16	D	120	5.05	125.05
7800		16	145	14	A	76	3.78	79.78
7860			103.30	16	D	115	3.78	118.78
8000		20	145	14	A	72	3.78	75.78

38

For expanded coverage of these items see *Means Concrete Cost Data 1988*

Figure 6.9d

COLUMNS & BEAMS	B3.1-114	C.I.P. Column, Square Tied

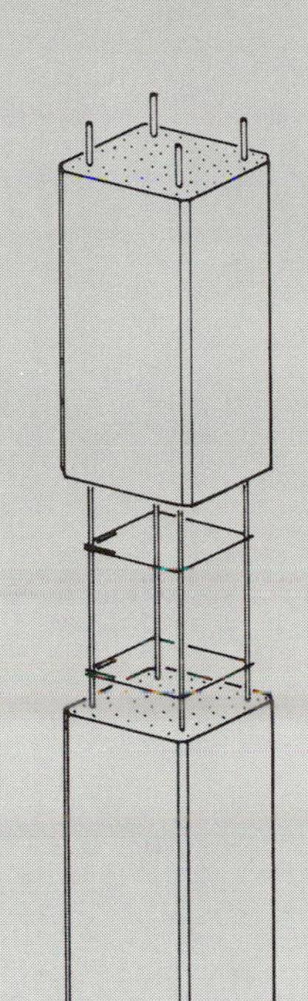

CONCRETE COLUMNS

General: It is desirable for purposes of consistency and simplicity to maintain constant column sizes throughout the building height. To do this, concrete strength may be varied (higher strength concrete at lower stories and lower strength concrete at upper stories), as well as varying the amount of reinforcing.

The first portion of the table provides probable minimum column sizes with related costs and weights per lineal foot of story height for bottom level columns.

The second portion of the table provides costs by column size for top level columns with minimum code reinforcement. Probable maximum loads for these columns are also given.

How to Use Table:

1. Enter the second portion (minimum reinforcing) of the table with the minimum allowable column size from the selected cast in place floor system.

   If the total load on the column does not exceed the allowable working load shown use the cost per L.F. multiplied by the length of columns required to obtain the column cost.

2. If the total load on the column exceeds the allowable working load shown in the second portion of the table enter the first portion of the table with the total load on the column and the minimum allowable column size from the selected cast in place floor system.

   Select a cost per L.F. for bottom level columns by total load or minimum allowable column size.

   Select a cost per L.F. for top level columns using the column size required for bottom level columns from the second portion of the table.

$$\frac{\text{Bottom + Top Col. Costs/L.F.}}{2} = \text{Average Column Cost/L.F.}$$

Column Cost = Average Col. Cost/L.F. × Length of Cols. Required.

See Example 3.1-112 in back of book to determine total loads.

Design and Pricing Assumptions:
Normal wt. concrete, f'c = 4 or 6 KSI, placed by pump
Steel, fy = 60 KSI, spliced every other level.
Minimum design eccentricity of 0.1t.
Assumed load level depth is 8" (weights prorated to full story basis).
Gravity loads only (no frame or lateral loads included).

Please see ⑬ in reference section for further design and cost information.

System Components	QUANTITY	UNIT	COST PER V.L.F. MAT.	INST.	TOTAL
SYSTEM 03.1-114-0640					
SQUARE COLUMNS, 100K LOAD,10' STORY, 10" SQUARE					
Forms in place, columns, plywood, 10" x 10", 4 uses	3.323	SFCA	2.03	14.43	16.46
Chamfer strip,wood, ¾" wide	4.000	L.F.	.36	1.92	2.28
Reinforcing in place, columns, #3 to #7	3.653	Lb.	1.50	1.90	3.40
Reinforcing in place, column ties	1.405	Lb.	.30	.38	.68
Concrete ready mix, regular weight, 4000 PSI	.026	C.Y.	1.56		1.56
Placing concrete, incl. vibrating, 12" sq./round columns, pumped	.026	C.Y.		1.30	1.30
Finish, break ties, patch voids, burlap rub w/grout	3.323	S.F.	.20	1.70	1.90
TOTAL			5.95	21.63	27.58

3.1-114 C.I.P. Column, Square Tied

	LOAD (KIPS)	STORY HEIGHT (FT.)	COLUMN SIZE (IN.)	COLUMN WEIGHT (P.L.F.)	CONCRETE STRENGTH (PSI)	COST PER V.L.F. MAT.	INST.	TOTAL
0640	100	10	10	96	4000	5.95	22	27.95
0680	⑬	12	10	97	4000	5.80	21	26.80
0700		14	12	142	4000	7.70	26	33.70
0710								

For expanded coverage of these items see *Means Concrete Cost Data 1988*

Figure 6.10a

COLUMNS & BEAMS	B3.1-114	C.I.P. Column, Square Tied

3.1-114 C.I.P. Column, Square Tied

	LOAD (KIPS)	STORY HEIGHT (FT.)	COLUMN SIZE (IN.)	COLUMN WEIGHT (P.L.F.)	CONCRETE STRENGTH (PSI)	COST PER V.L.F.		
						MAT.	INST.	TOTAL
0740	150	10	10	96	4000	5.95	22	27.95
0780		12	12	142	4000	7.75	26	33.75
0800		14	12	143	4000	7.70	26	33.70
0840	200	10	12	140	4000	7.80	26	33.80
0860		12	12	142	4000	7.75	26	33.75
0900		14	14	196	4000	9.70	30	39.70
0920	300	10	14	192	4000	9.95	31	40.95
0960		12	14	194	4000	9.85	31	40.85
0980		14	16	253	4000	11.15	33	44.15
1020	400	10	16	248	4000	12.30	35	47.30
1060		12	16	251	4000	12.15	35	47.15
1080		14	16	253	4000	12.05	35	47.05
1200	500	10	18	315	4000	15.95	42	57.95
1250		12	20	394	4000	16.10	44	60.10
1300		14	20	397	4000	16	44	60
1350	600	10	20	388	4000	19.15	48	67.15
1400		12	20	394	4000	18.85	47	65.85
1600		14	20	397	4000	18.70	47	65.70
1900	700	10	20	388	4000	19.35	48	67.35
2100		12	22	474	4000	19.10	48	67.10
2300		14	22	478	4000	18.95	47	65.95
2600	800	10	22	388	4000	28	59	87
2900		12	22	474	4000	28	59	87
3200		14	22	478	4000	28	59	87
3400	900	10	24	560	4000	28	60	88
3800		12	24	567	4000	27	59	86
4000		14	24	571	4000	27	59	86
4250	1000	10	24	560	4000	34	68	102
4500		12	26	667	4000	30	61	91
4750		14	26	673	4000	29	61	90
5600	100	10	10	96	6000	6.05	22	28.05
5800		12	10	97	6000	5.90	22	27.90
6000		14	12	142	6000	7.75	26	33.75
6200	150	10	10	96	6000	5.95	22	27.95
6400		12	12	98	6000	7.75	26	33.75
6600		14	12	143	6000	7.75	26	33.75
6800	200	10	12	140	6000	8.15	27	35.15
7000		12	12	142	6000	7.75	26	33.75
7100		14	14	196	6000	9.65	30	39.65
7300	300	10	14	192	6000	9.80	31	40.80
7500		12	14	194	6000	9.75	31	40.75
7600		14	14	196	6000	9.70	30	39.70
7700	400	10	14	192	6000	9.80	31	40.80
7800		12	14	194	6000	9.75	31	40.75
7900		14	16	253	6000	12	34	46
8000	500	10	16	248	6000	12.30	35	47.30
8050		12	16	251	6000	12.15	35	47.15
8100		14	16	253	6000	12.05	35	47.05
8200	600	10	18	315	6000	14.90	40	54.90
8300		12	18	319	6000	14.75	40	54.75
8400		14	18	321	6000	14.60	40	54.60
8500	700	10	18	315	6000	16.25	42	58.25
8600		12	18	319	6000	16.05	42	58.05
8700		14	18	321	6000	15.90	41	56.90
8800	800	10	20	388	6000	16.25	44	60.25
8900		12	20	394	6000	16.10	44	60.10

For expanded coverage of these items see *Means Concrete Cost Data 1988*

31

Figure 6.10b

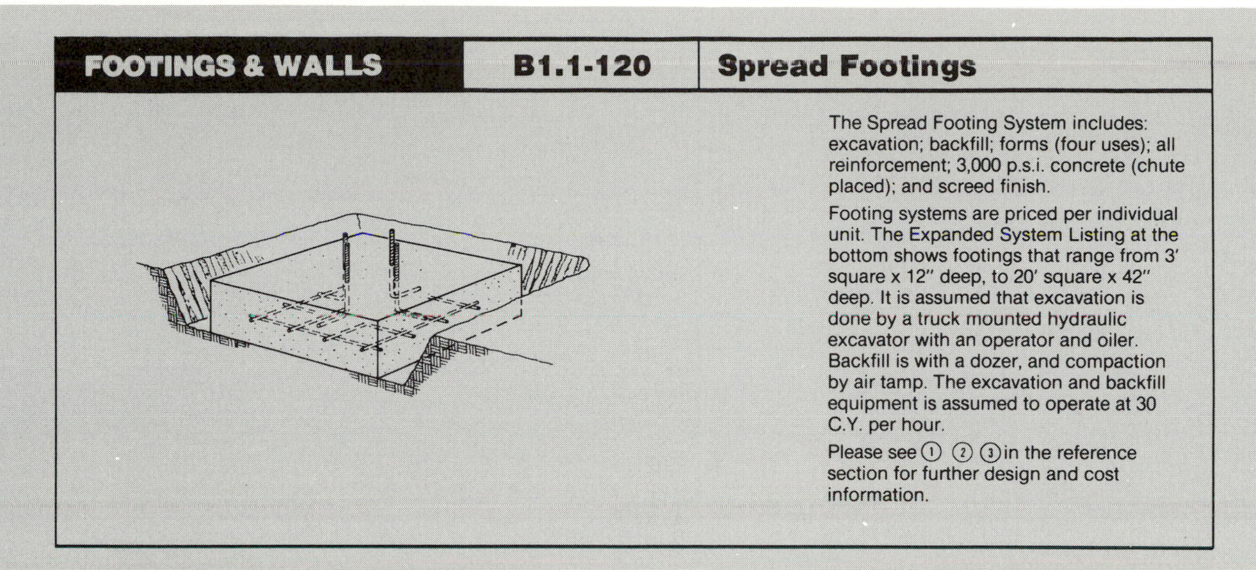

FOOTINGS & WALLS | **B1.1-120** | **Spread Footings**

The Spread Footing System includes: excavation; backfill; forms (four uses); all reinforcement; 3,000 p.s.i. concrete (chute placed); and screed finish.

Footing systems are priced per individual unit. The Expanded System Listing at the bottom shows footings that range from 3′ square x 12″ deep, to 20′ square x 42″ deep. It is assumed that excavation is done by a truck mounted hydraulic excavator with an operator and oiler. Backfill is with a dozer, and compaction by air tamp. The excavation and backfill equipment is assumed to operate at 30 C.Y. per hour.

Please see ① ② ③ in the reference section for further design and cost information.

Systems Components	QUANTITY	UNIT	COST EACH		
			MAT.	INST.	TOTAL
SYSTEM 01.1-120-7100					
SPREAD FOOTINGS, LOAD 25K, SOIL CAPACITY 3 KSF, 3′ SQ X 12″ DEEP					
Bulk excavation	.590	C.Y.		2.64	2.64
Hand trim	9.000	S.F.		3.78	3.78
Compacted backfill	.260	C.Y.		.44	.44
Formwork, 4 uses	12.000	S.F.	4.68	28.32	33
Reinforcing, fy = 60,000 psi	.006	Ton	3.47	3.29	6.76
Dowel or anchor bolt templates	6.000	L.F.	3.42	11.70	15.12
Concrete, f′c = 3,000 psi	.330	C.Y.	18.81		18.81
Place concrete, direct chute	.330	C.Y.		3.89	3.89
Screed finish	9.000	S.F.		2.34	2.34
TOTAL			30.38	56.40	86.78

1.1-120	Spread Footings	COST EACH		
		MAT.	INST.	TOTAL
7090	Spread footings, 3000 psi concrete, chute delivered			
7100	Load 25K, soil capacity 3 KSF, 3′-0″ sq. x 12″ deep	30	56	86
7150	Load 50K, soil capacity 3 KSF, 4′-6″ sq. x 12″ deep ①	65	98	163
7200	Load 50K, soil capacity 6 KSF, 3′-0″ sq. x 12″ deep	30	56	86
7250	Load 75K, soil capacity 3 KSF, 5′-6″ sq. x 13″ deep ②	105	140	245
7300	Load 75K, soil capacity 6 KSF, 4′-0″ sq. x 12″ deep	53	84	137
7350	Load 100K, soil capacity 3 KSF, 6′-0″ sq. x 14″ deep ③	130	165	295
7410	Load 100K, soil capacity 6 KSF, 4′-6″ sq. x 15″ deep	80	115	195
7450	Load 125K, soil capacity 3 KSF, 7′-0″ sq. x 17″ deep	205	235	440
7500	Load 125K, soil capacity 6 KSF, 5′-0″ sq. x 16″ deep	100	140	240
7550	Load 150K, soil capacity 3 KSF 7′-6″ sq. x 18″ deep	250	275	525
7610	Load 150K, soil capacity 6 KSF, 5′-6″ sq. x 18″ deep	135	175	310
7650	Load 200K, soil capacity 3 KSF, 8′-6″ sq. x 20″ deep	355	365	720
7700	Load 200K, soil capacity 6 KSF, 6′-0″ sq. x 20″ deep	180	215	395
7750	Load 300K, soil capacity 3 KSF, 10′-6″ sq. x 25″ deep	650	600	1,250
7810	Load 300K, soil capacity 6 KSF, 7′-6″ sq. x 25″ deep	340	360	700
7850	Load 400K, soil capacity 3 KSF, 12′-6″ sq. x 28″ deep	1,025	890	1,915
7900	Load 400K, soil capacity 6 KSF, 8′-6″ sq. x 27″ deep	470	470	940
7950	Load 500K, soil capacity 3 KSF, 14′-0″ sq. x 31″ deep	1,425	1,175	2,600
8010	Load 500K, soil capacity 6 KSF, 9′-6″ sq. x 30″ deep	650	610	1,260

1

Figure 6.11

General: Ground slabs are classified on the basis of use. Thickness is generally controlled by the heaviest concentrated load supported. If load area is greater than 80 sq. in., soil bearing may be important. The base granular fill must be a uniformly compacted material of limited capillarity, such as gravel or crushed rock. Concrete is placed on this surface or the vapor barrier on top of base.

Ground slabs are either single or two course floors. Single course are widely used. Two course floors have a subsequent wear resistant topping.

Reinforcement is provided to maintain tightly closed cracks.

Control joints limit crack locations and provide for differential horizontal movement only. Isolation joints allow both horizontal and vertical differential movement.

Use of Table: Determine appropriate type of slab (A, B, C, or D) by considering type of use or amount of abrasive wear of traffic type.

Determine thickness by maximum allowable wheel load or uniform load, opposite 1st column thickness. Increase the controlling thickness if details require, and select either plain or reinforced slab thickness and type.

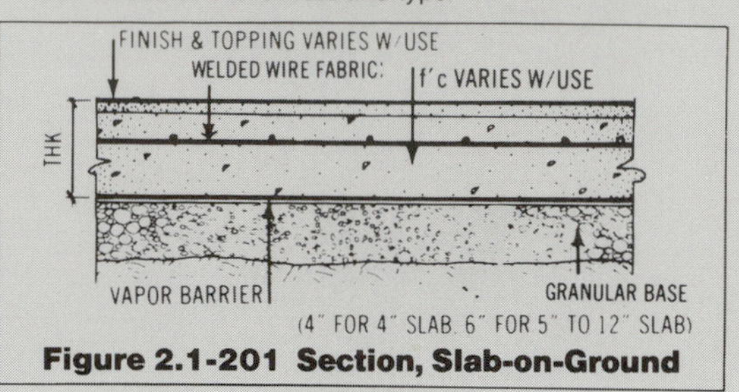

Figure 2.1-201 Section, Slab-on-Ground

Table 2.1-202 Thickness and Loading Assumptions by Type of Use

SLAB THICKNESS (IN.)	TYPE	A	B	C	D	◄ Slab I.D.
		Non	Light	Normal	Heavy	◄ Industrial
		Little	Light	Moderate	Severe	◄ Abrasion
		Foot Only	Pneumatic Wheels	Solid Rubber Wheels	Steel Tires	◄ Type of Traffic
		Load* (K)	Load* (K)	Load* (K)	Load* (K)	Max. Uniform Load to Slab ▼ (PSF)
4"	Reinf. Plain	4K				100
5"	Reinf. Plain	6K	4K			200
6"	Reinf. Plain		8K	6K	6K	500 to 800
7"	Reinf. Plain			9K	8K	1500
8"	Reinf. Plain				11K	*Max. Wheel Load in Kips (incl. impact)
10"	Reinf. Plain				14K	
12"	Reinf. Plain					
DESIGN ASSUMPTIONS	Concrete, Chuted	f'c = 3.5 KSI	4 KSI	4.5 KSI	Slab @ 3.5 KSI	ASSUMPTIONS BY SLAB TYPE
	Topping			1" Integral	1" Bonded	
	Finish	Steel Trowel	Steel Trowel	Steel Trowel	Screed & Steel Trowel	
	Compacted Granular Base	4" deep for 4" slab thickness 6" deep for 5" slab thickness & greater				ASSUMPTIONS FOR ALL SLAB TYPES
	Vapor Barrier	6 mil polyethylene				
	Forms & Joints	Allowances included				
	Rein-forcement	WWF As required ≥ 60,000 psi				

406

Figure 6.12

REFERENCE AIDS | **C15.2-200** | **Partition/Door Guidelines**

(101) **Table 15.2-201 Partition/Door Density**

Building Type		Stories	Partition/Density	Doors	Description of Partition
Apartments		1 story	9 SF/LF	90 SF/door	Plaster, wood doors & trim
		2 story	8 SF/LF	80 SF/door	Drywall, wood studs, wood doors & trim
		3 story	9 SF/LF	90 SF/door	Plaster, wood studs, wood doors & trim
		5 story	9 SF/LF	90 SF/door	Plaster, wood studs, wood doors & trim
		6-15 story	8 SF/LF	80 SF/door	Drywall, wood studs, wood doors & trim
Bakery		1 story	50 SF/LF	500 SF/door	Conc. block, paint, door & drywall, wood studs
		2 story	50 SF/LF	500 SF/door	Conc. block, paint, door & drywall, wood studs
Bank		1 story	20 SF/LF	200 SF/door	Plaster, wood studs, wood doors & trim
		2-4 story	15 SF/LF	150 SF/door	Plaster, wood studs, wood doors & trim
Bottling plant		1 story	50 SF/LF	500 SF/door	Conc. block, drywall, wood studs, wood trim
Bowling Alley		1 story	50 SF/LF	500 SF/door	Conc. block, wood & metal doors, wood trim
Bus Terminal		1 story	15 SF/LF	150 SF/door	Conc. block, ceramic tile, wood trim
Cannery		1 story	100 SF/LF	1000 SF/door	Drywall on metal studs
Car Wash		1 story	18 SF/LF	180 SF/door	Concrete block, painted & hollow metal door
Dairy Plant		1 story	30 SF/LF	300 SF/door	Concrete block, glazed tile, insulated cooler doors
Department Store		1 story	60 SF/LF	600 SF/door	Drywall, wood studs, wood doors & trim
		2-5 story	60 SF/LF	600 SF/door	30% concrete block, 70% drywall, wood studs
Dormitory		2 story	9 SF/LF	90 SF/door	Plaster, concrete block, wood doors & trim
		3-5 story	9 SF/LF	90 SF/door	Plaster, concrete block, wood doors & trim
		6-15 story	9 SF/LF	90 SF/door	Plaster, concrete block, wood doors & trim
Funeral Home		1 story	15 SF/LF	150 SF/door	Plaster on concrete block & wood studs, paneling
		2 story	14 SF/LF	140 SF/door	Plaster, wood studs, paneling & wood doors
Garage Sales & Service		1 story	30 SF/LF	300 SF/door	50% conc. block, 50% drywall, wood studs
Hotel		3-8 story	9 SF/LF	90 SF/door	Plaster, conc. block, wood doors & trim
		9-15 story	9 SF/LF	90 SF/door	Plaster, conc. block, wood doors & trim
Laundromat		1 story	25 SF/LF	250 SF/door	Drywall, wood studs, wood doors & trim
Medical Clinic		1 story	6 SF/LF	60 SF/door	Drywall, wood studs, wood doors & trim
		2-4 story	6 SF/LF	60 SF/door	Drywall, wood studs, wood doors & trim
Motel		1 story	7 SF/LF	70 SF/door	Drywall, wood studs, wood doors & trim
		2-3 story	7 SF/LF	70 SF/door	Concrete block, drywall on wood studs, wood paneling
Movie theater	200-600 seats	1 story	18 SF/LF	180 SF/door	Concrete block, wood, metal, vinyl trim
	601-1400 seats		20 SF/LF	200 SF/door	Concrete block, wood, metal, vinyl trim
	1401-2200 seats		25 SF/LF	250 SF/door	Concrete block, wood, metal, vinyl trim
Nursing Home		1 story	8 SF/LF	80 SF/door	Drywall, wood studs, wood doors & trim
		2-4 story	8 SF/LF	80 SF/door	Drywall, wood studs, wood doors & trim
Office		1 story	20 SF/LF	200-500 SF/door	30% concrete block, 70% drywall on wood studs
		2 story	20 SF/LF	200-500 SF/door	30% concrete block, 70% drywall on wood studs
		3-5 story	20 SF/LF	200-500 SF/door	30% concrete block, 70% movable partitions
		6-10 story	20 SF/LF	200-500 SF/door	30% concrete block, 70% movable partitions
		11-20 story	20 SF/LF	200-500 SF/door	30% concrete block, 70% movable partitions
Parking Ramp (Open)		2-8 story	60 SF/LF	600 SF/door	Stair and elevator enclosures only
Parking Garage		2-8 story	60 SF/LF	600 SF/door	Stair and elevator enclosures only
Pre-Engineered	Steel	1 story	0		
	Store	1 story	60 SF/LF	600 SF/door	Drywall on wood studs, wood doors & trim
	Office	1 story	15 SF/LF	150 SF/door	Concrete block, movable wood partitions
	Shop	1 story	15 SF/LF	150 SF/door	Movable wood partitions
	Warehouse	1 story	0		
Radio & TV Broadcasting		1 story	25 SF/LF	250 SF/door	Concrete block, metal and wood doors
& TV Transmitter		1 story	40 SF/LF	400 SF/door	Concrete block, metal and wood doors
Self Service Restaurant		1 story	15 SF/LF	150 SF/door	Concrete block, wood and aluminum trim
Cafe & Drive-In Restaurant		1 story	18 SF/LF	180 SF/door	Drywall, wood studs, ceramic & plastic trim
Restaurant with seating		1 story	25 SF/LF	250 SF/door	Concrete block, paneling, wood studs & trim
Supper Club		1 story	25 SF/LF	250 SF/door	Concrete block, paneling, wood studs & trim
Bar or Lounge		1 story	24 SF/LF	240 SF/door	Plaster or gypsum lath, wooded studs
Retail Store or Shop		1 story	60 SF/LF	600 SF/door	Drywall wood studs, wood doors & trim
Service Station	Masonry	1 story	15 SF/LF	150 SF/door	Concrete block, paint, door & drywall, wood studs
	Metal panel	1 story	15 SF/LF	150 SF/door	Concrete block, paint, door & drywall, wood studs
	Frame	1 story	15 SF/LF	150 SF/door	Drywall, wood studs, wood doors & trim
Shopping Center	(strip)	1 story	30 SF/LF	300 SF/door	Drywall, wood studs, wood doors & trim
	(group)	1 story	40 SF/LF	400 SF/door	50% concrete block, 50% drywall, wood studs
		2 story	40 SF/LF	400 SF/door	50% concrete block, 50% drywall, wood studs
Small Food Store		1 story	30 SF/LF	300 SF/door	Concrete block drywall, wood studs, wood trim
Store/Apt. above	Masonry	2 story	10 SF/LF	100 SF/door	Plaster, wood studs, wood doors, & trim
	Frame	2 story	10 SF/LF	100 SF/door	Plaster, wood studs, wood doors, & trim
	Frame	3 story	10 SF/LF	100 SF/door	Plaster, wood studs, wood doors, & trim
Supermarkets		1 story	40 SF/LF	400 SF/door	Concrete block, paint, drywall & porcelain panel
Truck Terminal		1 story	0		
Warehouse		1 story	0		

504

Figure 6.13

For the Chambers project, however, only the lobbies, rest rooms, elevators, and fire stairs are to receive finishes. To determine the door and partition requirements, rough floor plans are needed. An example showing the typical floor requirements and first floor entry lobby appears in Figure 6.14.

Exterior Closure

The exterior wall system used in low-rise office construction in the Chambers area is modular face brick over insulated drywall. For this project, the wall is to extend from the top of the basement wall to the top of a 2′ high parapet. Twenty percent of the exterior wall is assumed to be aluminum frame windows with 1/2″ insulated glass. Also included will be one aluminum and glass entry and exterior doors to the stairways.

Roofing

The roof system normally used for commercial construction in this location is a ballasted, single-ply membrane over 3″ of rigid insulation. The same system will be used in the Chambers office building project.

Interior Finishes

The interior finishes for the Chambers project are based on the levels of detail provided in the Project Program (see Chapter 5), with some additions required for the parking basement. In the basement area, walls of painted 8″ concrete block are assumed for the elevator lobby and stairs. The floor to floor height of 14′ less 1′ for depth of the first floor structure from Figure 6.5 results in a wall height of 13′. For the first through third floors, fire-resistant drywall is to be used. The wall height will be the full 14′ floor-to-floor height. The walls in the stairwells will be painted full height, while the lobby walls will have vinyl wall covering to a 10′ ceiling height. The rest room is to be finished with ceramic tile to an 8′ ceiling height. Quarry tile flooring will be used in the lobbies and ceramic tile in the rest rooms. A metal lay-in ceiling will be used in the lobbies, gypsum board in the rest rooms, and acoustical tile lay-in in the tenant spaces. Three-by-seven foot fire-rated doors are to be used throughout.

Specialties and Equipment

The only specialty and equipment items required for the Chambers project are cigarette urns and toilet accessories and partitions.

Conveying

Past experience suggests that two 3000 lb. capacity hydraulic elevators will be required.

Mechanical, Electrical, and Plumbing Criteria

A modeling exercise similar to that used for previous calculations could be used to establish mechanical, electrical, and plumbing criteria. However, square foot estimating methods generally prove just as accurate at this stage. The addition of basement parking to the Chambers project will have some effect on these costs. For example, an exhaust system may be needed to vent engine fumes, floor drains may be needed to remove rain or snow melted off of the cars, and lighting and additional electrical service will be needed. This extra cost is not as significant as it would be for an additional floor. It can be approximated by adding half of the gross square feet of basement floor area to the gross building floor area that is used in the Square Foot method. This area is determined as follows:

Building Area	56,700 S.F.
Basement: 18,900/2	9,450 S.F.
	66,150 S.F.

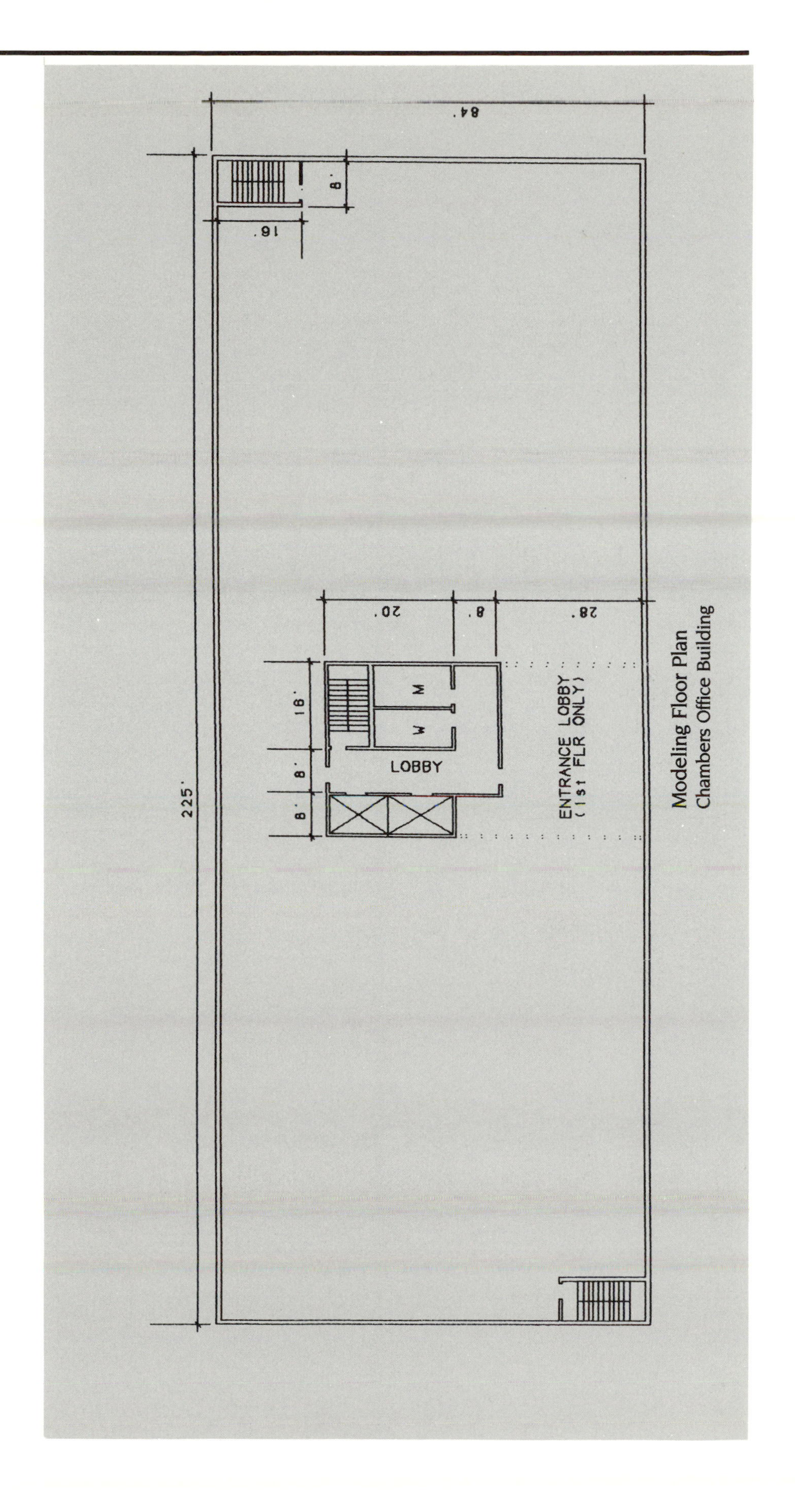

84'
16'
8'
20'
8'
28'
18'
8'
8'
225'
M
W
LOBBY
ENTRANCE LOBBY
(1st FLR ONLY)
Modeling Floor Plan
Chambers Office Building

Figure 6.14

The floor area of 66,150 S.F. is then used with square foot prices to determine the costs of the mechanical, plumbing, and electrical costs.

Site

The site study is conducted primarily to determine utility requirements and to finalize landscaping and parking requirements. Part of the information provided by the site plan attached as an exhibit to the program should be the location of adjacent utilities. It is assumed that all utilities are available from the street right of way. An 8" water line and an 8" sanitary sewer line will be extended from the edge of the road to a point one third of the depth of the site. Electrical service prices are included in the square foot price.

All of these assumptions must be recorded and kept as part of the record of this estimate. The Means *Preliminary Estimate Form* works well for this purpose. Such a form, filled in for our example project, is shown in Figure 6.15.

Estimate

An Assemblies/Square Foot estimate can now be drawn up based on the model. The estimate is prepared using quantities derived from the model and prices from *Means Assemblies Cost Data*. The Chambers estimate is shown in Figures 6.16 through 6.20. The prices for the spread footings, concrete columns, floor assemblies, and roof assemblies are obtained from the lines used in the structural analysis. The prices of the steel columns are an average of the costs for the top steel column and the bottom steel column as shown in the structural analysis procedure in Figure 6.6. The remaining general construction systems are priced using the appropriate assembly cost from *Means Assemblies Cost Data*.

The HVAC, plumbing, and electrical systems are priced using square foot costs. The prices for HVAC and plumbing are taken from lines 141-610-2770 and 141-610-2720 (Figure 6.21), respectively. There is, however, an alternate source for square foot electrical costs in *Means Assemblies Cost Data* that allows the electrical estimate to be tailored to the project. This data is shown in Figures 6.22a and 6.22b, a table listing square foot prices for components of electrical systems for a variety of building types. These components include service and distribution, lighting, devices, equipment connections, basic materials, and a number of special systems. The electrical estimate is tailored to the project by using only the components that are required for a specific project. The electrical system for the Chambers project requires service and distribution, lighting, devices, equipment connections, basic materials, and fire alarm and detection. To complete the estimate for the Chambers project, the square foot costs for HVAC, plumbing, and the electrical components are multiplied by 66,150 S.F., the gross area of the building plus half of the gross area of the parking basement previously determined. This estimate is shown in Figure 6.18.

A more detailed explanation of assemblies estimating techniques can be found in *Means Square Foot Estimating*.

Adjustments

The estimate Cost Summary, Figure 6.20, contains a recap and total of the UNIFORMAT division costs with adjustments for sales tax, general conditions, overhead and profit, bid conditions, location, and inflation. The information for sales tax, general conditions, and location is obtained from *Means Assemblies Cost Data*; the information for overhead and profit, inflation, and bid conditions is the product of past experience, together with inflation and construction trend studies.

PRELIMINARY ESTIMATE

PROJECT: Chambers Office Bldg
TOTAL SITE AREA: 310' x 365' = 113,150 sf = 2.6 ac
BUILDING TYPE: Office Building
OWNER: Chambers Investment Inc.
LOCATION: 103 Chambers Blvd. - Chambers, Ga.
ARCHITECT:
DATE OF CONSTRUCTION: ≈ 6-1-89
ESTIMATED CONSTRUCTION PERIOD: 12 mos.
BRIEF DESCRIPTION: approx 42,000 nsf lease space

PROGRAM ESTIMATE : 6-1-88

TYPE OF PLAN: model
TYPE OF CONSTRUCTION:
QUALITY: medium
BUILDING CAPACITY: 420 people

Floor		
Below Grade Levels:		
Area parking	≈ 18,900	S.F.
Area		S.F.
Total Area		S.F.
Ground Floor:		
Area 14,000 x 1.35	≈ 18,900 g	S.F.
Area		S.F.
Total Area		S.F.
Supported Levels:		
Area 1st	18,900 g	S.F.
Area 2nd	18,900 g	S.F.
Area		S.F.
Area		S.F.
Area		S.F.
Total Area		S.F.
Miscellaneous		
Area		S.F.
Area		S.F.
Area		S.F.
Area		S.F.
Total Area		S.F.
Net Finished Area		S.F.
Net Floor Area		S.F.
Gross Floor Area	75,600 g	S.F.
Roof		
Total Area	18,900	S.F.
Comments: single ply		
Volume		
Depth of Floor System:		
Minimum	12	in.
Maximum	18	in.
Foundation Wall Ht:	12	Ft.
Flr. to Flr. Ht:	14	Ft.
Flr. to Clg. Ht:	10	Ft.
Subgrade Volume:	226,800	C.F.
Above Grade Volume:	529,200	C.F.
Total Bldg. Volume:	755,000	C.F.

Wall Areas		
Foundation Walls:	600 L.F.,	8400 S.F.
Frost Walls:	48 L.F.,	192 S.F.
Exterior Closure: 600 lf x 44	Total	26,400 S.F.
Comment:		
Fenestration:	20 %	5280 S.F.
	%	S.F.
Exterior Wall: FB/DW	80 %	21,120 S.F.
	%	S.F.
Site Work 113,50 - 18,900 (site - bldg)	=	94,250
Parking: 82,469	S.F. (For	236 Cars)
Access Roads:	L.F. (X	Ft. Wide)
Sidewalk: 2456 sf	589 L.F. (X	4 Ft. Wide)
Landscaping: 9,425	S.F. (	10 % Unbuilt Site)
Building Codes		
City: Chambers, Ga.	County:	Fulton
National: SBC, NEC, NPC	Other:	ASHRAE
Loading		
Roof: 20 psf.,	Ground Flr.:	50 psf.
Supported Flrs: 75 avg psf.,	Corr:	psf.
Balcony: psf.,	Part'n. Allow:	psf.
Misc:		psf.
Live Load Reduction:		
Wind:		
Earthquake:	Zone:	
Comment:		
Soil: Type		
Bearing Capacity:		6 K.S.F.
Frost Depth:		.5 Ft.
Frame		
Type:	Bay Spacing:	25 x 22
Foundation: Standard	spread ftgs & strip ftgs	
Special		
Substructure:	12" Conc Walls & Conc cols	
Comment:		
Superstructure: Vert. steel	Horiz.	steel
Fireproofing: spray-on	☐ Columns	Hrs
☐ Girders	Hrs, ☐ Beams	Hrs
☐ Floor	Hrs, ☐ None	

Page 1 of 6

Figure 6.15

ASSEMBLY NUMBER	DESCRIPTION	QTY	UNIT	TOTAL COST UNIT	TOTAL COST TOTAL	COST PER S.F.
1.0	**Foundations**					
1.1-120-7900	Spread Ftgs 8.5' □ x 27" (int.)	16	ea	940	15040	
1.1-120-7700	✓ ✓ 6' □ x 20" (ext.)	20	ea	395	7900	
1.1-120-7610	✓ ✓ 5.5' □ x 18" (corner)	4	ea	310	1240	
1.1-140-2700	Strip Ftg - 24" x 12" - reinf.	600	LF	25.55	15330	
1.1-210-7260	CIP Wall - 12" x 12'	300	LF	115	34500	
1.1-210-5260	✓ ✓ 12" x 8'	✓	LF	76	22800	
1.1-210-1500	Elev. Pit Walls	48	LF	30.20	1450	
1.1-292-5800	Asphalt Bd & Mastic W.P. 1/4" x 12'	300	LF	13.09	3927	
1.1-292-5400	✓ ✓ x 8'	✓	LF	8.61	2583	
1.1-292-5000	✓ ✓ x 4'	48	LF	4.31	207	
1.1-294-1000	4" PVC Perimeter Drain	600	LF	4.00	2400	
1.9-100-4620	Bldg Exc - avg 8' - common earth	18,900	SF	1.58	29862	
	Total				137,239	
2.0	**Slab on Grade**					
2.1-200-3460	SOG @ Parking Level	18,900	SF	3.28	61992	
3.0	**Superstructure**					
3.1-114-7900	CIP Col. sq. tied - 16" □	240	LF	46	11040	
3.1-130-	Stl Col. @ P.H. (avg lines 1200, 5400)	224	LF	26.98	6044	
3.1-130-	✓ ✓ w/o P.H. (avg lines 1800, 5400)	1514	LF	29.11	44073	
3.5-160-5600	CIP Multispan Joist Slab @ 1st	18,900	SF	8.15	154035	
3.5-460-6050	Stl Joists & Bms on Cols @ 2,3,PH	38,550	SF	7.01	270236	
3.7-420-3700	✓ ✓ ✓ @ Roof	18,900	SF	3.14	59346	
3.9-100-0780	Stairs - conc/mtl pans	12	ea	5470	65640	
	Total				610414	
4.0	**Exterior Closure**					
4.1-252-5400	Ext. Wall - FB, 16ga @ 16"	21,120	SF	12.74	269069	
6.1-580-0700	✓ - 5/8" FR GB	✓	SF	.63	13306	
6.1-580-0960	✓ - tape & fin.	✓	SF	.27	5702	
4.6-100-6950	Entry - Alum & Glass - Dbl	1	opng	3065	3065	
4.6-100-3950	3070 HM Label door @ stairs	2	ea	975	1950	
4.7-582-1700	Windows - Alum Fram'g	5280	SF	12.50	66000	
4.7-584-1000	✓ - 1/2" Insul. Glass	✓	SF	11.35	59922	
	Total				419020	

Page 2 of 6

Figure 6.16

ASSEMBLY NUMBER	DESCRIPTION	QTY	UNIT	TOTAL COST UNIT	TOTAL COST TOTAL	COST PER S.F.
5.0	**Roofing**					
5.1-220-2100	Single Ply Roofing	18,900	sf	.86	16254	
5.7-101-2350	3" Polystyrene Insul.	✓	sf	1.19	22491	
5.8-100-0800	2'6" X 8'0" Roof Hatch	1	ea	960	960	
	Total				39705	
6.0	**Interior Construction**					
6.1-210-6000	8" CMU @ Bsmnt	2028	sf	4.92	9978	
6.1-510-5400	Drywall @ 1st - 3rd	10,416	sf	2.54	26457	
6.4-220-7220	3'0" 7'0" HM Labeled Doors/Frames	24	ea	429	10296	
	Hdwr - Allow	✓	ea	300	7200	
6.5-100-0320	Paint CMU	4056	sf	.80	3245	
6.5-100-0340	✓ add for block filler	✓	sf	.21	852	
6.5-100-0140	Paint ceiling @ stairs & RR	1984	sf	.35	694	
6.5-100-0080	Paint DW @ Stairs	6048	sf	.56	3387	
6.5-100-1820	VKC @ Lobbies	3072	sf	1.50	4608	
6.5-100-1940	Ceramic Tile @ RR Walls	1536	sf	3.69	5668	
6.6-100-1740	✓ ✓ Floors	576	sf	4.38	2523	
6.6-100-1800	Quarry Tile @ Lobbies	2064	sf	5.25	10836	
6.7-100-4800	DW Ceiling @ RR	576	sf	2.28	1313	
6.7-100-6300	Alum ✓ @ Lobbies	2064	sf	7.36	15191	
	Specialties - allow	56,700	sf	.50	28350	
6.7-100-5800	Acoustical lay-in @ tenant	51,948	sf	1.42	73766	
	Total				204364	
7.0	**Conveying**					
7.1-100-2300	Elev. - 89,000 ÷ 5 = 17,800 x 4 = 71,200	2	ea	71,200	142400	

Page 3 of 6

Figure 6.17

ASSEMBLY NUMBER	DESCRIPTION	QTY	UNIT	TOTAL COST UNIT	TOTAL COST TOTAL	COST PER S.F
8.0	**Mechanical System**					
141-610-2720	Plumbing	66,150	sf	2.48	164052	
141-610-2770	HVAC	✓	sf	4.90	324135	
	Total				488187	
9.0	**Electrical**					
C9.0-117	Service & Distribution	66,150	sf	1.39	91949	
	Lighting	✓	sf	3.20	211680	
	Devices	✓	sf	.13	8600	
	Equip. Connections	✓	sf	.51	33736	
	Basic Matls	✓	sf	2.06	136269	
	Fire Alarm & D.	✓	sf	.27	17861	
	Total				500095	

Page 4 of 6

Figure 6.18

ASSEMBLY NUMBER	DESCRIPTION	QTY	UNIT	TOTAL COST		COST PER S.F.
				UNIT	TOTAL	
10.0	**General Conditions**					
	See Summary					
11.0	**Special Construction**					
12.0	**Site Work**					
12.3-110-1840	Trenching for H_2O, SS – 2X 125 LF	250	LF	8.17	2042	
12.3-510-9060	SS piping	125	LF	8.65	1081	
12.3-540-2160	H_2O ✓	125	LF	17.35	2169	
12.5-510-1800	Parking	236	cars	415	97940	
12.7-500-5820	Mtl Halide Lighting	10	ea	1825	18250	
	Total				121482	
13.0	**Miscellaneous**					

Page 5 of 6

Figure 6.19

PRELIMINARY
ESTIMATE (Cost Summary)

PROJECT Chambers Office Bldg | TOTAL AREA 75,600 gsf | SHEET NO.

LOCATION 103 Chambers Blvd. - Chambers, Ga. | TOTAL VOLUME 755,000 | ESTIMATE NO.

ARCHITECT | COST PER S.F. | DATE

OWNER Chambers Investment Inc. | COST PER C.F. | NO. OF STORIES

QUANTITIES BY: RK | PRICES BY: RK | EXTENSIONS BY: RK | CHECKED BY:

NO.	DESCRIPTION	SUB TOTAL COST	COST/S.F.	%
1.0	**Foundation**	137239		
2.0	**Substructure**	61992		
3.0	**Superstructure**	610414		
4.0	**Exterior Closure**	419020		
5.0	**Roofing**	39705		
6.0	**Interior Construction**	204364		
7.0	**Conveying**	142400		
8.0	**Mechanical System**	488187		
9.0	**Electrical**	500095		
10.0	**General Conditions (Breakdown)**	—		
11.0	**Special Construction**	—		
12.0	**Site Work**	121482		
	Building Sub Total	$ 2,724,898		

$ 2,724,898

Sales Tax 4 % × Sub Total $ /2 = $ 54,474

General Conditions (%) 16 % × Sub Total $ =

General Conditions $ 435,984

Sub Total "A" $ 3,215,380

Overhead & Profit 10 % × Sub Total "A" $ = $ 321,538

Sub Total "B" $

~~Profit~~ Bid Conditions 8 % × Sub Total "B" $ = $ 257,230

Sub Total "C" $ 3,794,148

Location Factor Atlanta @ 89.2 % × Sub Total "C" $ =

Adjusted Building Cost $ 3,384,380

Architects Fee % × Adjusted Building Cost $ = $

~~Contingency~~ Inflation: 1-1-88 to 6-89 @ 4% % × Adjusted Building Cost $ = $ x 1.061

Total Cost 3,590,827

Square Foot Cost $ / S.F. = $/S.F.

Cubic Foot Cost $ / C.F. = $/C.F.

Page 6 of 6

Figure 6.20

141 | S.F. & C.F. and % of Total Costs

	141 200	S.F. & C.F. Costs	UNIT	UNIT COSTS ¼	MEDIAN	¾	% OF TOTAL ¼	MEDIAN	¾	
570	1800	Equipment	S.F.	1.78	3.52	5.05	3.20%	5.90%	7.20%	570
	2720	Plumbing		3.21	4.76	6.60	5.70%	6.90%	9.40%	
	2770	Heating, ventilating, air conditioning		3.81	5.60	7.20	6.60%	8%	10.40%	
	2900	Electrical		4.44	6.40	8.55	7.60%	9.70%	11.70%	
	3100	Total: Mechanical & Electrical	↓	10.35	14.70	19.30	17.30%	22.40%	27.20%	
590	0010	**MOTELS**	S.F.	50	51.55	76.45				590
	0020	Total project costs	C.F.	3.45	5.10	8.50				
	2720	Plumbing	S.F.	3.45	4.31	5.25	9.40%	10.50%	12.50%	
	2770	Heating, ventilating, air conditioning		1.82	3.14	5.65	4.90%	5.60%	10%	
	2900	Electrical		3.19	3.90	4.90	7.10%	8.10%	10.40%	
	3100	Total: Mechanical & Electrical	↓	7.55	9.45	13.30	18.50%	23.10%	26.10%	
	5000									
	9000	Per rental unit, total cost	Unit	14,200	26,100	36,300				
	9500	Total: Mechanical & Electrical	"	3,900	5,500	5,985				
600	0010	**NURSING HOMES**	S.F.	53.85	69.75	86				600
	0020	Total project costs	C.F.	4.27	5.60	7.30				
	1800	Equipment	S.F.	1.68	2.19	3.42	2%	3.70%	6%	
	2720	Plumbing		4.67	5.70	8.50	8.30%	10.30%	14.10%	
	2770	Heating, ventilating, air conditioning		5	7.25	8.50	10.60%	11.70%	11.80%	
	2900	Electrical		5.40	7.75	8.40	9.70%	10.90%	12.50%	
	3100	Total: Mechanical & Electrical	↓	12.30	16.40	24.15	22%	28.10%	33.20%	
	3200									
	9000	Per bed or person, total cost	Bed	20,900	27,100	36,300				
610	0010	**OFFICES** Low-Rise (1 to 4 story)	S.F.	43.15	55.40	72.95				610
	0020	Total project costs	C.F.	3.16	4.45	5.95				
	0100	Sitework	S.F.	3.17	5.35	8.15	5.20%	9.20%	13.70%	
	0500	Masonry		1.67	3.39	6.45	2.90	5.90%	8.60%	
	1800	Equipment		.60	1.01	2.69	1.30%	1.70%	4%	
	2720	Plumbing		1.65	2.48	3.54	3.60%	4.50%	6%	
	2770	Heating, ventilating, air conditioning		3.55	4.90	7.20	7.20%	10.40%	11.90%	
	2900	Electrical		3.61	5	6.90	7.40%	9.50%	11%	
	3100	Total: Mechanical & Electrical	↓	7.50	11.15	16.10	14.70%	20.50%	26.80%	
620	0010	**OFFICES** Mid-Rise (5 to 10 story)	S.F.	47.90	59.05	80.60				620
	0020	Total project costs	C.F.	3.33	4.25	6.05				
	2720	Plumbing	S.F.	1.47	2.23	3.21	2.80%	3.60%	4.50%	
	2770	Heating, ventilating, air conditioning		3.62	5.15	8.25	7.60%	9.30%	11%	
	2900	Electrical		3.08	4.41	7.45	6.50%	8.20%	10%	
	3100	Total: Mechanical & Electrical		8.20	11	18.65	16.70%	21.10%	25.70%	
630	0010	**OFFICES** High-Rise (11 to 20 story)	↓	57.50	73.70	92.80				630
	0020	Total project costs	C.F.	3.68	5.20	7.50				
	2900	Electrical	S.F.	2.87	4.18	6.45	5.80%	7%	10.50%	
	3100	Total: Mechanical & Electrical		9.75	13.35	24.35	16.50%	21.40%	29.40%	
640	0010	**POLICE STATIONS**	↓	71.90	92.30	117				640
	0020	Total project costs	C.F.	5.25	6.85	9.05				
	0500	Masonry	S.F.	7.55	11.70	14.45	6.80%	10.60%	13.20%	
	1140	Roofing		2.27	2.43	5.45	2%	2.20%	4.20%	
	1350	Glass & glazing		.65	.87	.93	.60%	.80%	.80%	
	1570	Floor covering		.32	.65	.72	.30%	.70%	.70%	
	1580	Painting		1.17	1.40	1.94	1.10%	1.50%	1.90%	
	1800	Equipment		1.13	5.90	9.90	2%	6%	13.30%	
	2720	Plumbing		4.06	5.75	9.45	5.70%	6.80%	10.60%	
	2770	Heating, ventilating, air conditioning		5.75	7.80	11.10	7%	10.50%	11.90%	
	2900	Electrical		7.20	11.20	14.45	9.40%	11.60%	14.50%	
	3100	Total: Mechanical & Electrical	↓	19.50	24.35	31.30	22.60%	27.50%	33%	

492

For expanded coverage of these items see *Means Square Foot Cost Data 1988*

Figure 6.21

ELECTRICAL | C9.0-110 | Estimating Procedure

A Conceptual Estimate of the costs for a building, when final drawings are not available, can be quickly figured by using **Table B9.0-117 Cost Per S.F. For Electrical Systems For Various Building Types.** The following definitions apply to this table.

1. **Service And Distribution:** This system includes the incoming primary feeder from the power company, the main building transformer, metering arrangement, switchboards, distribution panel boards, stepdown transformers, power and lighting panels. Items marked (*) include the cost of the primary feeder and transformer. In all other projects the cost of the primary feeder and transformer is paid for by the local power company.
2. **Lighting:** Includes all interior fixtures for decor, illumination, exit and emergency lighting. Fixtures for exterior building lighting are included but parking area lighting is not included unless mentioned. See also Section B9.2 for detailed analysis of lighting requirements and costs.
3. **Devices:** Includes all outlet boxes, receptacles, switches for lighting control, dimmers and cover plates.
4. **Equipment Connections:** Includes all materials and equipment for making connections for Heating, Ventilating and Air Conditioning, Food Service and other motorized items requiring connections.
5. **Basic Materials:** This category includes all disconnect power switches not part of service equipment, raceways for wires, pull boxes, junction boxes, supports, fittings, grounding materials, wireways, busways and cable systems.
6. **Special Systems:** Includes installed equipment only for the particular system such as fire detection and alarm, sound, emergency generator and others as listed in the table.

(56) Table 9.0-117 Cost Per S.F. for Electric Systems for Various Building Types

Type Construction	1. Service & Distrib.	2. Lighting	3. Devices	4. Equipment Connections	5. Basic Materials	6. Special Systems		
						Fire Alarm & Detection	Lighting Protection	Master TV Antenna
Apartment, luxury high rise	$ .95	$ .66	$ .47	$ .61	$1.62	$ .28		$ .18
Apartment, low rise	.54	.56	.41	.50	.94	.24		
Auditorium	1.23	3.41	.36	.90	1.97	.39		
Bank, branch office	1.44	3.72	.61	.90	1.84	1.09		
Bank, main office	1.10	2.03	.18	.40	1.99	.57		
Church	.70	2.00	.24	.19	.90	.57		
*College, science building	1.26	2.46	.75	.64	1.97	.48		
*College library	1.01	1.49	.15	.43	1.17	.57		
*College, physical education center	1.59	2.11	.25	.34	.90	.31		
Department store	.53	1.47	.15	.60	1.60	.24		
*Dormitory, college	.71	1.87	.16	.40	1.56	.42		.25
Drive-in donut shop	2.03	5.63	.88	.91	2.55	—		
Garage, commercial	.27	.68	.10	.27	.53	—		
*Hospital, general	3.92	2.89	1.03	.73	3.19	.35	$ .07	
*Hospital, pediatric	3.43	4.33	.86	2.64	5.92	.41		.31
*Hotel, airport	1.54	2.38	.16	.37	2.33	.31	.16	.28
Housing for the elderly	.43	.56	.25	.69	1.97	.41		.24
Manufacturing, food processing	.96	2.93	.13	1.33	2.15	.24		
Manufacturing apparel	.63	1.52	.19	.50	1.15	.20		
Manufacturing, tools	1.44	3.59	.17	.60	1.93	.25		
Medical clinic	.51	1.13	.29	.87	1.41	.40		
Nursing home	1.02	2.38	.31	.26	1.97	.54		.18
Office building	1.39	3.20	.13	.51	2.06	.27	.13	
Radio-TV studio	.92	3.11	.45	.91	2.29	.37		
Restaurant	3.68	3.13	.58	1.48	2.93	.20		
Retail store	.77	1.65	.16	.37	.90	—		
School, elementary	1.30	2.98	.36	.37	2.48	.34		.11
School, junior high	.78	2.49	.16	.66	1.97	.41		
*School, senior high	.87	1.96	.33	.86	2.19	.35		
Supermarket	.89	1.69	.22	1.42	1.89	.14		
*Telephone exchange	2.03	.66	.10	.57	1.21	.64		
Theater	1.68	2.26	.36	1.22	1.89	.48		
Town Hall	1.02	1.79	.36	.46	2.50	.31		
*U.S. Post Office	3.05	2.34	.37	.68	1.81	.31		
Warehouse, grocery	.55	1.00	.10	.37	1.34	.18		

* Includes cost of primary feeder and transformer. Cont'd on next page.

Figure 6.22a

ELECTRICAL	C9.0-110	Estimating Procedure

COST ASSUMPTIONS:

Each of the projects analyzed in Table C9.0-117 were bid within the last 10 years in the Northeastern part of the United States. Bid prices have been adjusted to Jan. 1, 1988 levels. The list of projects is by no means all-inclusive, yet by carefully examining the various systems for a particular building type, certain cost relationships will emerge. The use of Section C14 with the S.F. and C.F. electrical costs should produce a budget S.F. cost for the electrical portion of a job that is consistent with the amount of design information normally available at the conceptual estimate stage.

(56) Table 9.0-117 (Cont.) Cost Per S.F. for Electric Systems for Various Building Types

Type Construction	6. Special Systems, Cont'd.						
	Intercom Systems	Sound Systems	Closed Circuit TV	Snow Melting	Emergency Generator	Security	Master Clock Sys.
Apartment, luxury high rise	$ .40						
Apartment, low rise	.28						
Auditorium		$1.05	$ .49		$ .81		
Bank, branch office	.54		1.13			$ .95	
Bank, main office	.31		.23		.65	.52	$ .20
Church	.39						
* College, science building	.39				.81		.24
* College, library					.43		
* College, physical education center		.53					
Department store					.15		
* Dormitory, college	.52						
Drive-in donut shop							.07
Garage, commercial							.05
* Hospital, general	.41		.14		1.13		
* Hospital, pediatric	2.80	.28	.31		.71		
* Hotel, airport	.41				.42		
Housing for the elderly	.49						
Manufacturing food processing		.16			1.46		
Manufacturing, apparel		.24					
Manufacturing, tools		.31		$ .18			
Medical clinic							
Nursing home	.93				.36		
Office building		.12			.36	.14	.05
Radio-TV studio	.53				.92		.38
Restaurant		.24					
Retail store							
School, elementary		.14					.14
School, junior high		.45			.30		.31
* School, senior high	.37		.24		.43	.20	.21
Supermarket		.17			.38	.24	
* Telephone exchange					3.71	.10	
Theater		.36					
Town Hall							.14
* U.S. Post Office	.36			.05	.42		
Warehouse, grocery	.22						

*Includes cost of primary feeder and transformer. Cont'd on next page.

445

Figure 6.22b

Sales Tax

To raise revenues, many states charge a percentage tax on all sales within the state. Some states also allow municipalities to tax sales. These taxes must be added to the cost of the materials used in construction. It should be noted, however, that some projects are exempt from sales taxes. Included in the exempt category are projects built with public funds and, in some areas, projects for certain classes of non-profit organizations. Sales taxes and exemption requirements change from state to state and from city to city. Exact requirements must be determined for each location. Figure 6.23, a page from the 1988 *Means Assemblies Cost Data*, lists the percentages charged by those states that have sales taxes. The municipal rates are not included. The Chambers project is subject to both a state sales tax and a city sales tax. The table gives a 3% rate for Georgia; an additional 1% is assumed for the city of Chambers, for a total of 4%. Generally, materials are about one-half of the cost of a project. The sales tax determination for the Chambers project is:

$$\text{Total Tax Rate} \times \frac{\text{Building Subtotal}}{2} = \text{Sales Tax}$$

or

$$.04 \times \frac{\$2{,}723{,}698}{2} = \$54{,}474$$

General Conditions

The *Building Subtotal* contains the actual expenses to the general contractor for direct costs (his own company's labor, subcontractors, and materials), but does not include *General Conditions Costs* (supervision, insurance, job office, etc.). These costs represent a significant portion of the total cost of a construction project and must be estimated very carefully by the general contractor using the unit cost method. We can estimate General Conditions Costs as a percentage of the *Building Subtotal* using the information (from *Means Assemblies Cost Data*) shown in Figure 6.24. The table provides values for the various elements of a contractor's total overhead expense as percentages of the labor costs or as percentages of both the labor and material costs of the project. Of the items listed, all except *Main Office Expense* are attributable to the specific project and should be included in the *General Conditions* percentage. The *Main Office Expense* is included in *Overhead and Profit*, calculated below. Therefore, *General Conditions* costs for our project represent the grand total, 23.7%, less the *Main Office Expense*, 7.7%, or 16%. The result is:

$$\$2{,}723{,}698 \times .16 = \$435{,}792$$

Next, *Building Subtotal*, *Sales Tax*, and *General Conditions Costs* are added to obtain Subtotal A.

Overhead and Profit and Bid Conditions

Overhead and Profit includes a percentage for *Main Office Expense* and an amount for *Contractor's Profit* for the job. This total percentage has been found to vary widely according to the level of competition at the time of the bid. The author has had success using 10% of Subtotal A for overhead and profit and making an additional adjustment for bid conditions. This adjustment is 1% to 2% during high competition, around 5% during moderate competition, and approximately 8% during low competition. Business activity trends for both Chambers and Atlanta are high and climbing at the time of the estimate. Using this factor as a leading indicator, it is assumed that construction volume at bid time will be high. Consequently, the number of contractors who bid the project should be limited, and the level of competition should be low.

GENERAL CONDITIONS	C10.2-300	Unemployment & Social Security

(80) Unemployment Taxes and Social Security Taxes

Mass. State Unemployment tax ranges from 1.2% to 5.4% plus an experience rating assessment the following year, on the first $7,000 of wages. Federal Unemployment tax is 3.5% of the first $7,000 of wages. This is reduced by a credit for payment to the state. The minimum Federal Unemployment tax is .8% after all credits.

Combined rates in Mass. thus vary from 2.0% to 6.2% of the first $7,000 of wages. Combined average U.S. rate is about 5.5% of the first $7,000. Contractors with permanent workers will pay less since the average annual wages for skilled workers is $21.45 x 2,000 hours or about $42,900 per year. The average combined rate for U.S. would thus be 5.5% x 7,000 ÷ 42,900 = .90% of total wages for permanent employees.

Rates not only vary from state to state but also with the experience rating of the contractor.

Social Security (FICA) for 1988 is estimated at time of publication to be 7.51% of wages up to $45,000.

GENERAL CONDITIONS	C10.2-400	Overtime

Overtime

One way to improve the completion date of a project or eliminate negative float from a schedule, is to compress activity duration times. This can be achieved by increasing the crew size or working overtime with the proposed crew.

To determine the costs of working overtime to compress activity duration times, consider the following examples. Below is an overtime efficiency and cost chart based on a five, six, or seven day week with an eight through twelve hour day. Payroll percentage increases for time and one half and double time are shown for the various working days.

Days per Week	Hours per Day	Production Efficiency					Payroll Cost Factors	
		1 Week	2 Weeks	3 Weeks	4 Weeks	Average 4 Weeks	@ 1-1/2 Times	@ 2 Times
5	8	100%	100%	100%	100%	100%	100%	100%
	9	100	100	95	90	96.25	105.6	111.1
	10	100	95	90	85	91.25	110.0	120.0
	11	95	90	75	65	81.25	113.6	127.3
	12	90	85	70	60	76.25	116.7	133.3
6	8	100	100	95	90	96.25	108.3	116.7
	9	100	95	90	85	92.50	113.0	125.9
	10	95	90	85	80	87.50	116.7	133.3
	11	95	85	70	65	78.75	119.7	139.4
	12	90	80	65	60	73.75	122.2	144.4
7	8	100	95	85	75	88.75	114.3	128.6
	9	95	90	80	70	83.75	118.3	136.5
	10	90	85	75	65	78.75	121.4	142.9
	11	85	80	65	60	72.50	124.0	148.1
	12	85	75	60	55	68.75	126.2	152.4

GENERAL CONDITIONS	C10.3-100	General

Sales Tax

State sales tax on materials is tabulated below (5 states have no sales tax). Many states allow local jurisdictions, such as a county or city, to levy additional sales tax.

Some projects may be sales tax exempt, particularly those constructed with public funds.

State	Tax	State	Tax	State	Tax	State	Tax
Alabama	4%	Illinois	7%	Montana	0%	Rhode Island	6%
Alaska	0	Indiana	5	Nebraska	3.5	South Carolina	5
Arizona	5	Iowa	4	Nevada	5.75	South Dakota	4
Arkansas	4	Kansas	4	New Hampshire	0	Tennessee	5.5
California	6	Kentucky	5	New Jersey	6	Texas	6.25
Colorado	3	Louisiana	4	New Mexico	4.75	Utah	6.5
Connecticut	7.5	Maine	5	New York	4	Vermont	4
Delaware	0	Maryland	5	North Carolina	3	Virginia	4.5
District of Columbia	6	Massachusetts	5	North Dakota	4	Washington	6.5
Florida	5	Michigan	4	Ohio	5.5	West Virginia	5
Georgia	3	Minnesota	6	Oklahoma	3.25	Wisconsin	5
Hawaii	5	Mississippi	6	Oregon	0	Wyoming	3
Idaho	5	Missouri	4.225	Pennsylvania	6	Average	4.41%

470

Figure 6.23

GENERAL CONDITIONS | **C10.0-100** | **Overhead**

Table 10.0-101 General Contractor's Overhead

The table below shows a contractor's overhead as a percentage of direct cost in two ways. The figures on the right are for the overhead, markup based on both material and labor. The figures on the left are based on the entire overhead applied only to the labor. This figure would be used if the owner supplied the materials or if a contract is for labor only.

Items of General Contractor's Indirect Costs	% of Direct Costs	
	As a Markup of Labor Only	As a Markup of Both Material and Labor
Field Supervision	6.0%	2.4%
Main Office Expense (see details below)	9.2	7.7
Tools and Minor Equipment	1.0	0.4
Workers' Compensation & Employers' Liability. See (79)	11.4	4.6
Field Office, Sheds, Photos, Etc.	2.0	0.8
Performance and Payment Bond, .5% to .9%. See (78)	0.7	0.7
Unemployment Tax See (80) (Combined Federal and State)	5.5	2.2
Social Security and Medicare (7.51% of first $40,000)	7.5	3.0
Sales Tax — add if applicable 48/80 x % as markup of total direct costs including both material and labor. See (82)		
Sub Total	43.3%	21.8%
*Builder's Risk Insurance ranges from 0.215% to 0.586%. See (77)	0.4	0.4
*Public Liability Insurance	1.5	1.5
Grand Total	45.2%	23.7%

Paid by Owner or Contractor

Table 10.0-102 Main Office Expense

A General Contractor's main office expense consists of many items not detailed in the front portion of the book. The percentage of main office expense declines with increased annual volume of the contractor. Typical main office expense ranges from 2% to 20% with the median about 7.2% of total volume. This equals about 7.7% of direct costs. The following are approximate percentages of total overhead for different items usually included in a General Contractor's main office overhead. With different accounting procedures, these percentages may vary.

Item	Typical Range	Average
Managers', clerical and estimators' salaries	40% to 55%	48%
Profit sharing, pension and bonus plans	2 to 20	12
Insurance	5 to 8	6
Estimating and project management (not including salaries)	5 to 9	7
Legal, accounting and data processing	0.5 to 5	3
Automobile and light truck expense	2 to 8	5
Depreciation of overhead capital expenditures	2 to 6	4
Maintenance of office equipment	0.1 to 1.5	1
Office rental	3 to 5	4
Utilities including phone and light	1 to 3	2
Miscellaneous	5 to 15	8
Total		100%

Figure 6.24

Therefore, an adjustment of 8% will be used. The overhead and profit and bid conditions calculations are as follows.

Overhead & Profit	$3,213,964 × .10 = $321,396
Bid Conditions	$3,213,964 × .08 = $257,117

Subtotal A, Overhead and Profit, and Bid Conditions are all added to obtain Subtotal B.

Location

The prices from *Means Assemblies Cost Data* are nationwide averages and must be adjusted to Chambers, Georgia. Since Chambers is very close to Atlanta, Means City Cost Index for Atlanta will be used to adjust Subtotal B for location. (See Figure 6.25).

$3,792,477 × .892 = $3,382,889

Inflation

The prices in *Means Assemblies Cost Data* are dated January 1, 1988. These costs must be adjusted to include the amount of inflation between January 1, 1988 and the scheduled bid date. The scheduled bid date indicated on the project program is June, 1989. Inflation trend studies for Atlanta, like those described in Chapter 3, suggest that the inflation for the year and a half between January 1, 1988 and June, 1989 will average 4%. The total inflation factor is:

1.04 × 1.02 = 1.061

The inflation adjustment is:

$3,382,889 × 1.061 = $3,589,245

Program Estimate

The figure above is an estimate of the price that the contractor will charge to construct the Chambers office building. As stated in Chapter 1, to obtain the total project cost, prices for design fees, equipment and furnishings, financing, and planning costs must be added. As stated in Chapter 4, only prices for design fees and planning costs are needed for the Chambers project. The same percentages used in Chapter 4 (6.7% for design fees and 2% for planning costs) will be used here. It is common practice to round these numbers to the nearest $1000.

Construction Contract Cost	$3,589,000
Design Fees @ 6.7%	240,000
Total Construction Cost	$3,829,000
Planning Costs @ 2%	77,000
Total Project Costs	$3,906,000

These items—the construction contract cost, design fees, equipment and furnishings cost, financing, planning costs as required, and the total project costs comprise the program estimate. The *Total Project Cost* becomes the project budget.

Summary

As part of the detailed study expressing project needs, and to establish the project budget, a programming estimate is made using either the Square Foot method or a modeling exercise using a combination of the Assemblies and Square Foot methods. The modeling exercise procedure involves designing a model, preparing an Assemblies/Square Foot estimate based on the model, and performing the needed adjustments and additions to obtain a total project cost.

CITY COST INDEXES | C13.3-000 | Building Systems

DIV. NO.	BUILDING SYSTEMS	NEW HAVEN, CT MAT.	INST.	TOTAL	NORWALK, CT MAT.	INST.	TOTAL	STAMFORD, CT MAT.	INST.	TOTAL	WATERBURY, CT MAT.	INST.	TOTAL	WILMINGTON, DE MAT.	INST.	TOTAL
1-2	FOUND/SUBSTRUCTURES	101.8	103.3	102.7	106.4	103.3	104.5	118.9	103.9	109.6	112.4	101.5	105.6	102.5	109.0	106.6
3	SUPERSTRUCTURES	95.1	103.7	99.6	96.3	103.6	100.1	101.5	104.3	103.0	99.8	101.6	100.8	94.8	108.4	101.9
4	EXTERIOR CLOSURE	106.9	106.0	106.3	99.2	104.5	102.6	103.1	104.4	103.9	97.3	103.4	101.2	91.4	96.8	94.9
5	ROOFING	88.8	105.4	94.5	89.9	105.9	95.4	89.0	106.9	95.1	90.0	102.4	94.3	91.2	103.1	95.3
6	INTERIOR CONSTRUCTION	106.4	105.3	105.8	109.5	105.2	107.2	109.6	106.1	107.7	105.4	102.4	103.8	97.4	101.0	99.4
7	CONVEYING	100.0	105.9	101.9	100.0	105.9	101.9	100.0	105.9	101.9	100.0	105.9	101.9	100.0	105.1	101.6
8	MECHANICAL	101.9	103.2	102.6	102.4	103.2	102.8	101.3	107.8	104.6	100.6	97.9	99.2	100.0	105.3	102.7
9	ELECTRICAL	93.7	105.4	101.9	97.7	104.9	102.7	95.5	106.2	103.0	92.5	102.5	99.5	106.9	105.2	105.7
11	EQUIPMENT	100.0	105.3	101.4	100.0	104.8	101.3	100.0	106.1	101.6	100.0	102.4	100.6	100.0	105.1	101.3
12	SITEWORK	120.1	99.5	110.7	126.0	101.6	114.9	126.9	102.2	115.6	108.8	100.5	105.1	118.7	101.7	111.0
1-12	WEIGHTED AVERAGE	101.4	104.4	103.0	102.1	104.1	103.2	104.2	105.4	104.9	100.9	101.6	101.3	98.9	103.9	101.6

DIV. NO.	BUILDING SYSTEMS	WASHINGTON, D.C. MAT.	INST.	TOTAL	FT LAUDERDALE, FL MAT.	INST.	TOTAL	JACKSONVILLE, FL MAT.	INST.	TOTAL	MIAMI, FL MAT.	INST.	TOTAL	ORLANDO, FL MAT.	INST.	TOTAL
1-2	FOUND/SUBSTRUCTURES	105.3	87.9	94.6	96.8	92.7	94.3	96.6	82.0	87.5	94.7	94.1	94.3	97.4	83.2	88.6
3	SUPERSTRUCTURES	104.1	87.8	95.6	92.7	91.3	92.0	95.8	81.0	88.1	92.2	92.4	92.3	92.4	81.6	86.8
4	EXTERIOR CLOSURE	99.3	94.5	96.2	90.7	88.8	89.5	88.0	73.5	78.7	89.6	83.2	85.5	88.8	64.2	73.0
5	ROOFING	105.9	90.3	100.6	89.2	88.2	88.9	88.3	71.4	82.5	87.9	83.1	86.3	88.1	67.7	81.1
6	INTERIOR CONSTRUCTION	108.9	91.7	99.5	101.2	82.7	91.1	103.0	74.3	87.3	103.2	80.5	90.8	100.8	71.9	85.0
7	CONVEYING	100.0	91.9	97.4	100.0	85.7	95.4	100.0	74.5	91.8	100.0	84.6	95.0	100.0	73.6	91.5
8	MECHANICAL	101.6	84.7	92.9	100.7	85.8	93.0	100.0	78.8	89.1	97.6	89.4	93.4	97.0	78.3	87.4
9	ELECTRICAL	97.4	87.8	90.6	99.3	85.3	89.5	100.1	73.6	81.5	100.5	95.4	96.9	93.4	73.7	79.5
11	EQUIPMENT	100.0	90.5	97.4	100.0	85.7	96.1	100.0	74.5	93.1	100.0	84.6	95.8	100.0	73.6	92.9
12	SITEWORK	88.5	93.0	90.6	110.4	91.7	101.9	121.7	81.7	103.5	98.8	86.3	93.1	97.7	89.6	94.0
1-12	WEIGHTED AVERAGE	102.0	89.4	95.2	97.5	87.9	92.3	98.4	77.1	86.8	96.1	88.4	92.0	95.4	75.3	84.5

DIV. NO.	BUILDING SYSTEMS	TALLAHASSEE, FL MAT.	INST.	TOTAL	TAMPA, FL MAT.	INST.	TOTAL	ALBANY, GA MAT.	INST.	TOTAL	ATLANTA, GA MAT.	INST.	TOTAL	COLUMBUS, GA MAT.	INST.	TOTAL
1-2	FOUND/SUBSTRUCTURES	106.2	75.0	86.9	99.4	96.3	97.5	107.0	81.6	91.3	88.8	84.5	86.1	105.2	79.3	89.2
3	SUPERSTRUCTURES	97.4	73.2	84.8	99.2	93.3	96.1	104.2	79.6	91.3	99.6	82.3	90.6	99.9	77.8	88.4
4	EXTERIOR CLOSURE	88.2	66.3	74.1	97.7	78.6	85.5	85.8	70.0	75.6	91.8	77.6	82.7	89.6	54.4	67.0
5	ROOFING	87.7	59.9	78.2	104.3	69.2	92.3	92.8	68.9	84.6	97.1	71.1	88.2	93.3	66.6	84.1
6	INTERIOR CONSTRUCTION	99.4	60.0	77.9	98.5	79.4	88.1	92.9	70.1	80.5	106.1	82.3	93.1	93.9	62.9	77.0
7	CONVEYING	100.0	67.9	89.7	100.0	79.3	93.3	100.0	78.9	93.2	100.0	78.9	93.2	100.0	78.9	93.2
8	MECHANICAL	99.5	69.2	83.9	97.2	80.1	88.4	100.7	72.4	86.1	102.6	78.7	90.3	98.3	68.5	83.0
9	ELECTRICAL	92.1	69.2	76.0	93.4	80.9	84.6	102.4	70.8	80.2	95.7	80.3	84.9	97.7	61.6	72.4
11	EQUIPMENT	100.0	67.9	91.4	100.0	79.3	94.4	100.0	70.0	91.9	100.0	78.1	94.1	100.0	64.9	90.6
12	SITEWORK	121.0	85.7	104.9	112.3	91.9	103.0	120.1	88.9	105.9	107.0	92.2	100.3	126.2	88.3	108.9
1-12	WEIGHTED AVERAGE	98.4	69.2	82.6	99.2	84.0	91.0	99.7	74.4	86.0	99.2	80.7	89.2	98.9	68.0	82.2

DIV. NO.	BUILDING SYSTEMS	MACON, GA MAT.	INST.	TOTAL	SAVANNAH, GA MAT.	INST.	TOTAL	HONOLULU, HI MAT.	INST.	TOTAL	CEDAR RAPIDS, IA MAT.	INST.	TOTAL	DES MOINES, IA MAT.	INST.	TOTAL
1-2	FOUND/SUBSTRUCTURES	98.3	85.8	90.6	98.6	88.0	92.0	118.3	107.4	111.6	106.7	87.7	94.9	111.8	88.2	97.2
3	SUPERSTRUCTURES	93.7	83.5	88.4	99.7	85.1	92.1	114.8	107.2	110.9	98.9	85.8	92.1	99.6	87.1	93.1
4	EXTERIOR CLOSURE	89.8	69.3	76.6	89.3	70.7	77.4	117.7	115.5	116.3	95.8	79.4	85.2	93.5	82.0	86.1
5	ROOFING	93.0	68.1	84.4	90.7	71.3	84.0	109.5	114.5	111.2	89.9	80.4	86.6	89.1	82.0	86.6
6	INTERIOR CONSTRUCTION	93.5	69.4	80.4	106.9	70.7	87.1	134.9	116.8	125.1	106.1	80.3	92.1	105.4	82.0	92.6
7	CONVEYING	100.0	78.9	93.2	100.0	76.2	92.3	100.0	116.9	105.4	100.0	78.3	93.0	100.0	84.2	94.0
8	MECHANICAL	98.6	71.8	84.8	98.4	72.2	84.9	111.3	114.7	113.0	99.6	80.4	89.7	96.6	82.1	89.1
9	ELECTRICAL	108.5	70.8	82.0	101.8	76.7	84.2	106.2	105.5	105.7	97.6	80.4	85.5	99.4	82.0	87.2
11	EQUIPMENT	100.0	69.3	91.7	100.0	70.6	92.1	100.0	115.8	104.2	100.0	80.3	94.7	100.0	81.9	95.1
12	SITEWORK	119.2	88.1	105.1	121.9	87.2	106.1	135.6	105.9	122.1	103.3	92.6	98.4	102.5	92.4	97.9
1-12	WEIGHTED AVERAGE	97.6	75.1	85.4	99.8	76.8	87.4	115.6	111.7	113.5	100.0	82.3	90.4	99.6	84.0	91.1

DIV. NO.	BUILDING SYSTEMS	DAVENPORT, IA MAT.	INST.	TOTAL	SIOUX CITY, IA MAT.	INST.	TOTAL	WATERLOO, IA MAT.	INST.	TOTAL	BOISE, ID MAT.	INST.	TOTAL	POCATELLO, ID MAT.	INST.	TOTAL
1-2	FOUND/SUBSTRUCTURES	95.5	94.4	94.8	100.3	82.2	89.1	103.8	84.1	91.6	102.9	92.8	96.7	109.7	92.7	99.2
3	SUPERSTRUCTURES	93.1	93.9	93.5	95.5	80.2	87.5	106.3	82.6	94.0	98.0	92.0	94.8	100.6	91.9	96.1
4	EXTERIOR CLOSURE	101.2	90.1	94.0	98.2	70.8	80.6	102.5	73.3	83.8	104.6	86.8	93.2	108.3	88.1	95.4
5	ROOFING	90.4	90.1	90.3	87.4	70.8	81.7	87.7	73.2	82.7	105.5	88.3	99.6	103.6	88.1	98.3
6	INTERIOR CONSTRUCTION	102.5	90.0	95.6	98.3	70.8	83.3	110.5	73.2	90.1	98.8	88.3	93.0	103.5	88.2	95.2
7	CONVEYING	100.0	78.3	93.0	100.0	73.3	91.4	100.0	78.3	93.0	100.0	87.5	96.0	100.0	87.5	96.0
8	MECHANICAL	99.8	90.1	94.8	97.9	71.0	84.0	101.1	73.3	86.7	96.6	88.4	92.4	101.2	88.3	94.6
9	ELECTRICAL	98.1	90.1	92.4	99.7	81.6	87.0	93.0	73.2	79.1	96.7	88.3	90.8	93.7	88.2	89.8
11	EQUIPMENT	100.0	90.0	97.3	100.0	70.7	92.1	100.0	73.1	92.8	100.0	88.2	96.8	100.0	88.1	96.8
12	SITEWORK	110.4	88.4	100.4	115.9	89.9	104.1	113.1	90.2	102.7	92.8	95.8	94.2	102.8	95.0	99.3
1-12	WEIGHTED AVERAGE	98.8	90.8	94.4	98.5	75.8	86.2	102.9	76.8	88.8	99.3	89.4	93.9	102.5	89.6	95.5

DIV. NO.	BUILDING SYSTEMS	CHICAGO, IL MAT.	INST.	TOTAL	DECATUR, IL MAT.	INST.	TOTAL	JOLIET, IL MAT.	INST.	TOTAL	PEORIA, IL MAT.	INST.	TOTAL	ROCKFORD, IL MAT.	INST.	TOTAL
1-2	FOUND/SUBSTRUCTURES	98.8	102.4	101.0	103.4	95.2	98.3	100.8	102.1	101.6	97.5	88.1	91.7	105.2	95.5	99.2
3	SUPERSTRUCTURES	93.8	103.7	99.0	101.0	94.7	97.7	100.5	102.6	101.6	94.6	88.7	91.5	106.8	95.1	100.7
4	EXTERIOR CLOSURE	101.3	104.3	103.2	101.2	92.9	95.9	105.5	100.8	102.5	94.5	94.5	94.5	98.3	98.5	98.4
5	ROOFING	97.5	109.1	101.5	100.8	94.5	98.6	102.0	106.2	103.5	95.6	100.8	97.4	107.3	96.4	103.6
6	INTERIOR CONSTRUCTION	99.3	100.5	99.9	107.3	92.0	99.0	104.2	102.7	103.4	105.9	94.3	99.6	99.8	91.2	95.1
7	CONVEYING	100.0	103.9	101.2	100.0	88.7	96.3	100.0	103.9	101.2	100.0	86.3	95.6	100.0	100.2	100.0
8	MECHANICAL	96.4	96.5	96.4	95.6	93.1	94.3	97.2	95.0	96.1	96.1	92.4	94.2	101.7	99.7	100.7
9	ELECTRICAL	93.9	101.3	99.1	94.2	92.5	93.0	102.5	101.2	101.6	94.6	94.3	94.4	93.4	100.3	98.2
11	EQUIPMENT	100.0	102.6	100.7	100.0	92.4	97.9	100.0	101.1	100.3	100.0	94.2	98.4	100.0	100.2	100.0
12	SITEWORK	106.6	104.8	105.7	112.4	95.3	104.6	113.4	102.2	108.3	119.3	98.9	110.0	115.5	101.8	109.3
1-12	WEIGHTED AVERAGE	98.0	101.8	100.0	100.9	93.3	96.8	101.8	100.7	101.2	98.5	92.4	95.2	102.5	97.3	99.7

479

Figure 6.25

The limitations of using only the Square Foot method are demonstrated by the Chambers office building example, with its own particular requirements. In this case, the problem is meeting code parking requirements on the relatively small site. A parking basement is the solution. This addition causes the project budget to exceed the cost range used in preliminary planning.

7

DESIGN COST CONTROL ESTIMATES

7

DESIGN COST CONTROL ESTIMATES

Once the need analysis and funding processes are complete, the next step in the planning process is project design. Project design consists of two distinct phases: conceptual design and design development.

Conceptual Design

In the conceptual design phase, the designer prepares several design schemes for presentation to the owner. Each presentation consists of schematic drawings showing floor plans, elevations, massing studies, and site plans. The designer also presents a comparison of the schemes detailing the aesthetic and functional differences and the cost advantages and disadvantages of each. The owner then chooses one of the schemes for the designer to further develop. The designer works with the owner to finalize floor plans, elevations, and site plans, and to make initial materials choices. Rough structural, mechanical, and electrical studies are also drawn up and a refined estimate is prepared.

Design Development

In the design development phase, the finalized design and rough engineering studies from the conceptual design phase are used to develop the plans and specifications for building the project. During this phase, periodic checks must be made to confirm that design development is proceeding on schedule and that the project remains within the budget. These checks should occur when design development is scheduled to be 50%, 75%, 90%, and 100% complete. As the phase progresses, more and more information becomes available regarding the actual project design. Usually, most of the general construction details have been finalized by the 75% completion point, and most of the HVAC, plumbing, and electrical details have been finalized by the 90% completion point.

Estimates

Estimates are required at various intervals in both conceptual design and design development. During the conceptual design phase, estimates are needed to compare the costs of the initial design schemes and to check the cost of the scheme that is finally chosen. During the detailed design phase, estimates are prepared for the periodic checks conducted when the design is scheduled to be 50%, 75%, 90%, and 100% complete. The information available for these estimates varies from schematic plans and outline specifications in the conceptual design phase, to a complete set of project documents at the end of the design development phase. The methods used to prepare these estimates range from a combination of the assemblies/square foot methods during the conceptual design phase, to a complete assemblies estimate for all building systems at the end of the detailed design phase.

Assumptions and Adjustments

During the conceptual design and early design development phases, only schematic drawings and outline specifications are available for estimating. Materials, assemblies, and detailed dimensions cannot be figured into the estimate, since this information is not yet available. Consequently, it is necessary to make certain assumptions about the materials and assemblies that are to be used. These assumptions are based on experience and common construction practices, together with information from design aids such as the reference section of *Means Assemblies Cost Data*. As more information becomes available in the design development phase, the costs based on assumptions are replaced with assemblies estimates of the actual design. Usually, enough information is available for an accurate estimate of the cost of the general construction systems at the 75% completion point. Early HVAC, plumbing, and electrical estimates are prepared using the square foot method as shown in Chapter 6. At the 90% completion point, there is generally enough information to prepare a more accurate assemblies estimate of these systems.

It is not unusual for the estimates to indicate that the design has grown beyond the limits of the budget. In such cases, it is necessary to analyze the estimate to determine the reason for the cost overrun, and then identify possible changes in materials and/or design that will lower the cost. These possible changes are evaluated, and the most favorable ones incorporated into the design.

To determine the areas where assumptions are needed, the estimator must analyze the plans and specifications to discover what information is given, what may be inferred, and what must be assumed. To lend some structure to the process and to facilitate a smooth transition to the assemblies/square foot estimating method, the analysis is made according to the UNIFORMAT breakdown. The UNIFORMAT system of organization is a logical, sequential approach that reflects the order in which a building is constructed. The twelve UNIFORMAT divisions are as follows:

1. Foundations
2. Substructures
3. Superstructure
4. Exterior Closure
5. Roofing
6. Interior Construction
7. Conveying
8. Mechanical
9. Electrical
10. General Conditions
11. Special
12. Site Work

Estimating the Sample Project

The following pages demonstrate the assumption/estimate, and the cost overrun identification and correction procedures using the Chambers project as an example. The quantities shown in the sample estimate represent realistic conditions. Assumptions have also been made for items not shown that would normally be included in the plans and specifications of a project of this type.

1.0 Foundation

The foundation section in the UNIFORMAT Division 1 breakdown includes foundation systems, such as footings and pilings; substructure assemblies, such as grade beams and basement walls; foundation waterproofing and drainage systems; and building excavation and backfill. To estimate the costs of these systems, one must know the sizes and materials used for each system, the footprint area of the building, and the depth of the building excavation.

Plans for the Chambers project are shown in Figures 7.1 through 7.10. They illustrate the level of detail that can be expected in the schematic drawings at the end of the conceptual design phase. No outline specifications were submitted in this case. Figure 7.1, the Perspective Section, shows partially open basement walls on strip footings, spread footings supporting columns, and the suggestion of a perimeter drain. Figures 7.2 and 7.3, the Topographic Survey and the Finish Grading Plan, provide the grades that are needed to determine the depth of the building excavation and backfill. Figure 7.5, the Foundation Plan, specifies the dimensions of the basement, as well as the locations and dimensions of concrete strip footings and spread footings. It also shows an elevator pit. The size and depth of the elevator pit walls, the thickness of the basement walls, and the size of the perimeter drain must be assumed. The materials used for the pit and basement walls, the foundation waterproofing, and the perimeter drain must also be assumed. Elevator pits generally range from four to six feet in depth. For this project, a six foot depth will be used. Experience indicates that 6'' concrete walls are appropriate at the elevator pit. An asphalt-coated board and mastic basement waterproofing system is commonly used in the Atlanta area, along with a 4'' PVC perimeter drain.

2.0 Substructures (Slab on Grade)

UNIFORMAT Division 2 includes all building concrete slabs on grade. To estimate the costs of the slabs, one needs the slab dimensions and reinforcing requirements. The horizontal dimensions of the basement and elevator pit slabs may be obtained from Figure 7.5, the Foundation Plan. Slab thickness and reinforcing must be assumed. Experience suggests that an 8'' reinforced slab will be needed in the elevator pit and a 4" reinforced slab in the basement parking area.

3.0 Superstructure

UNIFORMAT Division 3 includes all building structural items such as columns and beams, metal siding subframing, floor assemblies, roof assemblies, and stairs. To estimate the costs of the columns and beams, one must know the material choice, the superimposed load or size of member, and the floor-to-floor height for the columns, or length of span for the beams. To estimate the costs of the metal siding subframing, one must have figures for building height, wind load, and column spacing. To estimate the costs of floor and roof systems, the material choice, superimposed load, and bay size must be known. For the cost of stairs, required information includes material choice and number of risers per flight.

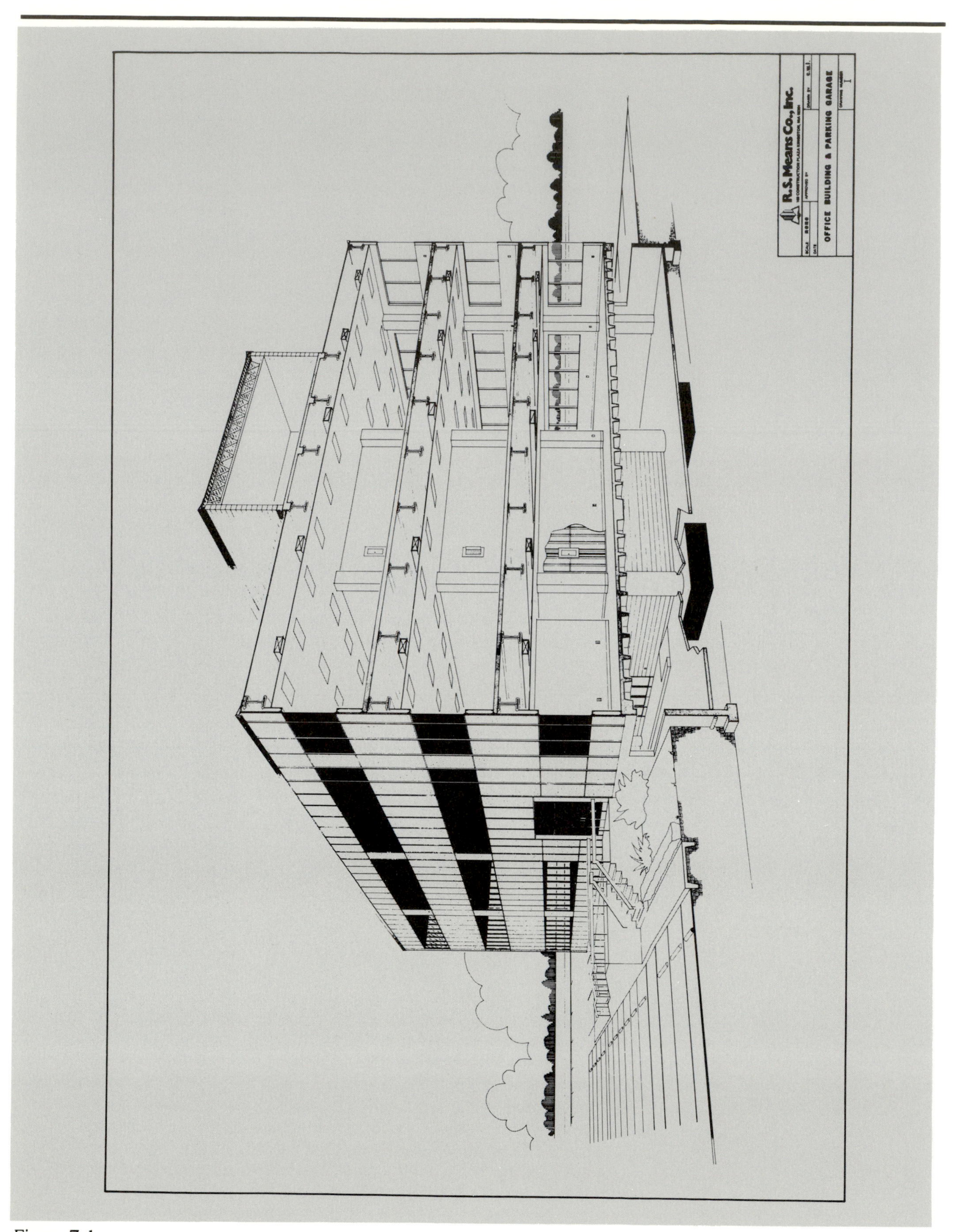

Figure 7.1

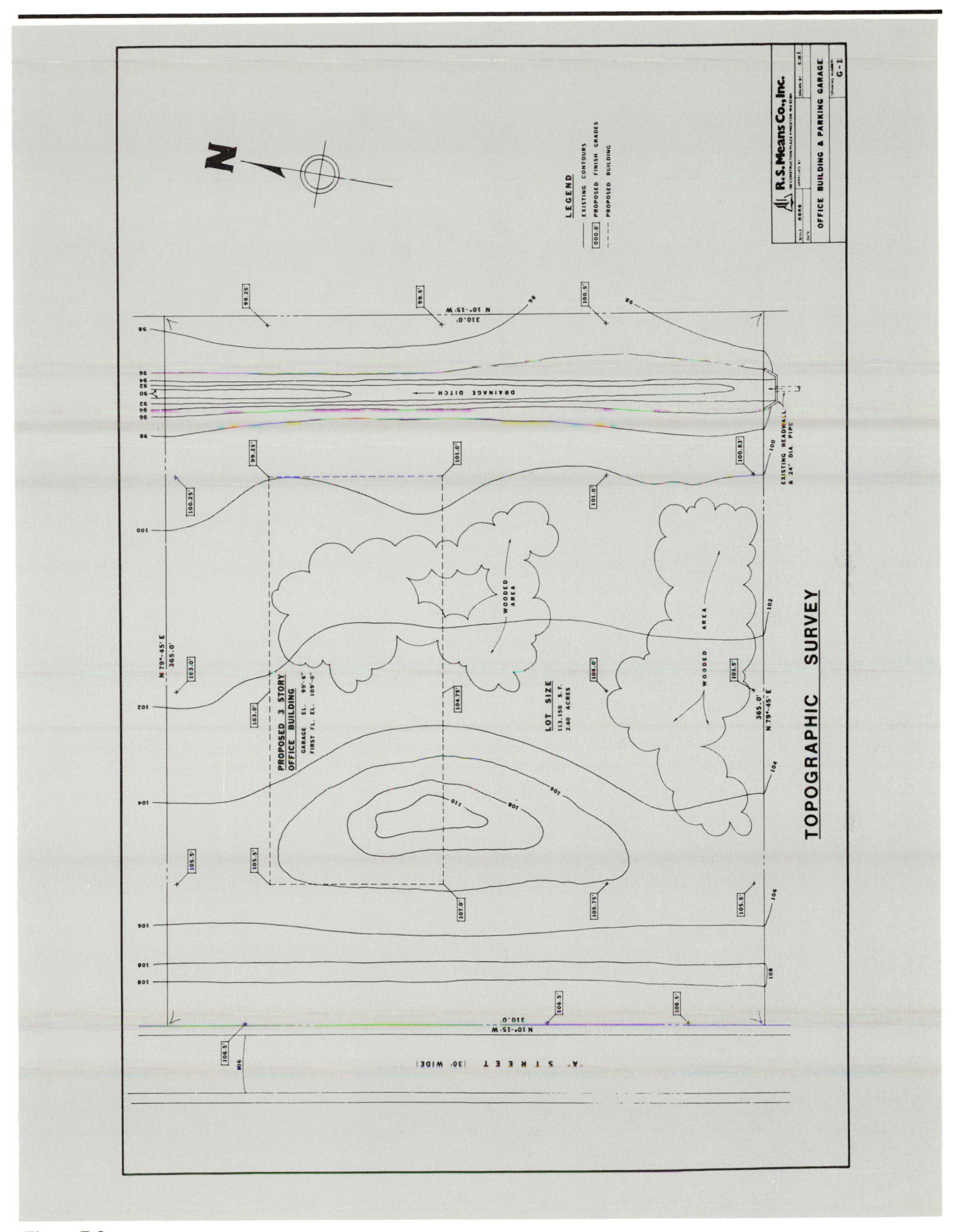

N
LEGEND
EXISTING CONTOURS
PROPOSED FINISH GRADES
PROPOSED BUILDING
R.S. Means Co., Inc.
OFFICE BUILDING & PARKING GARAGE
G-1
DRAINAGE DITCH
EXISTING HEADWALL & 24" DIA. PIPE
WOODED AREA
PROPOSED 3 STORY OFFICE BUILDING
GARAGE EL. 99'-6"
FIRST FL. EL. 109'-0"
LOT SIZE
113,150 S.F.
2.60 ACRES
N 79°-45' E
365.0'
N 10°-15' W
310.0'
"A" STREET (30' WIDE)
TOPOGRAPHIC SURVEY

Figure 7.2

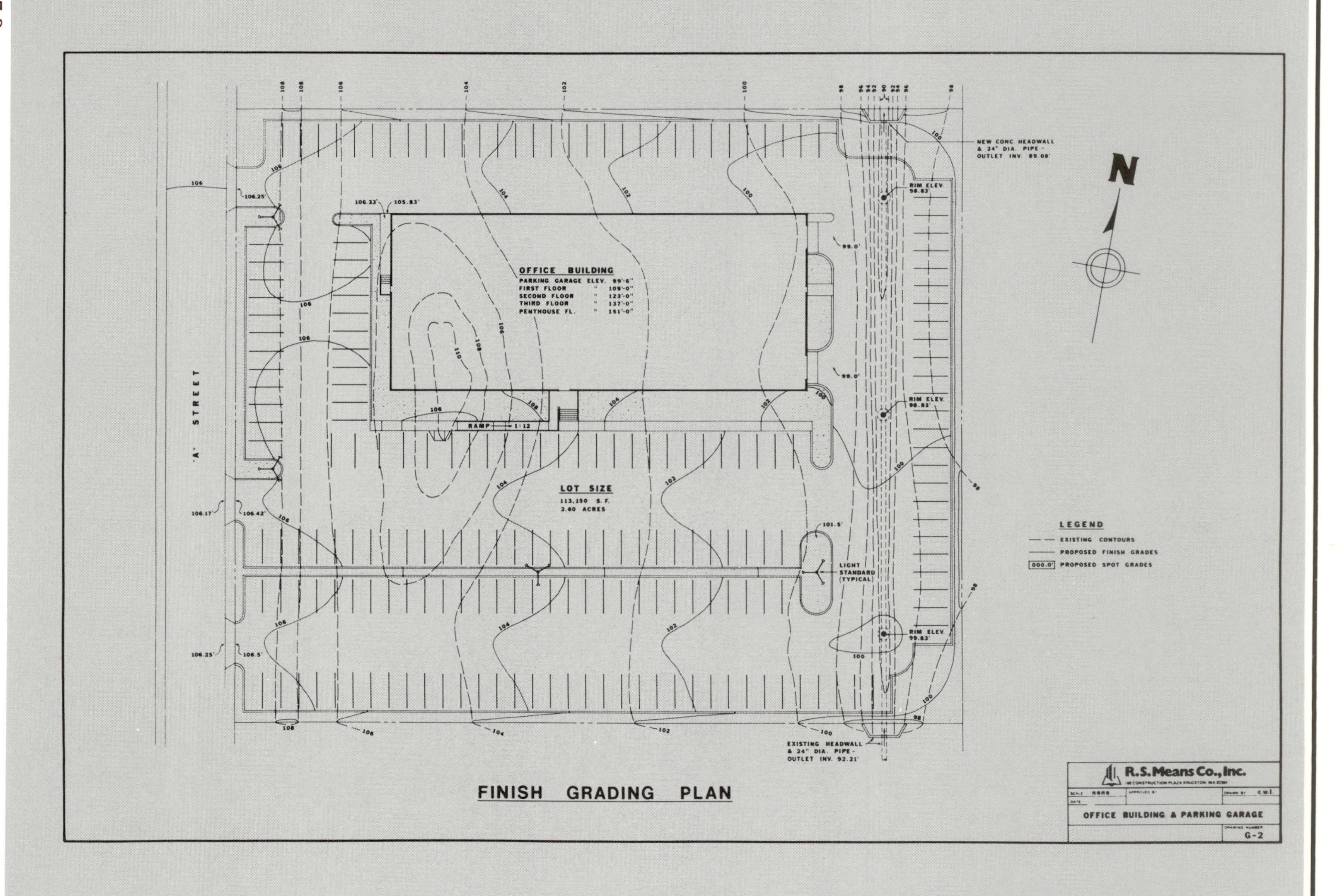

FINISH GRADING PLAN
R.S. Means Co., Inc.
OFFICE BUILDING & PARKING GARAGE
G-2
N
LEGEND
EXISTING CONTOURS
PROPOSED FINISH GRADES
000.0' PROPOSED SPOT GRADES
OFFICE BUILDING
PARKING GARAGE ELEV. 99'-6"
FIRST FLOOR " 109'-0"
SECOND FLOOR " 123'-0"
THIRD FLOOR " 137'-0"
PENTHOUSE FL. " 151'-0"
LOT SIZE
113,150 S.F.
2.60 ACRES
RAMP 1:12
'A' STREET
NEW CONC. HEADWALL & 24" DIA. PIPE - OUTLET INV. 89.08'
EXISTING HEADWALL & 24" DIA. PIPE - OUTLET INV. 92.21'
RIM ELEV. 98.83'
RIM ELEV. 98.83'
RIM ELEV. 99.83'
LIGHT STANDARD (TYPICAL)

Figure 7.3

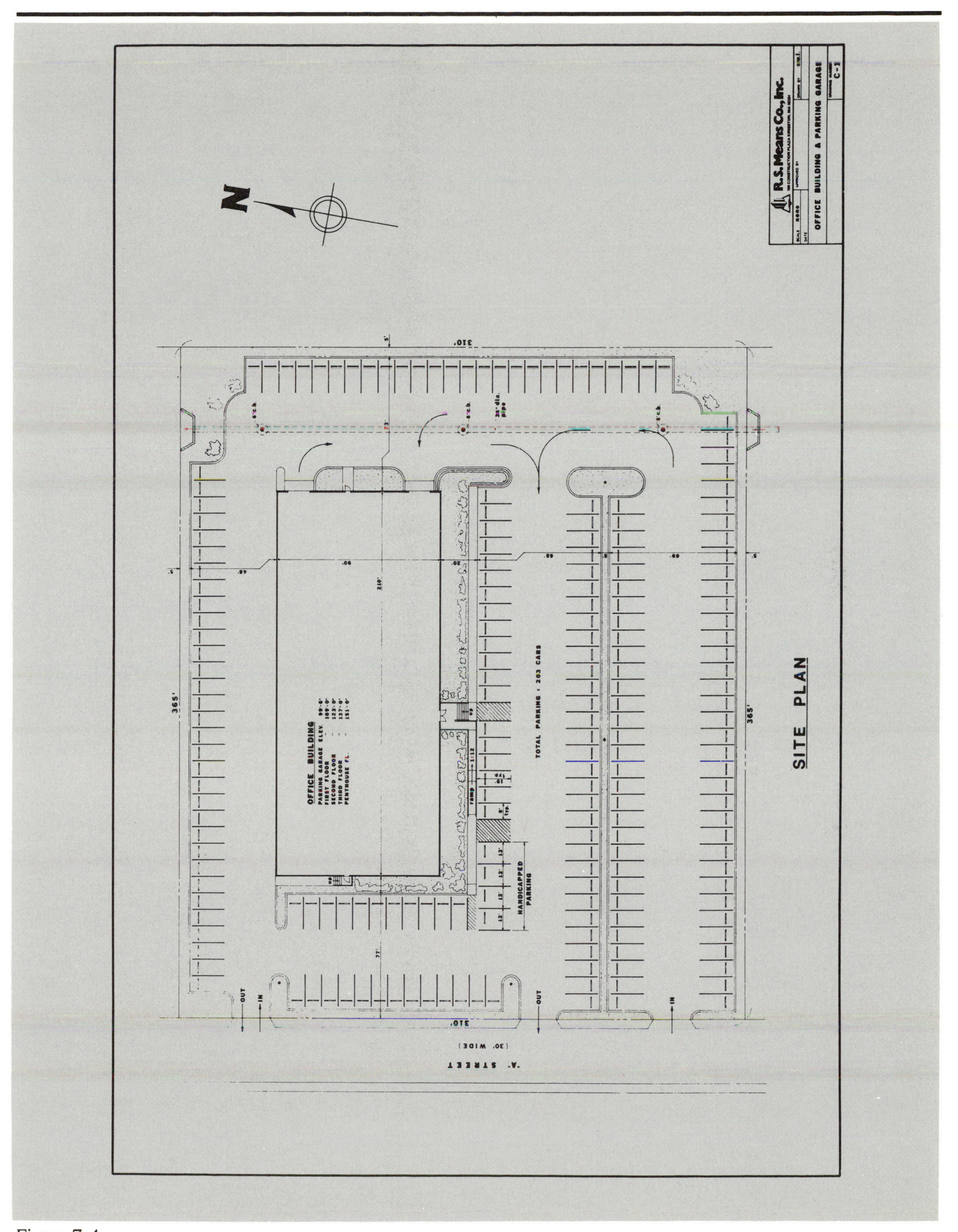
R.S. Means Co., Inc.
OFFICE BUILDING & PARKING GARAGE
C-1
N
310'
365'
210'
77'
OFFICE BUILDING
PARKING GARAGE ELEV. 99'-6"
FIRST FLOOR 109'-0"
SECOND FLOOR 123'-0"
THIRD FLOOR 137'-0"
PENTHOUSE FL. 151'-0"
TOTAL PARKING : 203 CARS
HANDICAPPED PARKING
ramp 1:12
OUT
IN
'A' STREET
(30' WIDE)
SITE PLAN

Figure 7.4

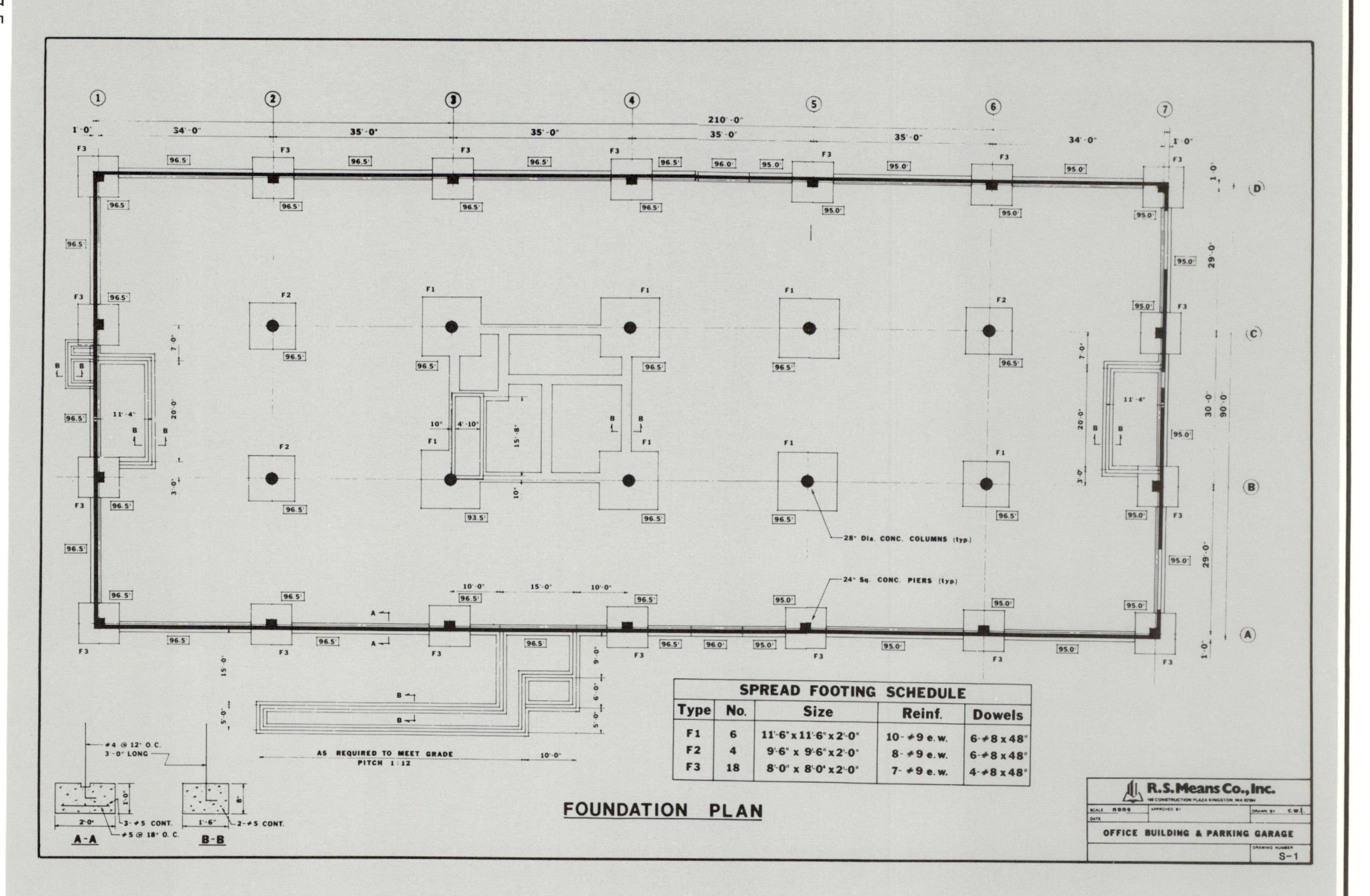

SPREAD FOOTING SCHEDULE				
Type	No.	Size	Reinf.	Dowels
F1	6	11'-6" x 11'-6" x 2'-0"	10-#9 e.w.	6-#8 x 48"
F2	4	9'-6" x 9'-6" x 2'-0"	8-#9 e.w.	6-#8 x 48"
F3	18	8'-0" x 8'-0" x 2'-0"	7-#9 e.w.	4-#8 x 48"

Figure 7.5

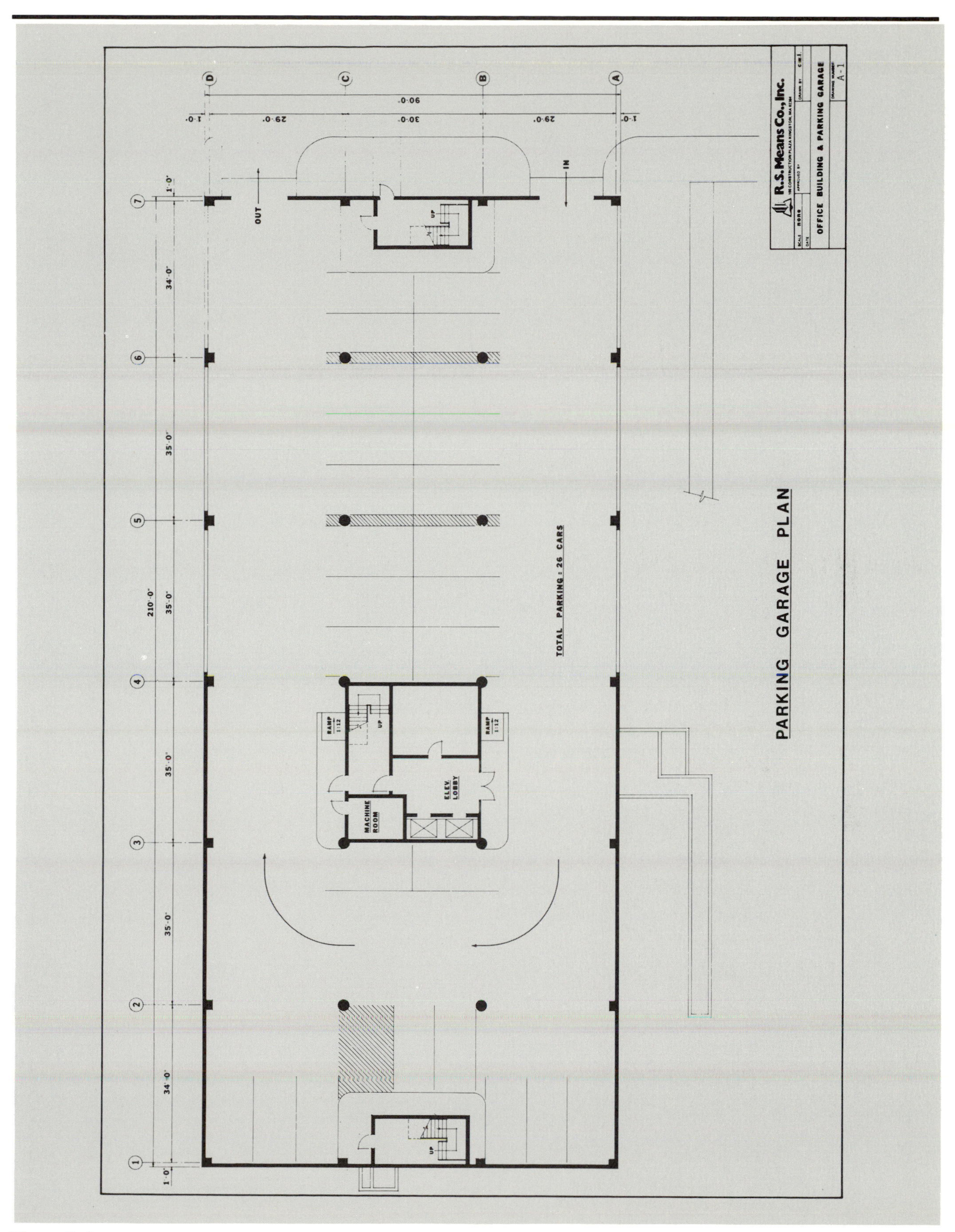
PARKING GARAGE PLAN
TOTAL PARKING: 26 CARS
R.S. Means Co., Inc.
OFFICE BUILDING & PARKING GARAGE
A-1
MACHINE ROOM
ELEV LOBBY
RAMP 1:12
UP
OUT
IN
210'-0"
90'-0"
34'-0"
35'-0"
29'-0"
30'-0"
1'-0"

Figure 7.6

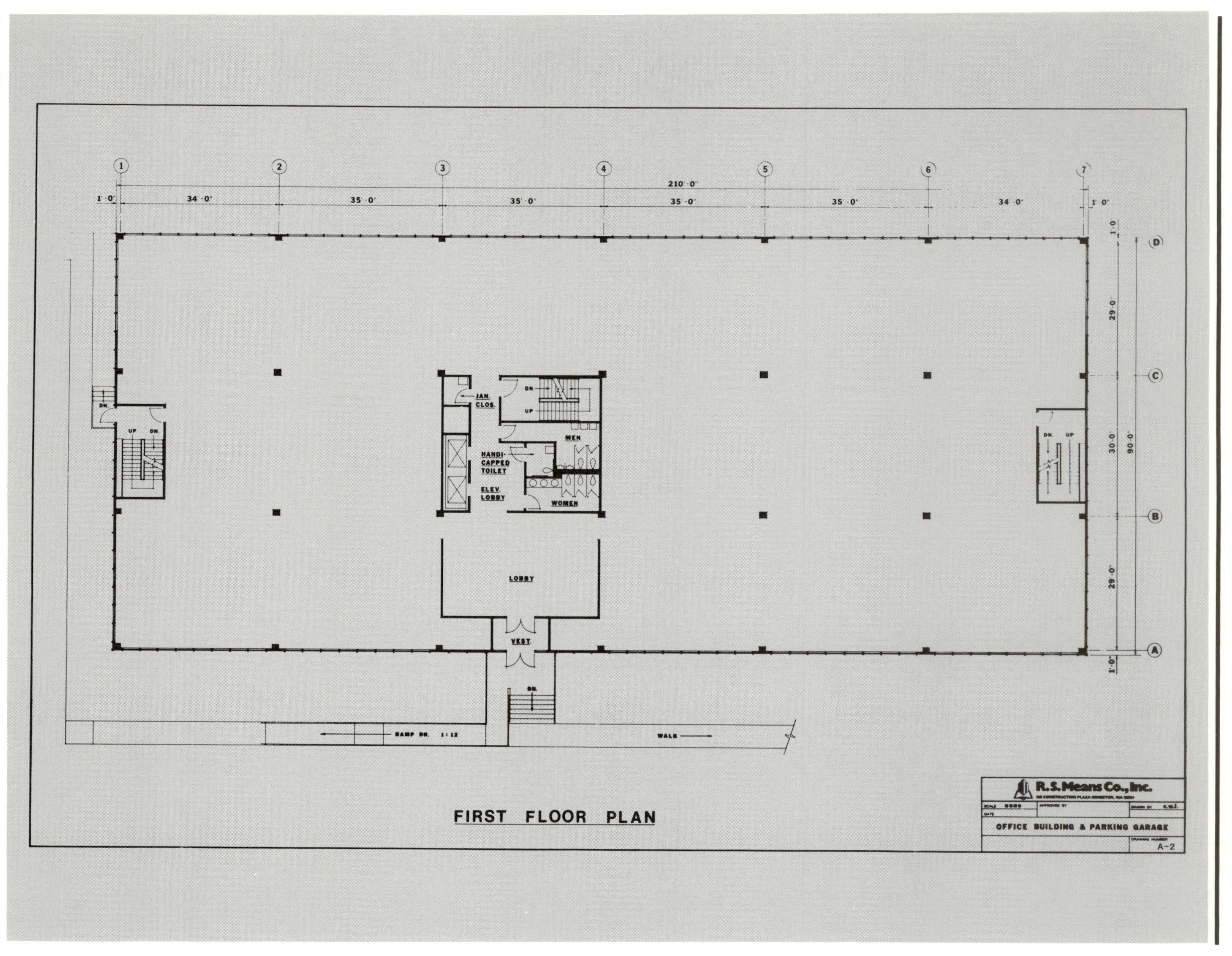
FIRST FLOOR PLAN
R.S. Means Co., Inc.
OFFICE BUILDING & PARKING GARAGE
A-2
LOBBY
VEST.
MEN
WOMEN
HANDI-CAPPED TOILET
ELEV. LOBBY
JAN. CLOS.
UP
DN.
WALK
RAMP DN. 1:12
210'-0"
1'-0"
34'-0"
35'-0"
35'-0"
35'-0"
35'-0"
34'-0"
1'-0"
90'-0"
29'-0"
30'-0"
29'-0"
1'-0"

Figure 7.7

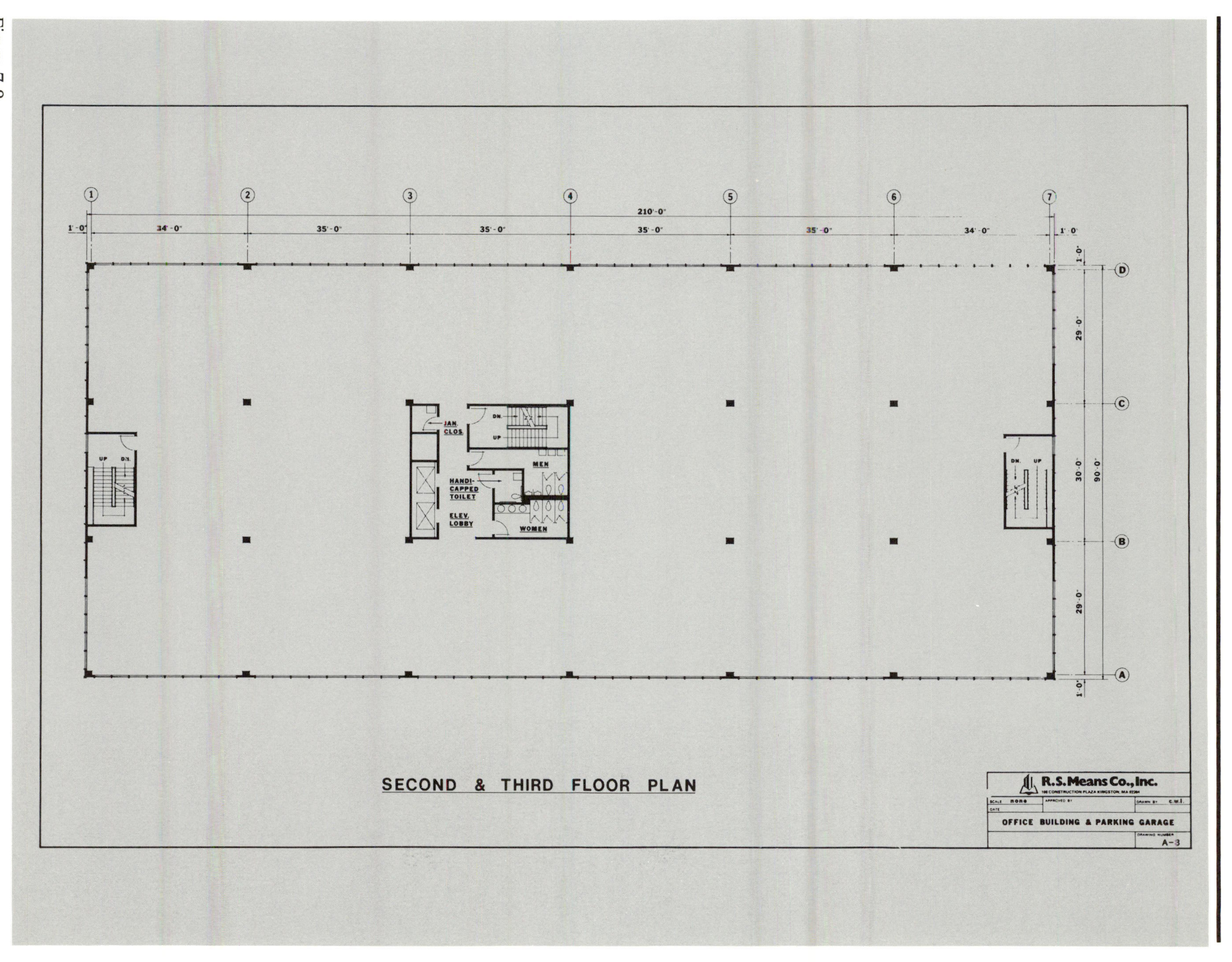
SECOND & THIRD FLOOR PLAN
R.S. Means Co., Inc.
OFFICE BUILDING & PARKING GARAGE
A-3
210'-0"
1'-0"
34'-0"
35'-0"
90'-0"
29'-0"
30'-0"
JAN. CLOS.
HANDI-CAPPED TOILET
ELEV. LOBBY
MEN
WOMEN
UP
DN.

Figure 7.8

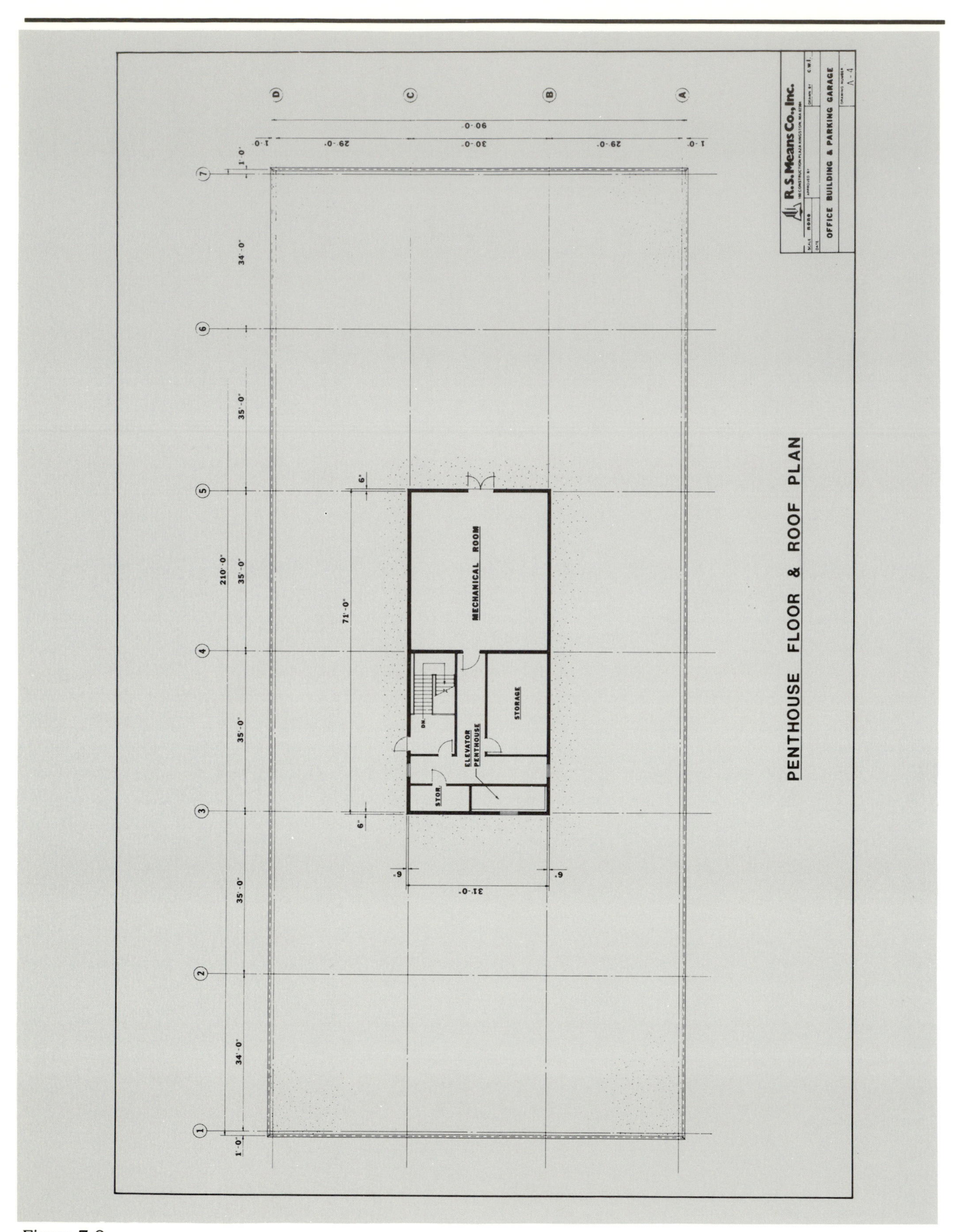
PENTHOUSE FLOOR & ROOF PLAN
MECHANICAL ROOM
STORAGE
ELEVATOR PENTHOUSE
STOR.
210'-0"
71'-0"
31'-0"
90'-0"
R.S. Means Co., Inc.
OFFICE BUILDING & PARKING GARAGE

Figure 7.9

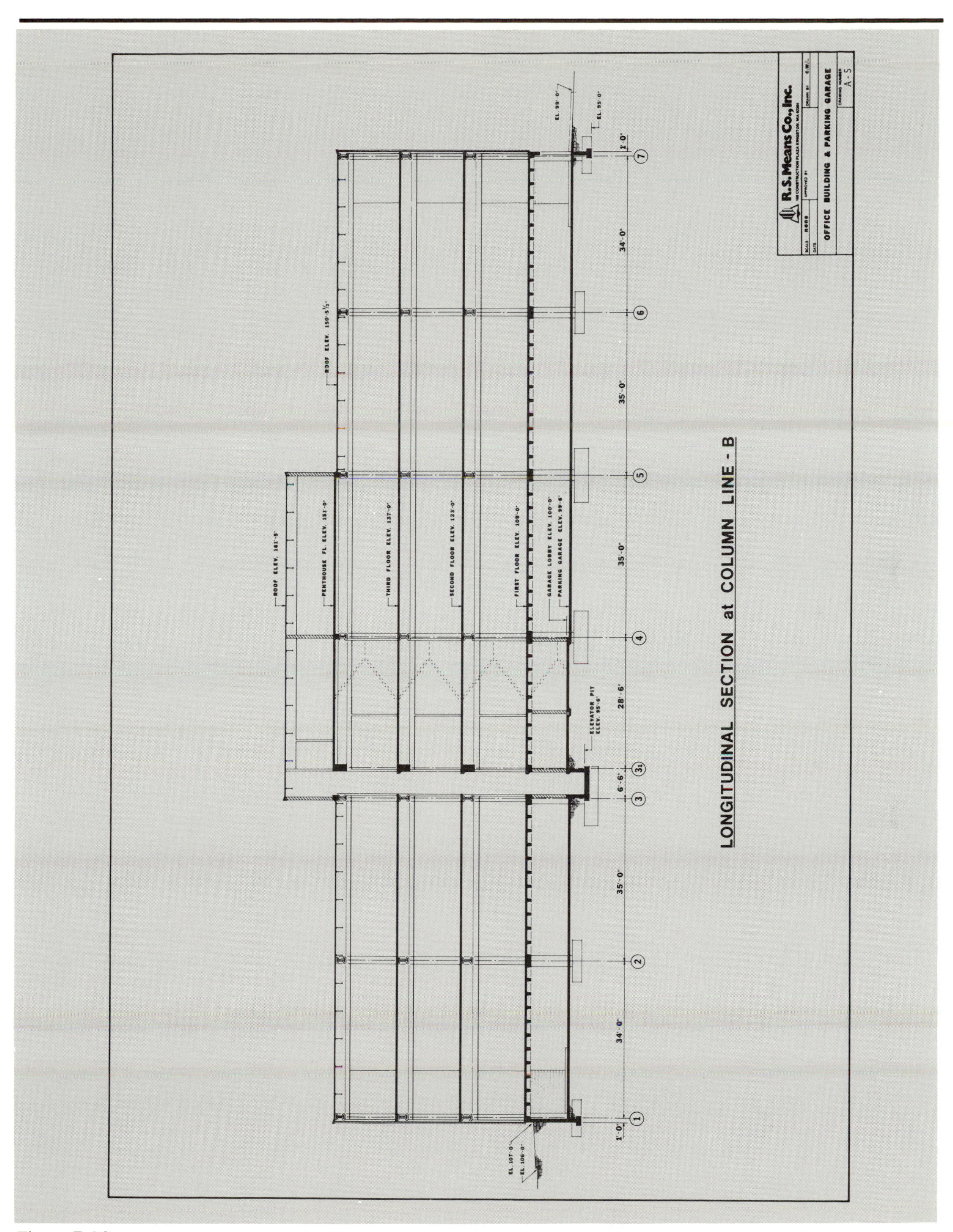
LONGITUDINAL SECTION at COLUMN LINE - B
R.S. Means Co., Inc.
OFFICE BUILDING & PARKING GARAGE
A-5
34'-0"
35'-0"
6'-6"
28'-6"
1'-0"

Figure 7.10

The Perspective Section, Figure 7.1, shows a concrete multispan joist slab system at the first floor, composite steel beam and cast-in-place slab systems for the second, third, and penthouse floors, and a steel joist on wall system at the penthouse roof. The Foundation Plan, Figure 7.5, identifies the material and sizes of the columns supporting the first floor. The floor and roof plans, Figures 7.6 through 7.9, provide the bay sizes. Floor elevations can be found on the Longitudinal Section, Figure 7.10, which also shows a steel joists and beams on columns system at the roof. The floor-to-floor heights of all columns is also given. The columns supporting the first floor are shown to be concrete and the sizes specified. The rest of the columns are assumed to be steel because of the steel frame floor and roof systems. An average size for these columns will have to be determined by a structural analysis (as shown in Chapter 6). Beams are included in the costs of the floor assemblies (in *Means Assemblies Cost Data*) and, therefore, do not have to be priced separately. Metal siding subframing is not required.

The systems used for floor and roof structures may be inferred from the drawings: multispan joist slab at the first floor, composite steel beam and cast-in-place slabs at the second and third floors and the penthouse, steel joists and beams on columns at the main roof, and steel joists on walls at the penthouse roof. The bay sizes are provided. Experience suggests that a 75 P.S.F. superimposed load will be required for the tenant floors, a 125 P.S.F. load for the penthouse to support the HVAC equipment, and a 20 P.S.F. for the roofs. A concrete fill pan stair system is commonly used in this type of construction. The number of risers is determined by dividing the floor-to-floor height by an average riser height of 7″.

4.0 Exterior Closure

UNIFORMAT Division 4 includes exterior wall systems, doors, and windows. Materials and size information are needed to estimate the costs of these systems. The Perspective Section, Figure 7.1, shows a curtain wall exterior wall system at the tenant spaces, a concrete masonry unit exterior wall at the penthouse, and a storefront entry at the first floor. The floor plans, Figures 7.6 through 7.9, provide the horizontal dimensions for the exterior walls, and show exterior doors at the stairs and penthouse. They also show that the penthouse walls are 6″ thick. The Longitudinal Section, Figure 7.10, lists the vertical dimensions for the exterior walls. The materials used in the curtain wall and the exact sizes and types of doors must be assumed.

Experience suggests that the entry storefront units will be aluminum and glass, the curtain wall system will be aluminum frame with insulated glass vision panels and insulated glass and drywall spandrel units, and that fire-rated hollow metal doors and frames will be required for the exterior doors. It is assumed that rolling metal grilles will be used to control access to the parking basement and that the penthouse walls will require a perlite insulating fill.

5.0 Roofing

UNIFORMAT Division 5 includes roof coverings, roof insulation, and specialty items such as skylights and gravel stop. The horizontal dimensions of the main and penthouse roof are provided in the Penthouse Floor and Roof Plan, Figure 7.9. The type of roofing, edging, and base flashing must be assumed. A ballasted, single-ply EPDM roofing system with 3″ of polystyrene insulation is commonly used in the Atlanta area. Compatible edging and base flashing systems must also be estimated.

6.0 Interior Construction

UNIFORMAT Division 6 includes partitions and doors, and wall, floor, and ceiling finishes. The dimensions and materials must be known in order to prepare the estimate. The floor plans, Figures 7.6 through 7.9, and the Longitudinal Section, Figure 7.10, provide the horizontal and vertical dimensions of the partitions and the door locations. The materials and door sizes must be assumed. Based on past experience, it is assumed that painted 8" CMU walls will be used in the basement, painted 6" CMU walls in the penthouse, and painted drywall in the tenant areas. Plastic laminate toilet partitions and urinal screens will be used in the rest rooms. Ceramic tile flooring is assumed for the rest rooms, and quarry tile flooring in the lobbies and vestibules. Acoustical tile ceilings are to be used in the tenant areas, lobbies, and vestibules, and gypsum wall board for the rest room ceilings. Allowances will be included for finish hardware and specialties.

7.0 Conveying

UNIFORMAT Division 7 includes elevators, moving stairs, and miscellaneous conveying systems, such as conveyors or pneumatic tube systems. Two elevators are shown in the floor plans, and the number of stops is given in the Longitudinal Section. Their weight capacity, speed, and type must be assumed. Hydraulic elevators are commonly used in low-rise construction. A 2500 pound capacity with a speed of 100 feet per minute is assumed.

8.0 Mechanical System

UNIFORMAT Division 8 includes plumbing fixtures and piping, fire protection, heating, and air conditioning. Although the plans show plumbing fixtures and HVAC ductwork, there is not enough information available to prepare an assemblies estimate for the plumbing and HVAC systems.

9.0 Electrical System

UNIFORMAT Division 9 includes electrical service, lighting and power, and equipment connections. The plans contain no information about the electrical system.

10.0 General Conditions

General Conditions costs are to be determined by use of a percentage on the summary sheet.

11.0 Special Construction

Special construction includes Architectural Specialties, such as bathroom accessories and partitions; Architectural Equipment, such as bank or kitchen equipment; furnishings like blinds or furniture, and Special Construction items like domes or greenhouses. Some Specialties items are required; their cost will be covered by an allowance.

12.0 Site Work

UNIFORMAT Division 12 includes utilities, roadways and walks, site lighting, and site excavation. The Finish Grading Plan, Figure 7.3, and the Site Plan, Figure 7.4, identify the number of parking spaces, dimensions for the sidewalk, curb, and gutter; and dimensions and sizes for the storm sewer system. The pavement and sidewalk materials, the curb and gutter size and materials, the steps, water, sewer, and electrical service must be assumed. Experience suggests that a 3" asphalt over 6" gravel base paving system, 4" concrete sidewalks and ramps, 6" × 18" concrete curb, 100 L.F. of 8" clay sanitary pipe, 100 L.F. of 8" P.V.C. water pipe, metal halide pole lights, and a 1200 amp electrical service with 100 L.F. of feeder will be required.

13.0 Miscellaneous

The miscellaneous section is a catch-all for items not covered elsewhere. None is required for this project.

The Cost Control Estimate

A combination of the assemblies and square foot methods is used to prepare the Cost Control Estimate. An example based on the preceding project analysis is shown in Figures 7.11 through 7.15. The prices are from the 1988 *Means Assemblies Cost Data*.

For the Need Analysis estimate, national average prices from *Means Assemblies Cost Data* can be used for the assembly. They provide a close approximation of the assembly size and materials used in the modeling exercise. This approach is well suited to the modeling exercise because the goal of the estimate is simply to set an adequate budget for the project. A little extra "cushion" is actually desirable to cover items that may be omitted or under-priced. The cost control estimate, on the other hand, requires greater accuracy and adherence to a specific, developed design. The Means prices must, therefore, be adjusted to match the sizes and materials actually used. For example, the footing schedule in Figure 7.5 shows footing F1 to be 11′6″ square by 2′0″ deep; F2 is 9′6″ square by 2′ deep; and F3 is 8′ square by 2′ deep. The spread footing table in Figure 7.16 does not provide costs for assemblies of these specific sizes. However, prices of similar assemblies from the table can be adapted to fit. The 10′6″ square by 25″ deep footing in line 1.1-120-7750 can be adapted to price the interior footings and the 7′6″ square by 25″ deep footing in line 1.1-120-7810 can be adapted for the exterior footings. The prices are adjusted according to a percentage based on the difference in volumes of the plan footing and the table footing. The procedure is as follows.

$$\text{Table Price} \times \frac{\text{Plan Size}}{\text{Table Size}} = \text{Adapted Price}$$

$$\text{F1: } \$1{,}250 \times \frac{11.5' \times 11.5' \times 2'}{10.5' \times 10.5' \times 2.08} = \$1{,}442$$

$$\text{F2: } \$1{,}250 \times \frac{9.5' \times 9.5' \times 2'}{10.5' \times 10.5' \times 2.08'} = \$\ 984$$

$$\text{F3: } \$\ 700 \times \frac{8' \times 8' \times 2'}{7.5' \times 7.5' \times 2.08'} = \$\ 766$$

The adjustment is indicated by a plus added to the assembly number. The other prices in *Means Assemblies Cost Data* can also be adjusted as needed to match the actual design.

At this stage, there is not enough information available to estimate mechanical, electrical, and plumbing costs using the assemblies method. For these systems, square foot costs from *Means Assemblies Cost Data* are used, just as they were applied to the program estimate.

Estimate Analysis

Next, the cost control estimate is compared with the project budget. During the conceptual design phase, this comparison provides some guidance as to which schemes should be discarded as too expensive, and which are economically feasible and thereby merit further consideration. For the final conceptual design scheme, such a comparison points up the following: Are modifications needed to reduce costs? Is the scheme within budget, or should it be discarded altogether in favor of a less expensive scheme? In the design development phase,

ASSEMBLY NUMBER	DESCRIPTION	QTY	UNIT	TOTAL COST UNIT	TOTAL COST TOTAL	COST PER S.F.
1.0	**Foundations**					
1.1-120-7750+	Ftg F1 11^{5}□ x 2^{0}	6	ea	1442	8652	
1.1-120-7750+	✓ F2 9^{6}□ x 2^{0}	4	ea	984	3936	
1.1-120-7810+	✓ F3 8^{0}□ x 2^{0}	18	ea	766	13788	
1.1-140-2700	✓ AA 2^{0} x 1^{0}	580^{42}	LF	25^{55}	14830	
1.1-140-2100+	✓ BB 1^{5} x 0^{67}	215	LF	15^{41}	3313	
1.1-210-3000	Elev Pit Walls 6'x6"	41	LF	44^{30}	1816	
1.1-210-7220	Bsmnt Walls 12'x8"	300	LF	100	30000	
1.1-210-5020	✓ ✓ 8'x8"	186	LF	65	12090	
1.1-292-6400	WP perimeter 8' avg ht	600	LF	12^{34}	7404	
1.9-100-4620	Bldg Exc & BF 8' avg depth	18,900	sf	1^{19}	22491	
1.1-294-1000	4" PVC Perimeter Drain	600	LF	4	2400	
	Total				120720	
2.0	**Slab on Grade**					
2.1-200-4520	Elev. Pit Slab - 6", reinf.	137^{29}	sf	3^{77}	518	
2.1-200-2280	Bsmnt Slab - 4", reinf.	18,228	sf	3^{06}	55778	
	Total				56296	
3.0	**Superstructure**					
3.1-112-1490	Int. Cols 28" ⌀	130	LF	68	8190	
3.1-114-4000+	Pilasters @ Ext. Wall 24" x 12"	234	LF	43	10062	
3.5-160-7000	CIP Multispan Joist Slab @ 1st	18,900	sf	8^{54}	161406	
3.5-520-6000	Composit Bm & CIP Slab @ 2 & 3	37,800	sf	11^{15}	421470	
3.1-130-4800	Stl Cols	1176	LF	29^{78}	35021	
3.5-520-6250	Composit Bm & CIP Slab @ PH	2201	sf	13^{50}	29714	
3.7-420-4700	Stl Joists & Bms on Cols @ Roof	16,679	sf	3^{17}	52936	
3.7-430-1700	Stl Joists on Walls @ PH	2201	sf	1^{92}	4226	
3.9-100-0780	Stairs - Conc in stl pans	10	ea	5470	54700	
	Total				777725	
4.0	**Exterior Closure**					
4.1-211-3300	6" Lt. Wt. CMU w/ perlite	2244	sf	4^{77}	10704	
4.6-100-7150	Entry	2	opng	3825	7650	
4.6-100-4200	Dbl dr @ bsmnt & PH	2	opng	1940	3880	
4.6-100-3950	Sgl Dr. ✓	7	opng	975	6825	
4.7-582-1850	Curtain Wall Frame	25,800	sf	10^{75}	277350	
4.7-584-1100	1/2" insul, tinted glass	9,840	sf	13^{10}	128904	
4.7-584-3100	1/2" ℄ w/ 2" insul. @ spandrel	15,960	sf	12^{26}	195670	
4.6-100-6150	O.H. Rolling Grilles - motor op.	2	opng	4325	8650	
	Total				639633	

Page 2 of 6

Figure 7.11

ASSEMBLY NUMBER	DESCRIPTION	QTY	UNIT	TOTAL COST UNIT	TOTAL COST TOTAL	COST PER S.F.
5.0	**Roofing**					
5.1-220-2100	EPDM - .045, balasted	18,900	sf	.86	16254	
5.1-520-1200	Alum. Roof edge	804	lf	13.50	10854	
5.1-510-1100	Base Flashing	204	lf	15.10	3080	
5.7-101-2350	3" polystyrene insul.	18,900	sf	1.19	22491	
	Total				52679	
6.0	**Interior Construction**					
6.1-210-6000	8" CMU @ Bsmnt	3406	sf	4.92	16758	
6.1-210-5500	6" ✓ @ PH	2282	sf	4.40	10041	
6.1-510-5400	DW @ Core 1-3	15,162	sf	2.54	38511	
6.1-510-5300	✓ @ Ext Walls	13,620	sf	1.57	21383	
6.1-870-0420	T. Partitions	15	ea	465	6975	
6.1-870-1300	U. Screens	3	ea	216	648	
6.4-260-6000	Wd. Dr. - 1 hr. flush birch	27	ea	384	10368	
	Hdwr - allow	✓	ea	250	6750	
	Specialties - allow	1	LS	5000	5000	
6.5-100-0080	Paint DW	14,391	sf	.56	8059	
6.5-100-0320/40	Paint & Fill CMU	8532	sf	1.01	8617	
6.6-100-1720	CT Flooring	768	sf	4.30	3302	
6.6-100-1820	QT Flooring	1625	sf	9.40	15275	
6.7-100-5800	Acou Ceiling	55,803	sf	1.42	79240	
	Total				230927	
7.0	**Conveying**					
7.1-100-2090t	Elev. - 2500#, 100 FPM, 4 Flrs	2	ea	70,800	141600	

Figure 7.12

ASSEMBLY NUMBER	DESCRIPTION	QTY	UNIT	TOTAL COST UNIT	TOTAL COST TOTAL	COST PER S.F.
8.0	**Mechanical System**					
14.1-610-2720	Plumbing	68,351	sf	2.48	169510	
14.1-610-2770	HVAC	✓	sf	4.90	334920	
	Total				504430	
9.0	**Electrical**					
C9.0-117	Service & Dist.	68,351	sf	1.39	95008	
	Lighting	✓	sf	3.20	218723	
	Devices	✓	sf	.13	8886	
	Equip. Conn.	✓	sf	.51	34859	
	Mat'ls	✓	sf	2.06	140803	
	Fire Alarm & Detec.	✓	sf	.27	18455	
	Total				516734	

Page 4 of 6

Figure 7.13

ASSEMBLY NUMBER	DESCRIPTION	QTY	UNIT	TOTAL COST UNIT	TOTAL COST TOTAL	COST PER S.F.
10.0	**General Conditions**					
	see summary					
11.0	**Special Construction**					
12.0	**Site Work**					
12.5-510-1500	Parking	205	cars	365	74825	
12.7-140-1600	4' SW	260	LF	7.26	1888	
12.7-610-2000	Curb & Gutter	3120	LF	6.40	19968	
12.3-710-5860	Catch Basins	3	ea	1405	4215	
12.3-510-4650	SS - 24" RCP	310	LF	24	7440	
12.3-110-3900	Trenching	570	LF	29.20	16644	
12.3-510-9060	Sanitary	100	LF	8.65	865	
12.3-540-2160	H_2O	100	LF	17.35	1735	
12.7-500-5820	MH Halide Lighting	10	ea	1825	18250	
	Site Exc. - Allow	11,883	SY	1.00	11883	
~~10.0~~	~~Miscellaneous~~					
9.1-210-0480	1200 Amp Service	1	ea	8725	8725	
9.1-310-0480	1200 Amp Feeder	100	LF	128	12800	
	Conc Steps - allow	1	LS	3000	3000	
1.1-140-2100+	Ftg - BB @ Ramp & steps	120	LF	15.41	1849	
1.1-210-1500	Ret'g Wall 4' x 6"	✓	LF	30.20	3624	
	Total				187711	

Page 5 of 6

Figure 7.14

PRELIMINARY
ESTIMATE (Cost Summary)

PROJECT Chambers Office Bldg — TOTAL AREA — SHEET NO.
LOCATION 103 Chambers Blvd., Chambers, Ga. — TOTAL VOLUME — ESTIMATE NO.
ARCHITECT — COST PER S.F. — DATE
OWNER Chambers Investment Inc. — COST PER C.F. — NO. OF STORIES
QUANTITIES BY: RK — PRICES BY: RK — EXTENSIONS BY: RK — CHECKED BY:

NO.	DESCRIPTION	SUB TOTAL COST	COST/S.F.	%
1.0	**Foundation**	120720		
2.0	**Substructure**	56296		
3.0	**Superstructure**	777725		
4.0	**Exterior Closure**	639633		
5.0	**Roofing**	52679		
6.0	**Interior Construction**	230927		
7.0	**Conveying**	141600		
8.0	**Mechanical System**	504430		
9.0	**Electrical**	516734		
10.0	**General Conditions (Breakdown)**	—		
11.0	**Special Construction**	—		
12.0	**Site Work**	187711		

Building Sub Total $ 3,228,455 → $ 3,228,455

Sales Tax 4 % × Sub Total $ ______ /2 = $ 64,569

General Conditions (%) 16 % × Sub Total $ ______ = ______

General Conditions $ 516,553

Sub Total "A" $ 3,809,577

Overhead & Profit 10 % × Sub Total "A" $ ______ = $ 380,958

Sub Total "B" $ ______

~~Profit~~ Bid Conditions - low @ 8 % × Sub Total "B" $ ______ = $ 304,766

Sub Total "C" $ 4,495,301

Location Factor Atlanta @ 89.2 % × Sub Total "C" $ ______ =

Adjusted Building Cost $ 4,009,808

Architects Fee ______ % × Adjusted Building Cost $ ______ = $ ______

~~Contingency~~ Inflation: 1-1-88 to 6-87 - 4 % × Adjusted Building Cost $ ______ = $ x 1.061

Total Cost 4,254,406

Square Foot Cost $ ______ / ______ S.F. = ______ $/S.F.

Cubic Foot Cost $ ______ / ______ C.F. = ______ $/C.F.

Page 6 of 6

Figure 7.15

FOOTINGS & WALLS	B1.1-120	Spread Footings

The Spread Footing System includes: excavation; backfill; forms (four uses); all reinforcement; 3,000 p.s.i. concrete (chute placed); and screed finish.

Footing systems are priced per individual unit. The Expanded System Listing at the bottom shows footings that range from 3' square x 12" deep, to 20' square x 42" deep. It is assumed that excavation is done by a truck mounted hydraulic excavator with an operator and oiler. Backfill is with a dozer, and compaction by air tamp. The excavation and backfill equipment is assumed to operate at 30 C.Y. per hour.

Please see ① ② ③ in the reference section for further design and cost information.

Systems Components	QUANTITY	UNIT	COST EACH MAT.	INST.	TOTAL
SYSTEM 01.1-120-7100					
SPREAD FOOTINGS, LOAD 25K, SOIL CAPACITY 3 KSF, 3' SQ X 12" DEEP					
Bulk excavation	.590	C.Y.		2.64	2.64
Hand trim	9.000	S.F.		3.78	3.78
Compacted backfill	.260	C.Y.		.44	.44
Formwork, 4 uses	12.000	S.F.	4.68	28.32	33
Reinforcing, fy = 60,000 psi	.006	Ton	3.47	3.29	6.76
Dowel or anchor bolt templates	6.000	L.F.	3.42	11.70	15.12
Concrete, f'c = 3,000 psi	.330	C.Y.	18.81		18.81
Place concrete, direct chute	.330	C.Y.		3.89	3.89
Screed finish	9.000	S.F.		2.34	2.34
TOTAL			30.38	56.40	86.78

1.1-120	Spread Footings		COST EACH MAT.	INST.	TOTAL
7090	Spread footings, 3000 psi concrete, chute delivered				
7100	Load 25K, soil capacity 3 KSF, 3'-0" sq. x 12" deep		30	56	86
7150	Load 50K, soil capacity 3 KSF, 4'-6" sq. x 12" deep	①	65	98	163
7200	Load 50K, soil capacity 6 KSF, 3'-0" sq. x 12" deep		30	56	86
7250	Load 75K, soil capacity 3 KSF, 5'-6" sq. x 13" deep	②	105	140	245
7300	Load 75K, soil capacity 6 KSF, 4'-0" sq. x 12" deep		53	84	137
7350	Load 100K, soil capacity 3 KSF, 6'-0" sq. x 14" deep	③	130	165	295
7410	Load 100K, soil capacity 6 KSF, 4'-6" sq. x 15" deep		80	115	195
7450	Load 125K, soil capacity 3 KSF, 7'-0" sq. x 17" deep		205	235	440
7500	Load 125K, soil capacity 6 KSF, 5'-0" sq. x 16" deep		100	140	240
7550	Load 150K, soil capacity 3 KSF 7'-6" sq. x 18" deep		250	275	525
7610	Load 150K, soil capacity 6 KSF, 5'-6" sq. x 18" deep		135	175	310
7650	Load 200K, soil capacity 3 KSF, 8'-6" sq. x 20" deep		355	365	720
7700	Load 200K, soil capacity 6 KSF, 6'-0" sq. x 20" deep		180	215	395
7750	Load 300K, soil capacity 3 KSF, 10'-6" sq. x 25" deep		650	600	1,250
7810	Load 300K, soil capacity 6 KSF, 7'-6" sq. x 25" deep		340	360	700
7850	Load 400K, soil capacity 3 KSF, 12'-6" sq. x 28" deep		1,025	890	1,915
7900	Load 400K, soil capacity 6 KSF, 8'-6" sq. x 27" deep		470	470	940
7950	Load 500K, soil capacity 3 KSF, 14'-0" sq. x 31" deep		1,425	1,175	2,600
8010	Load 500K, soil capacity 6 KSF, 9'-6" sq. x 30" deep		650	610	1,260

1

Figure 7.16

the same kind of comparison shows the effect that the materials and assemblies choices are having on project cost.

In the example project, the estimate of the final conceptual design scheme is $4,254,406, some $664,000 above the project budget of $3,590,827. This discrepancy suggests that the scheme should either be modified to reduce costs or discarded in favor of a less expensive one.

Assuming that this estimate was run late in the design development phase, changes would have to be made to reduce costs to meet the budget. It is very tempting to the architect and engineer at this point to reduce costs by deleting items that have little or no effect on the design of the major building systems, such as certain pieces of equipment, or the air conditioning of a floor or wing. These kinds of changes have the least effect on the plans and specifications, and are, therefore, the easiest and least expensive to the architect and engineers. If the image or appearance of the facility is one of the most important design considerations, then this approach may have some merit. If, however, facility and functional requirements are paramount, and the deletion of equipment or air conditioning would adversely affect the usefulness of the facility, then the substitution of less expensive systems and finishes and/or redesigning the project is the approach that should be taken.

For the Chambers project, image is important, but the program budget already represents the maximum that can be funded. Therefore, changes must be made to reduce project costs. The first step is to compare the design with the program to determine if there are items that can be deleted and still meet program requirements. An analysis of the proposed systems and materials is then made to determine if less expensive systems can be substituted. If these steps do not produce enough savings, then the project should be redesigned.

A comparison of the design with the program shows that space has been added to the penthouse to house HVAC equipment. This addition increases the square footage of the building by 2201 square feet. While adding square footage increases the amount of leasable space, the efficiency ratio used in the program had already allowed adequate space for both equipment and leasable area to make the project feasible. Consequently, the scheme can be made some 2201 square feet smaller and still meet program requirements. The effect of this reduction can be quickly approximated by multiplying the square foot cost of the design scheme by 2201 square feet.

$$(\$4{,}254{,}406/68{,}351 \text{ S.F.}) \times 2201 \text{ S.F.} = \$136{,}998$$

To further cut costs, a change in materials and systems should be investigated. To begin with, experience suggests that a steel joists and beams on columns structural system is less expensive than the composite steel beams and cast-in-place concrete slabs in the design. The assembly shown in Figures 7.17a and b gives a price of $11.15/S.F. for the composite system. Figures 7.18a and b show a cost of $8.47/S.F. for a similar joist system. The cost savings resulting from a change in the floor system of the second and third floors is as follows.

$$\begin{aligned} 37{,}800 \times \$11.15 &= \$421{,}470 \\ 37{,}800 \times \$\ 8.47 &= \underline{\ 320{,}166} \\ &\ \$101{,}304 \end{aligned}$$

FLOORS	B3.5-520	Composite Beam & C.I.P. Slab

General: Composite construction of wide flange beams and concrete slabs is most efficiently used when loads are heavy and spans are moderately long. It is stiffer with less deflection than non-composite construction of similar depth and spans.

In practice, composite construction is typically shallower in depth than non-composite would be.

Design Assumptions:

Steel, fy = 36 KSI
Beams unshored during construction
Deflection limited to span/360
Shear connectors, welded studs
Concrete, f'c = 3 KSI

System Components	QUANTITY	UNIT	COST PER S.F. MAT.	INST.	TOTAL
SYSTEM 03.5-520-3800					
20X25 BAY, 40 PSF S. LOAD, 4" THICK SLAB, 20" TOTAL DEPTH					
Structural steel	3.820	Lb.	1.83	.76	2.59
Welded shear connectors ¾" diameter 3-⅜" long	.140	Ea.	.06	.14	.20
Forms in place, floor slab forms hung from steel beams, 4 uses	1.000	S.F.	.54	2.67	3.21
Edge forms to 6" high on elevated slab, 4 uses	.050	L.F.	.01	.10	.11
Reinforcing in place, elevated slabs #4 to #7	1.190	Lb.	.36	.24	.60
Concrete ready mix, regular weight, 3000 PSI	.330	C.F.	.70		.70
Place and vibrate concrete, elevated slab less than 6", pumped	.330	C.F.		.25	.25
Finishing floor, monolithic steel trowel finish for resilient tile	1.000	S.F.		.42	.42
Curing with sprayed membrane curing compound	.010	S.F.	.02	.04	.06
Spray mineral fiber/cement for fire proof, 1" thick on beams	.520	S.F.	.18	.30	.48
TOTAL			3.70	4.92	8.62

3.5-520 Composite Beam & Cast In Place Slab

	BAY SIZE (FT.)	SUPERIMPOSED LOAD (P.S.F.)	SLAB THICKNESS (IN.)	TOTAL DEPTH (FT. - IN.)	TOTAL LOAD (P.S.F.)	COST PER S.F. MAT.	INST.	TOTAL
3800	20x25	40	4	1 - 8	94	3.70	4.92	8.62
3900		75	4	1 - 8	130	4.29	5.25	9.54
4000	(18)	125	4	1 - 10	181	4.98	5.60	10.58
4100		200	5	2 - 2	272	6.45	6.35	12.80
4200	25x25	40	4-½	1 - 8-½	99	3.89	4.97	8.86
4300		75	4-½	1 - 10-½	136	4.68	5.40	10.08
4400		125	5-½	2 - -½	200	5.55	5.85	11.40
4500		200	5-½	2 - 2-½	278	6.85	6.50	13.35
4600	25x30	40	4-½	1 - 8-½	100	4.33	5.20	9.53
4700		75	4-½	1 - 10-½	136	5.05	5.55	10.60
4800		125	5-½	2 - 2-½	202	6.20	6.10	12.30
4900		200	5-½	2 - 5-½	279	7.35	6.70	14.05
5000	30x30	40	4	1 - 8	95	4.29	5.20	9.49
5200		75	4	2 - 1	131	4.94	5.50	10.44
5400		125	4	2 - 4	183	6.10	6.10	12.20
5600		200	5	2 - 10	274	7.75	6.90	14.65
5800	30x35	40	4	2 - 1	95	4.53	5.30	9.83
6000		75	4	2 - 4	139	5.40	5.75	11.15
6250		125	4	2 - 4	185	7	6.50	13.50
6500		200	5	2 - 11	276	8.55	7.30	15.85

For expanded coverage of these items see *Means Concrete Cost Data 1988*

Figure 7.17a

FLOORS	B3.5-520	Composite Beam & C.I.P. Slab

3.5-520 Composite Beam & Cast In Place Slab

	BAY SIZE (FT.)	SUPERIMPOSED LOAD (P.S.F.)	SLAB THICKNESS (IN.)	TOTAL DEPTH (FT. - IN.)	TOTAL LOAD (P.S.F.)	COST PER S.F.		
						MAT.	INST.	TOTAL
7000	35x30	40	4	2-1	95	4.65	5.35	10
7200		75	4	2-4	139	5.50	5.80	11.30
7400		125	4	2-4	185	7	6.50	13.50
7600		200	5	2-11	276	8.55	7.30	15.85
8000	35x35	40	4	2-1	96	4.77	5.40	10.17
8250		75	4	2-4	133	5.65	5.85	11.50
8500		125	4	2-10	185	6.75	6.40	13.15
8750		200	5	3-5	276	8.80	7.35	16.15
9000	35x40	40	4	2-4	97	5.30	5.65	10.95
9250		75	4	2-4	134	6.15	6	12.15
9500		125	4	2-7	186	7.40	6.70	14.10
9750		200	5	3-5	278	9.65	7.75	17.40

90

For expanded coverage of these items see *Means Concrete Cost Data 1988*

Figure 7.17b

FLOORS	B3.5-460	Steel Joists & Beams On Cols.

Table 3.5-460 lists costs for a floor system on steel columns and beams using open web steel joists, galvanized steel slab form, and 2-1/2" concrete slab reinforced with welded wire fabric.

Design and Pricing Assumptions:
Structural Steel is A36.
Concrete f'c = 3 KSI placed by pump.
WWF 6 x 6 #10/#10 (W1.4/W1.4).
Columns are 12' high.
Building is 4 bays long by 4 bays wide.
Joists are 2' O.C. ± and span the long direction of the bay.

Joists at columns have bottom chords extended and are connected to columns.

Slab form is 28 gauge galvanized. Columns costs in table are for columns to support 1 floor plus roof loading in a 2-story building; however, column costs are from ground floor to 2nd floor only. Joist costs include appropriate bridging. Deflection is limited to 1/360 of the span. Screeds and steel trowel finish.

Design Loads	Min.	Max.
S.S. & Joists	6.3 PSF	15.3 PSF
Slab Form	1.0	1.0
2-1/2" Concrete	27.0	27.0
Ceiling	3.0	3.0
Misc.	5.7	1.7
	43.0 PSF	48.0 PSF

System Components	QUANTITY	UNIT	COST PER S.F.		
			MAT.	INST.	TOTAL
SYSTEM 03.5-460-2350					
15'X20'BAY 40 PSF S. LOAD, 17" DEPTH, 83 PSF TOTAL LOAD					
Structural steel	1.974	Lb.	.93	.39	1.32
Open web joists	3.140	Lb.	1.22	.53	1.75
Slab form, galvanized steel 9/16" deep, 28 gauge	1.020	S.F.	.43	.23	.66
Welded wire fabric rolls, 6 x 6, #10/10 (w1.4/w1.4) 21 lb/csf	1.000	S.F.	.09	.16	.25
Concrete ready mix, regular weight, 3000 PSI	.210	C.F.	.44		.44
Place and vibrate concrete, elevated slab less than 6", pumped	.210	C.F.		.16	.16
Finishing floor, monolithic steel trowel finish for finish floor	1.000	S.F.		.48	.48
Curing with sprayed membrane curing compound	.010	S.F.	.02	.04	.06
TOTAL			3.13	1.99	5.12

3.5-460	Steel Joists, Beams & Slab On Columns							
	BAY SIZE (FT.)	SUPERIMPOSED LOAD (P.S.F.)	DEPTH (IN.)	TOTAL LOAD (P.S.F.)	COLUMN ADD	COST PER S.F.		
						MAT.	INST.	TOTAL
2350	15x20 (18)	40	17	83		3.13	1.99	5.12
2400					column	.44	.19	.63
2450	15x20	65	19	108		3.43	2.12	5.55
2500					column	.44	.19	.63
2550	15x20	75	19	119		3.59	2.19	5.78
2600					column	.49	.21	.70
2650	15x20	100	19	144		3.81	2.29	6.10
2700					column	.49	.21	.70
2750	15x20	125	19	170		4.02	2.58	6.60
2800					column	.65	.28	.93
2850	20x20	40	19	83		3.35	2.09	5.44
2900					column	.36	.15	.51
2950	20x20	65	23	109		3.70	2.24	5.94
3000					column	.49	.21	.70
3100	20x20	75	26	119		3.89	2.32	6.21
3200					column	.49	.21	.70
3400	20x20	100	23	144		4.03	2.38	6.41
3450					column	.49	.21	.70
3500	20x20	125	23	170		4.49	2.58	7.07
3600					column	.58	.25	.83

86

For expanded coverage of these items see *Means Concrete Cost Data 1988*

Figure 7.18a

FLOORS	B3.5-460	Steel Joists & Beams On Cols.

3.5-460 Steel Joists, Beams & Slab On Columns

	BAY SIZE (FT.)	SUPERIMPOSED LOAD (P.S.F.)	DEPTH (IN.)	TOTAL LOAD (P.S.F.)	COLUMN ADD	COST PER S.F.		
						MAT.	INST.	TOTAL
3700	20x25	40	44	83		3.64	2.41	6.05
3800					column	.39	.17	.56
3900	20x25	65	26	110		3.97	2.55	6.52
4000					column	.39	.17	.56
4100	20x25	75	26	120		4.13	2.42	6.55
4200					column	.47	.20	.67
4300	20x25	100	26	145		4.37	2.53	6.90
4400					column	.47	.20	.67
4500	20x25	125	29	170		4.90	2.75	7.65
4600					column	.54	.23	.77
4700	25x25	40	23	84		3.90	2.51	6.41
4800					column	.37	.16	.53
4900	25x25	65	29	110		4.12	2.62	6.74
5000					column	.37	.16	.53
5100	25x25	75	26	120		4.54	2.60	7.14
5200					column	.43	.18	.61
5300	25x25	100	29	145		5.10	2.83	7.93
5400					column	.43	.18	.61
5500	25x25	125	32	170		5.35	2.95	8.30
5600					column	.48	.20	.68
5700	25x30	40	29	84		4.01	2.59	6.60
5800					column	.36	.15	.51
5900	25x30	65	29	110		4.18	2.70	6.88
6000					column	.36	.15	.51
6050	25x30	75	29	120		4.53	2.48	7.01
6100					column	.40	.17	.57
6150	25x30	100	29	145		4.91	2.62	7.53
6200					column	.40	.17	.57
6250	25x30	125	32	170		5.20	3.20	8.40
6300					column	.46	.20	.66
6350	30x30	40	29	84		4.23	2.36	6.59
6400					column	.33	.14	.47
6500	30x30	65	29	110		4.80	2.59	7.39
6600					column	.33	.14	.47
6700	30x30	75	32	120		4.89	2.63	7.52
6800					column	.38	.16	.54
6900	30x30	100	35	145		5.45	2.84	8.29
7000					column	.45	.19	.64
7100	30x30	125	35	172		5.85	3.50	9.35
7200					column	.50	.21	.71
7300	30x35	40	29	85		4.77	2.57	7.34
7400					column	.29	.12	.41
7500	30x35	65	29	111		5.25	3.22	8.47
7600					column	.37	.16	.53
7700	30x35	75	32	121		5.25	3.22	8.47
7800					column	.38	.16	.54
7900	30x35	100	35	148		5.75	2.95	8.70
8000					column	.46	.20	.66
8100	30x35	125	38	173		6.35	3.19	9.54
8200					column	.47	.20	.67
8300	35x35	40	32	85		4.89	2.63	7.52
8400					column	.33	.14	.47
8500	35x35	65	35	111		5.50	3.35	8.85
8600					column	.40	.17	.57
9300	35x35	75	38	121		5.65	3.41	9.06
9400					column	.40	.17	.57

For expanded coverage of these items see *Means Concrete Cost Data 1988*

87

Figure 7.18b

The $101,304 figure does not include sales tax, General Conditions, overhead, etc., as shown on the Estimate Summary Sheet in Figure 7.15. These costs are added as follows:

Structural change savings	$101,304
Sales Tax @ 4% × .5	2,026
General Conditions @ 16%	16,209
Subtotal "A"	$119,539
Overhead & Profit @ 10%	11,954
Bid Conditions @ 8%	9,563
Subtotal "B"	$141,056
Location Factor @ 89.2%	× .892
Subtotal "C"	$125,822
Inflation @ 4%/yr	× 1.061
Savings	**$133,497**

This figure does not represent the total potential savings from the change because the difference in floor system weight of 9 P.S.F. also means that the supporting structure can be lighter and less expensive. A rough structural analysis similar to the one in the last chapter would be needed to change the supporting structure by re-sizing columns and footings. This simplified example does, however, illustrate how easily cost comparisons for different systems can be made.

By reducing the building size and changing the floor system, we have identified potential savings of approximately $271,000. Further savings might be found by using a different exterior closure system, less expensive interior finishes, and a less expensive paving system for the parking lot.

Summary

Drawing up cost control estimates in the conceptual and detailed design phases can present something of a challenge because, until late in the detailed design phase, design information is limited. Those plans and specifications that are available must be carefully analyzed to determine what information they contain, what can be inferred, and what must be assumed. The product of this analysis is used to create an assemblies/square foot estimate. This estimate is used to determine if the scheme or current project development is still within the budget. If corrective measures are needed, the assemblies/square foot estimate can also be used to determine the effect of potential cost savings changes.

As more information becomes available, the assemblies costs for the actual materials/assemblies used can be substituted for the earlier assumptions in order to refine the cost control estimates. *Means Square Foot Estimating* is a good source of techniques for the assemblies estimating method. Chapters 8, 9, and 10 present short-cut methods for estimating mechanical, electrical, and plumbing—using assemblies costs.

8

ESTIMATING PLUMBING COSTS

8

ESTIMATING PLUMBING COSTS

One of the challenges of effective design cost control is obtaining accurate estimates of plumbing, HVAC, and electrical costs. Early estimate accuracy is limited by the fact that little or no information about these systems is available at this stage, and the square foot method shown in Chapter 6 must be used. Sufficient information for highly accurate estimates of these systems is usually available only when the plans and specifications are about 90% complete, and that type of detailed estimate requires that the estimator be a specialist in the field.

The solution for an early design cost control estimate is a simplified system using a combination of assemblies and unit price methods. This approach is easy enough for the non-specialist, and can provide check estimates accurate enough for effective design cost control.

Applying this combined assemblies/unit price method requires some explanation. We will begin, in this chapter, with these procedures for estimating plumbing costs. In assemblies estimating, plumbing is divided into *site systems* and *building systems*. Estimating site systems is described in the analysis by UNIFORMAT division in Chapter 7. Here we will address the issue of building systems. First, we will provide a necessary background in the different components of a building plumbing system. Next, the appropriate plumbing assemblies prices from *Means Assemblies Cost Data* will be presented to show which components can be estimated using the assemblies method, and which must be estimated using the unit price method. Finally, we will outline a system synthesizing the two methods, and demonstrate the system by preparing a plumbing estimate for the Chambers office building.

The Plumbing System

A building plumbing system consists of a number of components which provide the means to supply, use, and dispose of fluids. These components include pipe and fittings, fixtures, drains, carriers, pumps, stand pipes, and fire sprinkler systems. Each of these categories is described below.

Pipe and Fittings

The piping system transports fluids throughout the building. These fluids include potable water, domestic hot water, storm and sanitary waste, and gas. The system includes pipe, couplings, fittings, valves, and hangers and supports. Pipe comes in a variety of sizes and materials. Sizes can range from 1/2″ to 12″ in diameter. Materials include cast iron, steel, copper, plastic, and clay.

Couplings and fittings are used to connect lengths of pipe. Fittings may also provide a change in the direction of flow. Tees, elbows, traps, and wyes are all types of fittings. Some examples of plastic couplings and fittings are shown in Figure 8.1. Valves are used to control the flow, pressure, or level of the fluid in the piping system. Examples are shown in Figure 8.2.

Hangers and supports include a variety of devices that support the pipe and control pipe expansion and contraction. These devices include inserts, hanger rods, clamps, u-bolts, and brackets. Examples are shown in Figure 8.3.

Fixtures

Plumbing fixtures are the receptacles attached to and served by piping systems. Included are such items as drinking fountains, lavatories, water closets, urinals, sinks, showers, hot water heaters, and dishwashers. (See examples in Figure 8.4a and b).

Drains

Drains are devices to collect and transport waste fluids to the waste systems. Examples are shown in Figure 8.5.

Carriers

Carriers are used to support plumbing fixtures and attendant piping. Examples are shown in Figure 8.6.

Pumps

Pumps, such as sewage ejectors and sump pumps, are occasionally required in plumbing systems to move liquids out of a building.

Stand Pipe

A stand pipe system includes piping, fire hose valves, hose cabinets, and fire department and pumper connections. It provides a manual fire fighting system in the building. Examples of stand pipe equipment are shown in Figure 8.7.

Fire Sprinkler Systems

Fire sprinkler systems provide a means of automatic fire protection in the building. The fire suppression medium may be water or various types of chemicals, foams, or gasses. Such a system includes all piping, fittings, fixtures, and pumps that are required. Fire sprinkler systems are installed by a specially licensed subcontractor.

More detailed information about plumbing systems and unit price plumbing estimating can be found in *Means Mechanical Estimating*.

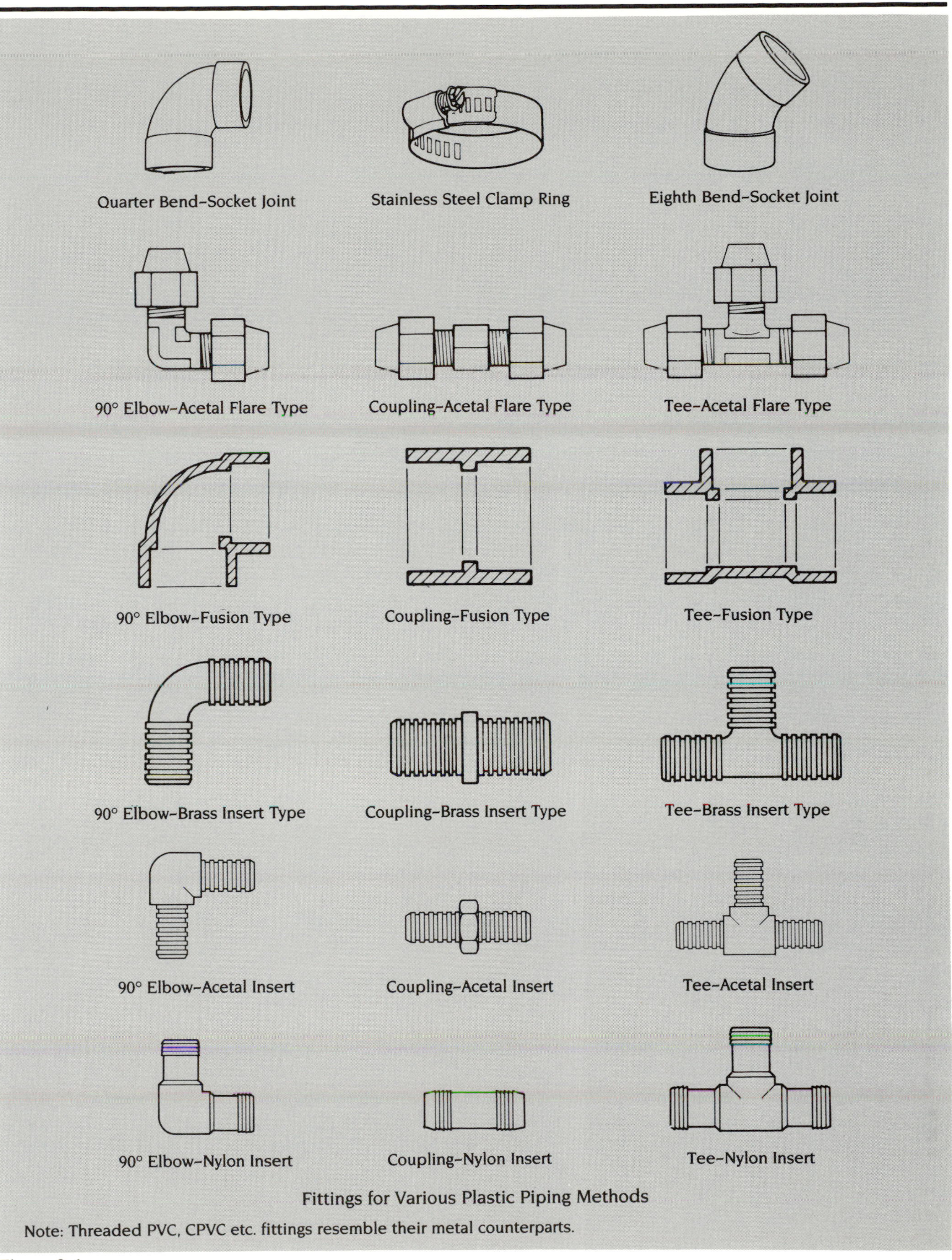

Fittings for Various Plastic Piping Methods

Note: Threaded PVC, CPVC etc. fittings resemble their metal counterparts.

Figure 8.1

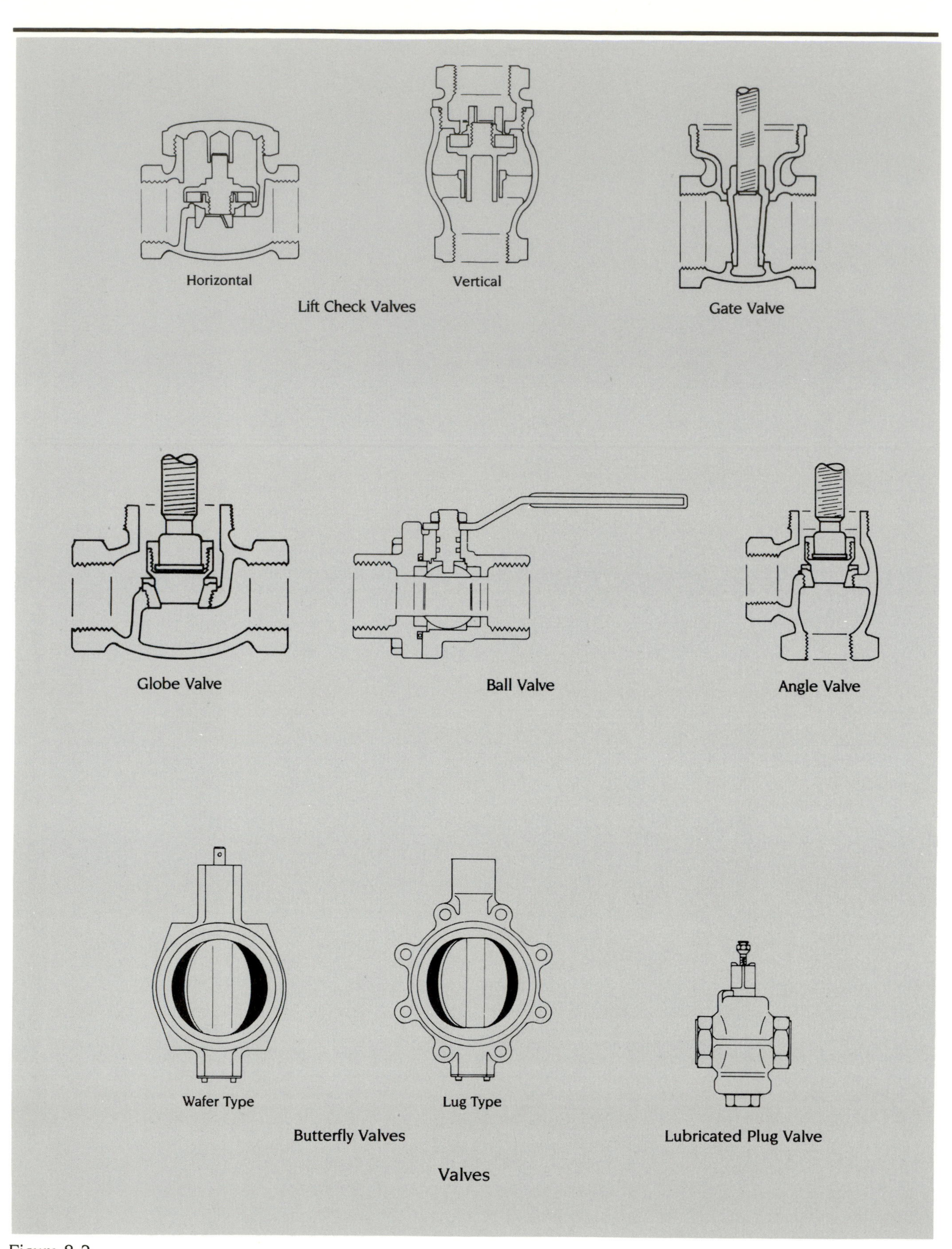
Horizontal
Vertical
Lift Check Valves
Gate Valve
Globe Valve
Ball Valve
Angle Valve
Wafer Type
Lug Type
Butterfly Valves
Lubricated Plug Valve
Valves

Figure 8.2

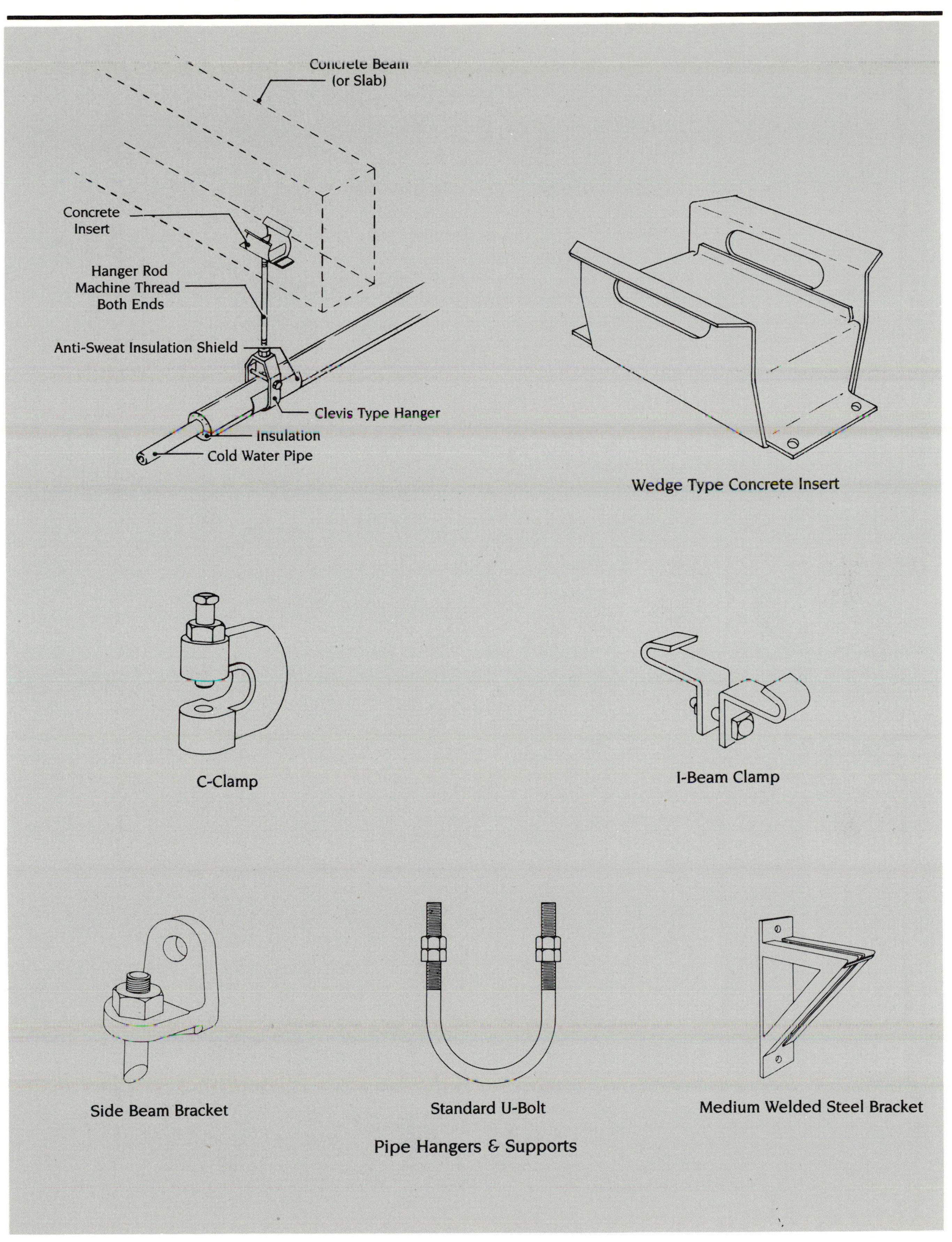
Concrete Beam
(or Slab)
Concrete
Insert
Hanger Rod
Machine Thread
Both Ends
Anti-Sweat Insulation Shield
Clevis Type Hanger
Insulation
Cold Water Pipe
Wedge Type Concrete Insert
C-Clamp
I-Beam Clamp
Side Beam Bracket
Standard U-Bolt
Medium Welded Steel Bracket
Pipe Hangers & Supports

Figure 8.3

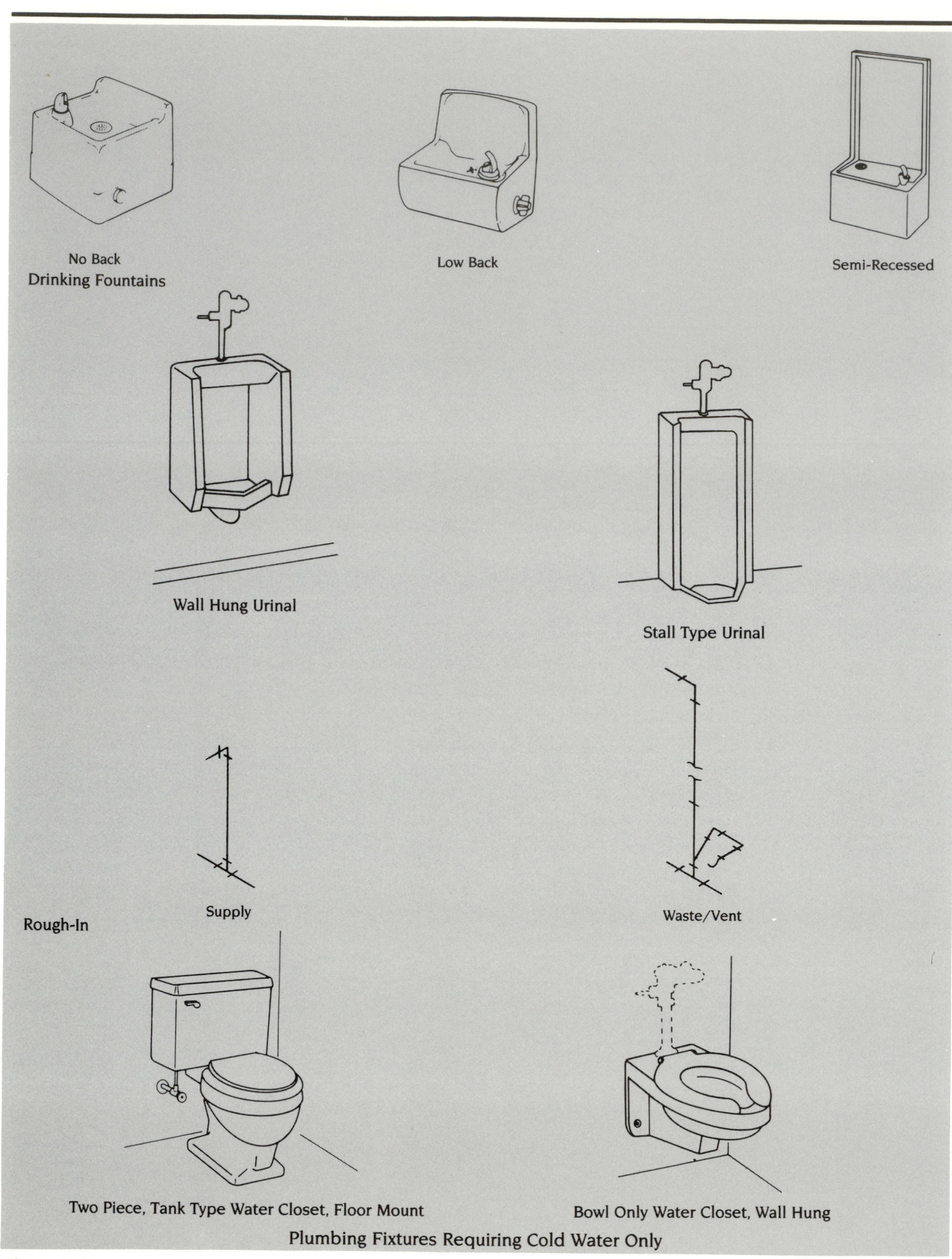

Plumbing Fixtures Requiring Cold Water Only

Figure 8.4a

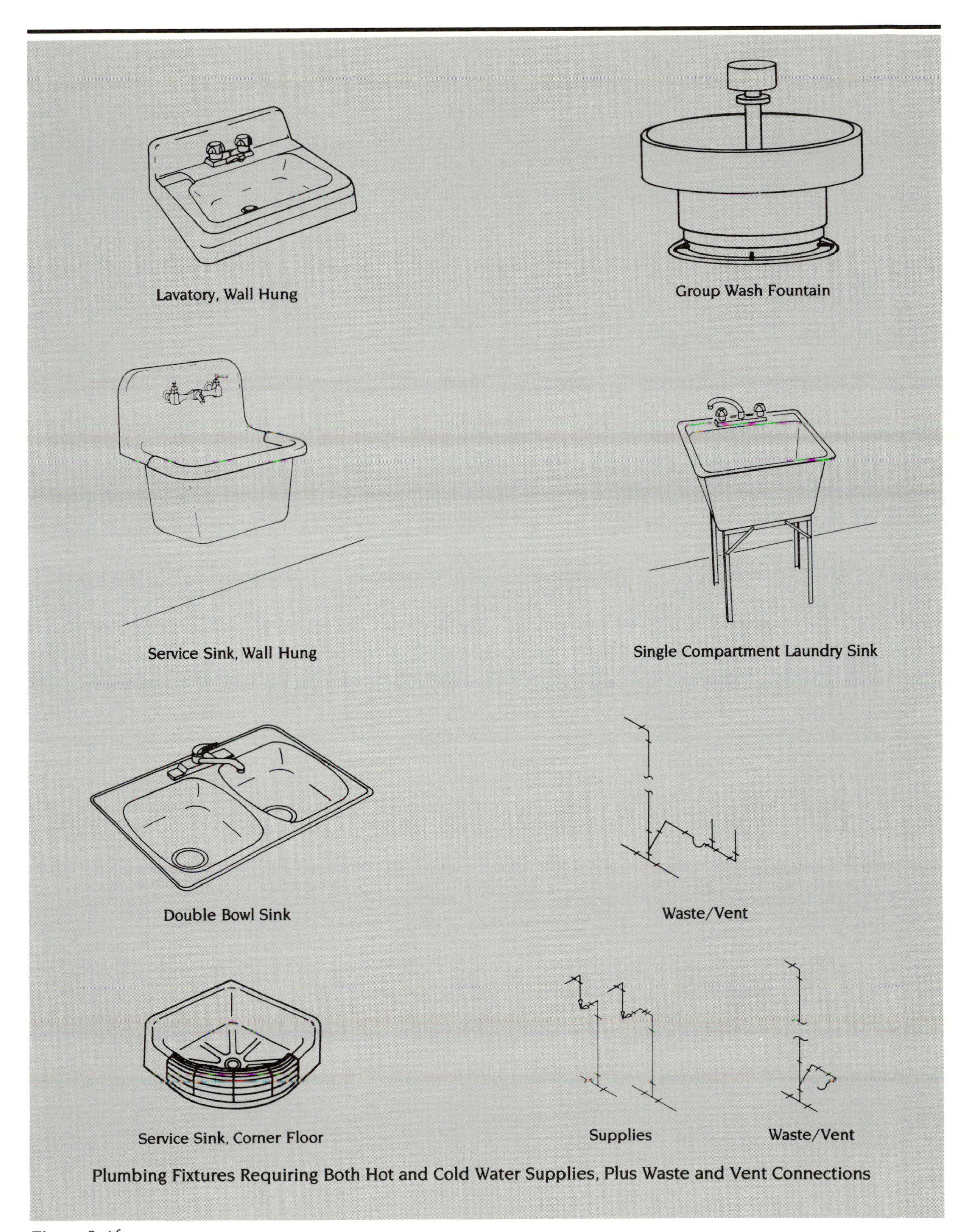
Lavatory, Wall Hung
Group Wash Fountain
Service Sink, Wall Hung
Single Compartment Laundry Sink
Double Bowl Sink
Waste/Vent
Service Sink, Corner Floor
Supplies
Waste/Vent
Plumbing Fixtures Requiring Both Hot and Cold Water Supplies, Plus Waste and Vent Connections

Figure 8.4b

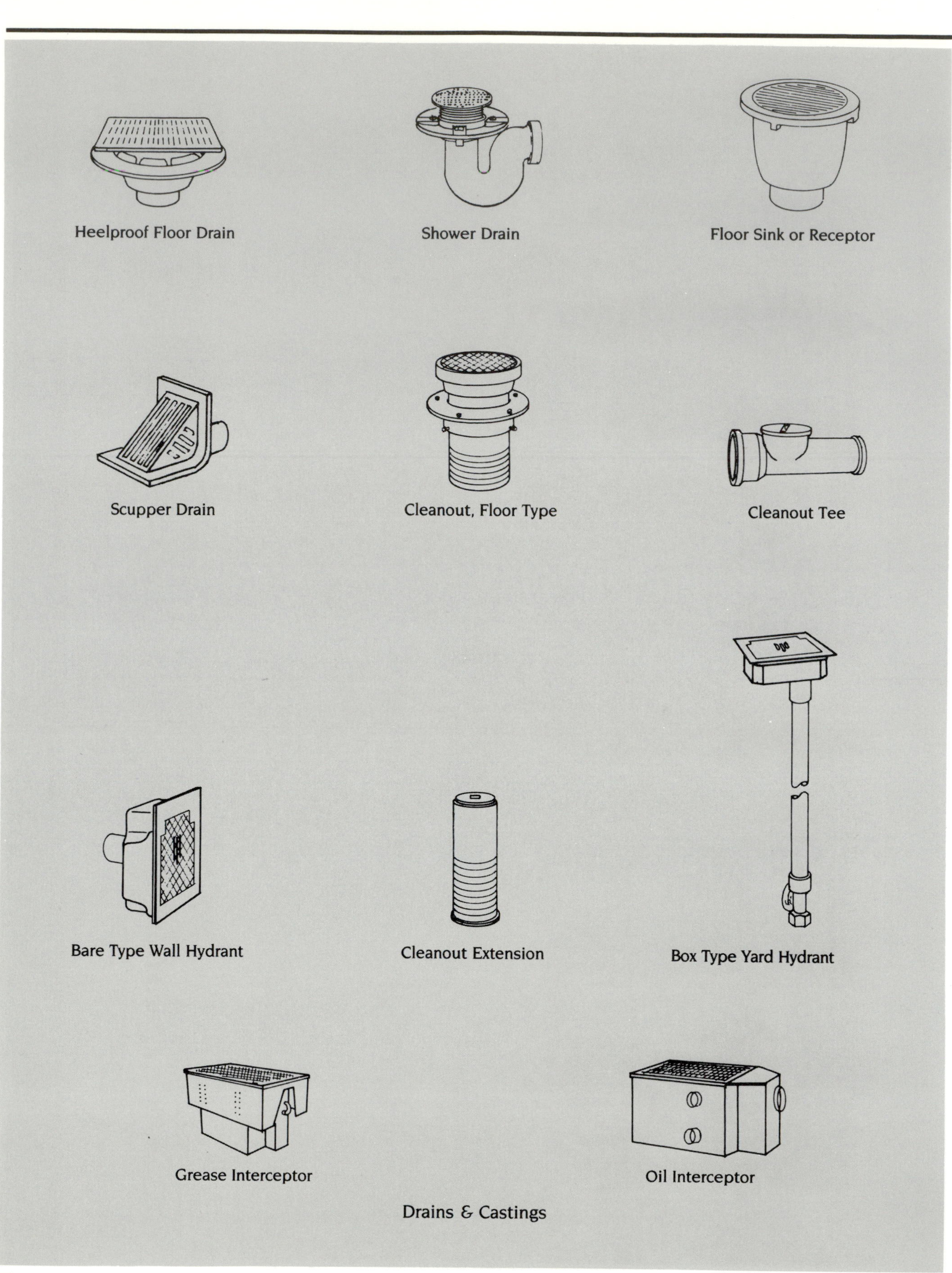

Drains & Castings

Figure 8.5

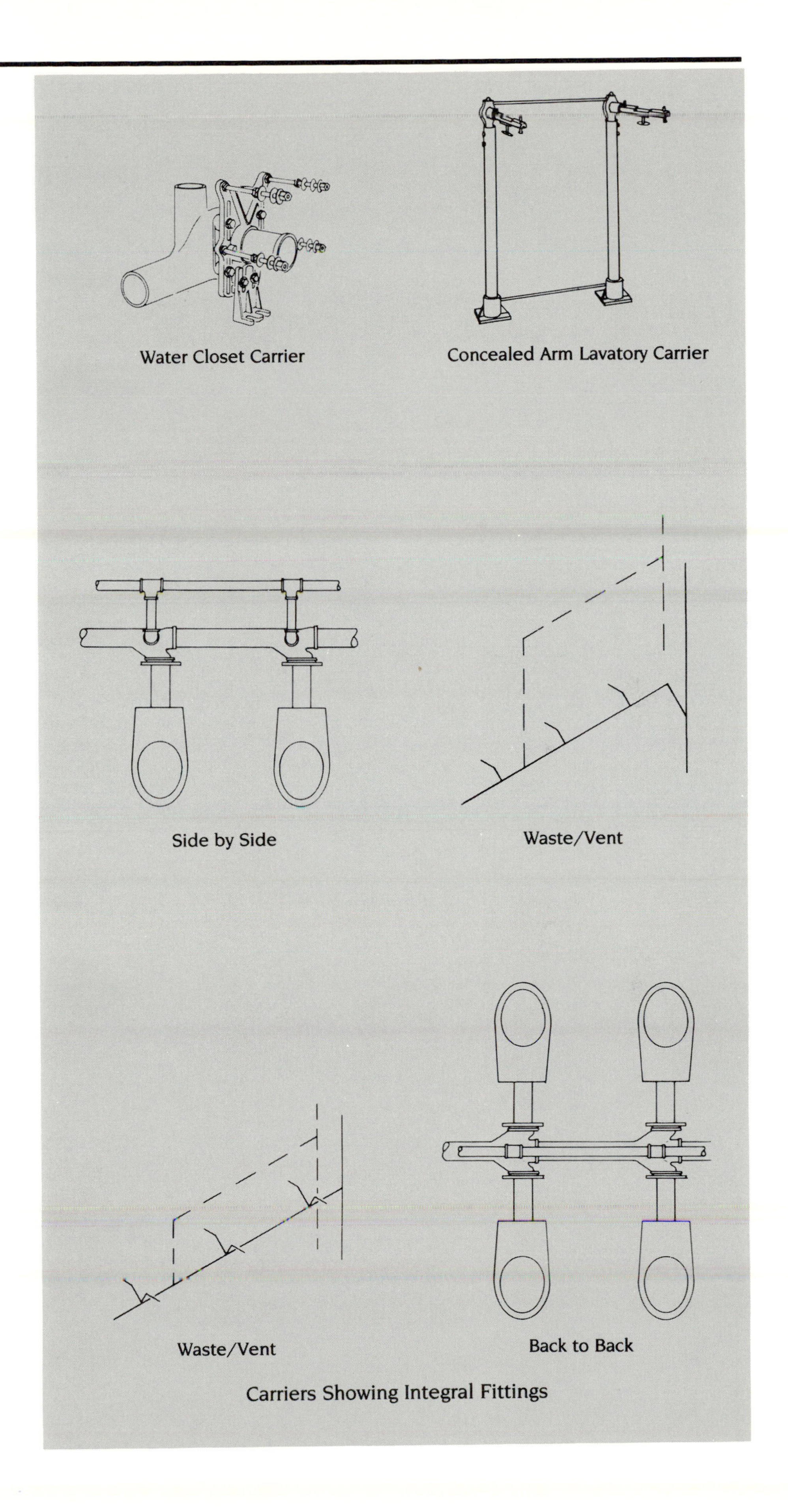

Carriers Showing Integral Fittings

Figure 8.6

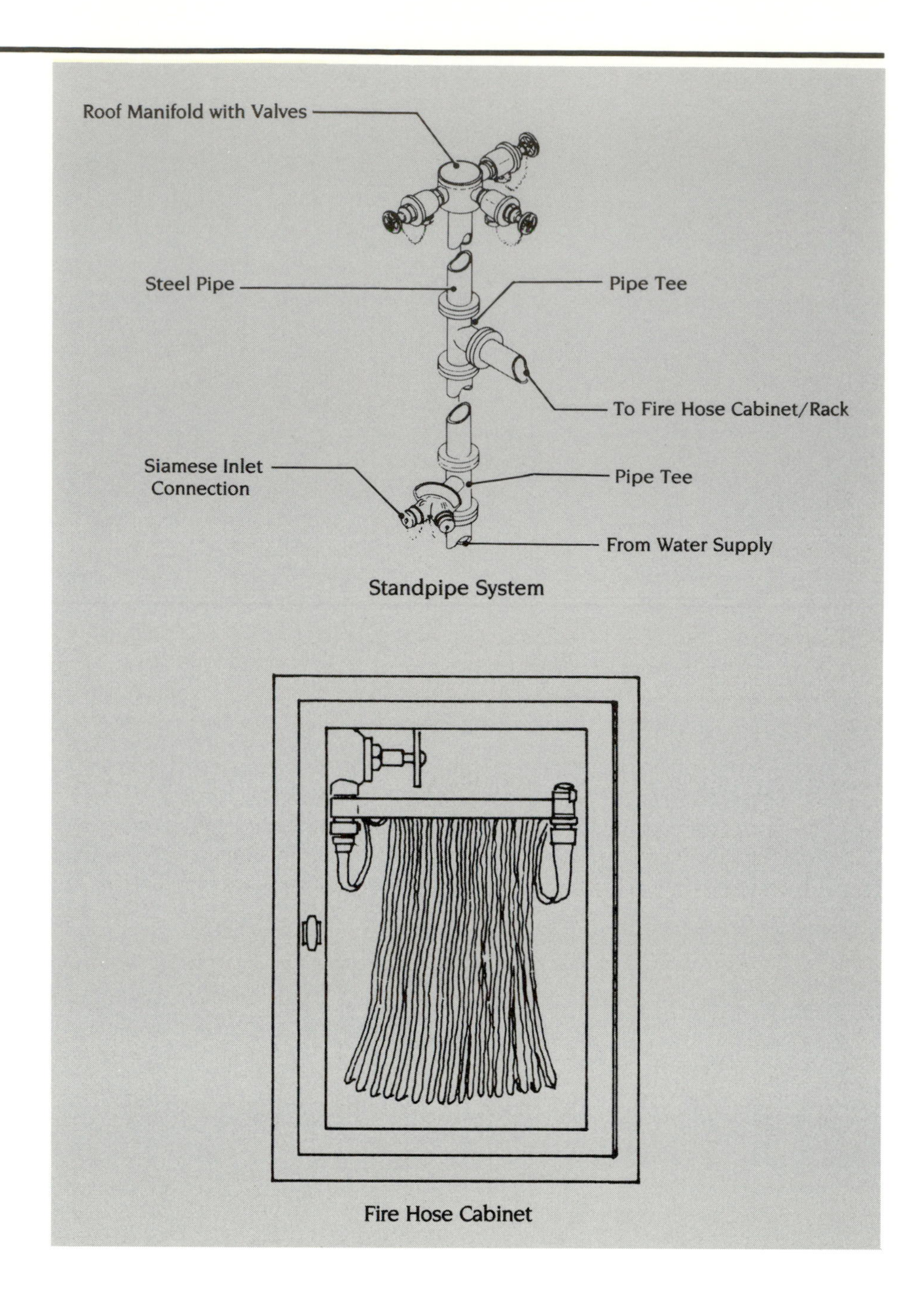
Roof Manifold with Valves
Steel Pipe
Pipe Tee
To Fire Hose Cabinet/Rack
Siamese Inlet
Connection
Pipe Tee
From Water Supply
Standpipe System
Fire Hose Cabinet

Figure 8.7

Assemblies and Unit Cost Applications

An accurate estimate for plumbing equipment may be obtained by using a combination of assemblies and unit price methods. Where assemblies cost data is available, it saves time, since individual items do not have to be priced. For items that cannot be made into assemblies and are therefore unavailable in this form, the unit price method is used.

Means Assemblies Cost Data provides assemblies prices for a number of plumbing systems. Among them are single and multi-unit fixture assemblies, roof drains, fire sprinkler systems, and stand pipe systems. The fixture assembly prices include the fixtures, rough-in piping, and carriers as needed. Figure 8.8 is an example of a single unit assembly; a multi-unit assembly appears in Figure 8.9. Roof drain assemblies, Figure 8.10, include the drain and rough-in piping.

Fire sprinkler systems (Figure 8.11a and b) are priced by square foot of floor area. Separate tables are given for each type of sprinkler system. System prices include all pipe, fittings, valves, heads, and accessories.

The cost of stand pipe systems, Figure 8.12a and b, is given per floor for the size of pipe and type of system. The price includes pipe, fittings, valves, and fire department and pumper connections. Associated items, such as hose valves, hose cabinets, and nozzles, are shown as unit prices in Figure 8.13.

Of these four types of assemblies, only the price for fire sprinklers represents a *total system cost*. Fixture assemblies, roof drains, and stand pipes require additional piping to supply and/or take away fluids. Also excluded from the assemblies are pumps and floor drains.

A sample estimate from *Means Assemblies Cost Data* (Figure 8.14) shows a method for estimating the cost of the supply/waste piping needed for the roof drains, fixtures, and associated piping. In the example, the roof drain and plumbing fixture assemblies for an apartment building are counted and priced; percentages of the total cost of these assemblies are added for water control devices, and pipe and fittings. The total is modified by a multiplier to adjust for the relative quality and complexity of the project.

The table in Figure 8.15 provides some explanation of the adjustment percentages. The water control adjustment includes water meters, backflow preventers, shock absorbers, vacuum breakers, and mixers, and ranges from 10% to 15%. Although not specifically stated, it is assumed that the pipe and fittings adjustment (30% - 60%) includes pipe, couplings, fittings, valves, and hangers. The lower end of the pipe and fittings range is for buildings with compact plumbing systems, and the higher end is buildings with widely spread plumbing systems. It is also noted that the pipe and fittings adjustment may range as high as 100% in rare cases, and that special purpose and process piping is excluded. A quality/complexity modifier is added to adjust the Means prices, which are based on an economy, low-complexity installation. This modifier is 0% to 5% for economy installations, 5% to 15% for good quality and medium complexity installations, and 15% to 25% for above average quality and complexity installations.

Individual costs for pumps and drains are not included in *Means Assemblies Cost Data*, but can be estimated using unit prices from *Means Building Construction Cost Data*. (For expanded coverage of plumbing systems and individual items, see the 1988 edition of *Means Mechanical Cost Data*, and as of 1989, *Means Plumbing Cost Data*.) Prices for floor drains are shown in Figure 8.16, and for sump pumps, in Figure 8.17. These prices do not include piping, but the percentages (from Figure 8.15) can be added to the cost of the pumps and drains to cover the cost of the waste piping.

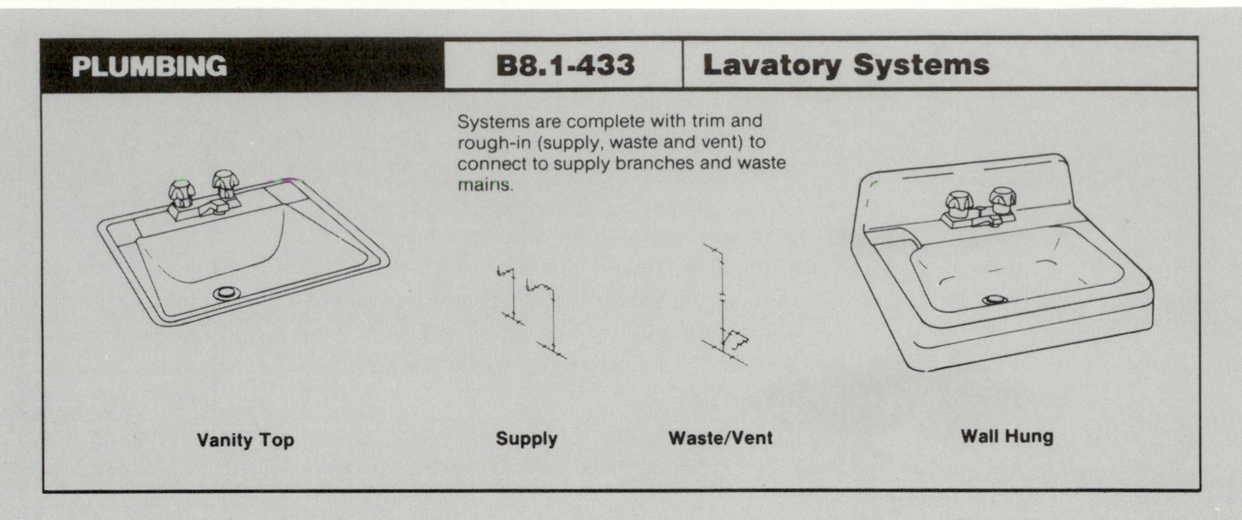

System Components	QUANTITY	UNIT	COST EACH MAT.	COST EACH INST.	COST EACH TOTAL
SYSTEM 08.1-433-1560					
LAVATORY W/TRIM, VANITY TOP, P.E. ON C.I., 20" X 18"					
Lavatory w/trim, PE on CI, white, vanity top, 20" x 18" oval	1.000	Ea.	121	79	200
Pipe, steel, galvanized, schedule 40, threaded, 1-¼" diam	4.000	L.F.	9.16	22.64	31.80
Copper tubing type DWV, solder joint, hanger 10'OC 1-¼" diam	4.000	L.F.	7.48	37.32	44.80
Wrought copper DWV, Tee, sanitary, 1-¼" diam	1.000	Ea.	2.30	30.70	33
P trap w/cleanout, 20 ga, 1-¼" diam	1.000	Ea.	9.85	15.15	25
Copper tubing type L, solder joint, hanger 10' OC ½" diam	10.000	L.F.	8.30	52.70	61
Wrought copper 90° elbow for solder joints ½" diam	2.000	Ea.	.24	28.06	28.30
Wrought copper Tee for solder joints, ½" diam	2.000	Ea.	.44	43.56	44
Stop, chrome, angle supply, ½" diam	2.000	Ea.	29.70	26.30	56
TOTAL			188.47	335.43	523.90

8.1-433	Lavatory Systems	COST EACH MAT.	COST EACH INST.	COST EACH TOTAL
1560	Lavatory w/trim, vanity top, PE on CI, 20" x 18"	190	335	525
1600	19" x 16" oval	180	335	515
1640	18" round	175	335	510
1680	Cultured marble, 19" x 17"	145	335	480
1720	25" x 19"	160	335	495
1760	Stainless, self-rimming, 25" x 22"	205	335	540
1800	17" x 22"	140	335	475
1840	Steel enameled, 20" x 17"	135	340	475
1880	19" round	135	345	480
1920	Vitreous china, 20" x 16"	250	350	600
1960	19" x 16"	195	350	545
2000	22" x 13"	200	350	550
2040	Wall hung, PE on CI, 18" x 15"	370	365	735
2080	19" x 17"	335	370	705
2120	20" x 18"	260	370	630
2160	Vitreous china, 18" x 15"	260	375	635
2200	19" x 17"	210	375	585
2240	24" x 20"	310	375	685

(40)

For expanded coverage of these items see *Means Mechanical Cost Data 1988*

259

Figure 8.8

PLUMBING	B8.1-630	Three Fixture Bathrooms

Three fixture bathroom systems consisting of a lavatory, water closet, bathtub or shower and service piping.
- Prices for plumbing and fixtures only.

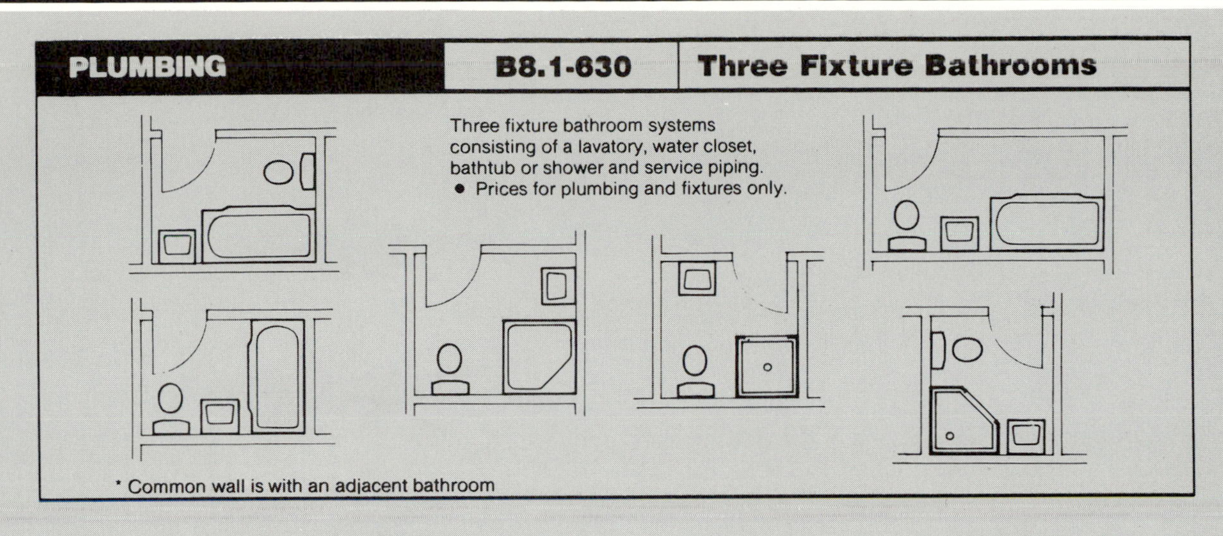

* Common wall is with an adjacent bathroom

System Components	QUANTITY	UNIT	COST EACH MAT.	COST EACH INST.	COST EACH TOTAL
SYSTEM 08.1-630-1170					
BATHROOM, LAVATORY, WATER CLOSET & BATHTUB					
ONE WALL PLUMBING, STAND ALONE					
Wtr closet, 2 pc close cpld vit china flr mntd w/seat supply & stop	1.000	Ea.	126.50	93.50	220
Water closet, rough-in waste & vent	1.000	Set	91.45	258.55	350
Lavatory w/ftngs, wall hung, white, PE on CI, 20" x 18"	1.000	Ea.	139.70	65.30	205
Lavatory, rough-in waste & vent	1.000	Set	119.60	305.40	425
Bathtub, white PE on CI, w/ftgs, mat bottom, recessed, 5' long	1.000	Ea.	288.20	116.80	405
Baths, rough-in waste and vent	1.000	Set	74.58	262.92	337.50
TOTAL			840.03	1,102.47	1,942.50

8.1-630	Three Fixture Bathroom, One Wall Plumbing	COST EACH MAT.	COST EACH INST.	COST EACH TOTAL
1150	Bathroom, three fixture, one wall plumbing			
1160	Lavatory, water closet & bathtub			
1170	Stand alone	840	1,100	1,940
1180	Share common plumbing wall *	735	835	1,570

8.1-630	Three Fixture Bathroom, Two Wall Plumbing	COST EACH MAT.	COST EACH INST.	COST EACH TOTAL
2130	Bathroom, three fixture, two wall plumbing			
2140	Lavatory, water closet & bathtub			
2160	Stand alone	845	1,125	1,970
2180	Long plumbing wall common *	765	920	1,685
3610	Lavatory, bathtub & water closet			
3620	Stand alone	905	1,275	2,180
3640	Long plumbing wall common *	855	1,150	2,005
4660	Water closet, corner bathtub & lavatory			
4680	Stand alone	1,550	1,125	2,675
4700	Long plumbing wall common *	1,450	840	2,290
6100	Water closet, stall shower & lavatory			
6120	Stand alone	895	1,425	2,320
6140	Long plumbing wall common *	855	1,325	2,180
7060	Lavatory, corner stall shower & water closet			
7080	Stand alone	1,050	1,300	2,350
7100	Short plumbing wall common *	980	930	1,910

268

For expanded coverage of these items see *Means Mechanical Cost Data 1988*

Figure 8.9

PLUMBING | B8.1-310 | Storm Drainage - Roof Drains

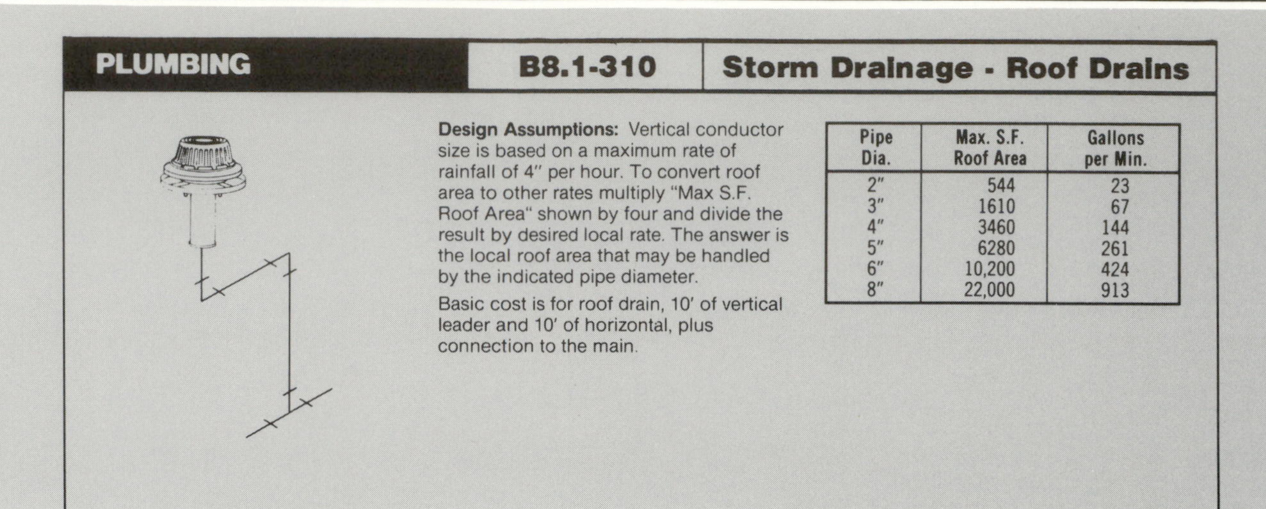

Design Assumptions: Vertical conductor size is based on a maximum rate of rainfall of 4" per hour. To convert roof area to other rates multiply "Max S.F. Roof Area" shown by four and divide the result by desired local rate. The answer is the local roof area that may be handled by the indicated pipe diameter.

Basic cost is for roof drain, 10' of vertical leader and 10' of horizontal, plus connection to the main.

Pipe Dia.	Max. S.F. Roof Area	Gallons per Min.
2"	544	23
3"	1610	67
4"	3460	144
5"	6280	261
6"	10,200	424
8"	22,000	913

System Components	QUANTITY	UNIT	COST EACH MAT.	COST EACH INST.	COST EACH TOTAL
SYSTEM 081-310-1880					
ROOF DRAIN, DWV PVC PIPE, 2" DIAM., 10' HIGH					
Drain, roof, main, PVC, dome type 2" pipe size	1.000	Ea.	36.85	36.15	73
Clamp, roof drain, underdeck	1.000	Ea.	9.74	21.26	31
Pipe, Tee, PVC DWV, schedule 40, 2" pipe size	1.000	Ea.	1.33	29.67	31
Pipe, PVC, DWV, schedule 40, 2" diam.	20.000	L.F.	25.80	171.20	197
Pipe, elbow, PVC schedule 40, 2" diam.	2.000	Ea.	1.96	36.04	38
TOTAL			75.68	294.32	370

8.1-310	Roof Drain Systems	COST PER SYSTEM MAT.	COST PER SYSTEM INST.	COST PER SYSTEM TOTAL
1880	Roof drain, DWV PVC, 2" diam., piping, 10' high	76	295	371
1920	For each additional foot add	1.29	8.55	9.84
1960	3" diam., 10' high	105	355	460
2000	For each additional foot add	2.39	9.50	11.89
2040	4" diam., 10' high	140	395	535
2080	For each additional foot add	3.28	10.50	13.78
2120	5" diam., 10' high	305	430	735
2160	For each additional foot add	4.77	11.75	16.52
2200	6" diam., 10' high	395	510	905
2240	For each additional foot add	5.95	12.95	18.90
2280	8" diam., 10' high	630	810	1,440
2320	For each additional foot add	9.30	16.70	26
3940	C.I., soil, single hub, service wt., 2" diam. piping, 10' high	175	305	480
3980	For each additional foot add	2.67	8	10.67
4120	3" diam., 10' high	215	330	545
4160	For each additional foot add	3.75	8.40	12.15
4200	4" diam., 10' high	255	365	620
4240	For each additional foot add	4.87	9.20	14.07
4280	5" diam., 10' high	335	405	740
4320	For each additional foot add	6.75	10.30	17.05
4360	6" diam., 10' high	435	430	865
4400	For each additional foot add	7.95	10.75	18.70
4440	8" diam., 10' high	730	870	1,600
4480	For each additional foot add	13.15	17.85	31
6040	Steel galv. sch 40 threaded, 2" diam. piping, 10' high	200	295	495
6080	For each additional foot add	3.65	7.90	11.55

For expanded coverage of these items see *Means Mechanical Cost Data 1988*

253

Figure 8.10

FIRE PROTECTION	B8.2-150	Firecycle Sprinkler Systems

Firecycle is a fixed fire protection sprinkler system utilizing water as its extinguishing agent. It is a time delayed, recycling, preaction type which automatically shuts the water off when heat is reduced below the detector operating temperature and turns the water back on when that temperature is exceeded.

The system senses a fire condition through a closed circuit electrical detector system which controls water flow to the fire automatically. Batteries supply up to 90 hour emergency power supply for system operation. The piping system is dry (until water is required) and is monitored with pressurized air. Should any leak in the system piping occur, an alarm will sound, but water will not enter the system until heat is sensed by a Firecycle detector.

All areas are assumed to be open.

System Components	QUANTITY	UNIT	COST EACH MAT.	COST EACH INST.	COST EACH TOTAL
SYSTEM 08.2-150-0580					
FIRECYCLE SPRINKLER SYSTEM, STEEL BLACK SCH. 40 PIPE					
LIGHT HAZARD, ONE FLOOR, 2000 S.F.					
Valve, gate, iron body 125 lb, OS&Y, flanged, 4" pipe size	1.000	Ea.	136.13	126.38	262.51
Valve, angle, bronze, 150 lb, rising stem, threaded, 2" pipe size	1.000	Ea.	78.38	19.13	97.51
Valve, swing check, bronze, 125 lb, regrinding disc, 2-½" pipe size	1.000	Ea.	61.05	25.20	86.25
*Alarm valve, 2-½" pipe size	1.000	Ea.	400.13	128.63	528.76
Alarm, water motor, complete with gong	1.000	Ea.	79.61	55.39	135
Pipe, steel, black, schedule 40, 4" diam	10.000	L.F.	57.45	107.55	165
Fire alarm, horn, electric	1.000	Ea.	21.45	31.05	52.50
Pipe, steel, black, schedule 40, threaded, cplg & hngr 10'OC 2-½" diam	20.000	L.F.	77.70	151.05	228.75
Pipe, steel, black, schedule 40, threaded, cplg & hngr 10'OC 2" diam	12.500	L.F.	29.91	73.69	103.60
Pipe, steel, black, schedule 40, threaded, cplg & hngr 10'OC 1-¼" diam	37.500	L.F.	56.25	158.91	215.16
Pipe, steel, black, schedule 40, threaded, cplg & hngr 10'OC 1" diam	112.000	L.F.	136.92	442.68	579.60
Pipe, Tee, malleable iron, black, 150 lb threaded, 4" pipe size	2.000	Ea.	52.80	187.20	240
Pipe, Tee, malleable iron, black, 150 lb threaded, 2-½" pipe size	2.000	Ea.	18.24	83.76	102
Pipe, Tee, malleable iron, black, 150 lb threaded, 2" pipe size	1.000	Ea.	3.89	34.36	38.25
Pipe, Tee, malleable iron, black, 150 lb threaded, 1-¼" pipe size	5.000	Ea.	10.76	135.49	146.25
Pipe, Tee, malleable iron, black, 150 lb threaded, 1" pipe size	4.000	Ea.	5.31	105.69	111
Pipe, 90° elbow, malleable iron, black, 150 lb threaded, 1" pipe size	6.000	Ea.	5.13	98.37	103.50
Sprinkler head std spray, brass 135°-286°F ½" NPT, ⅜" orifice	12.000	Ea.	30.24	158.76	189
Firecycle controls, incls panel, battery, solenoid valves, press switches	1.000	Ea.	3,885.75	820.50	4,706.25
Detector, firecycle system	2.000	Ea.	280.50	27	307.50
Firecycle pkg, swing check & flow control valves w/trim 4" pipe size	1.000	Ea.	1,130.25	388.50	1,518.75
Air compressor, auto, complete, 200 Gal sprinkler sys cap, ⅓ HP	1.000	Ea.	367.13	165.38	532.51
*Standpipe connection, wall, flush, brass w/plug & chain 2-½"x2-½"	1.000	Ea.	70.95	79.05	150
Valve, gate, bronze, 300 psi, NRS, class 150, threaded, 1" diam	1.000	Ea.	12.29	10.96	23.25
TOTAL			7,008.22	3,614.68	10,622.90
COST PER S.F.			3.50	1.81	5.31

*Not included in systems under 2000 S.F.

8.2-150	Firecycle Sprinkler Systems	COST PER S.F. MAT.	COST PER S.F. INST.	COST PER S.F. TOTAL
0520	Firecycle sprinkler systems, steel black sch. 40 pipe			
0530	Light hazard, one floor, 500 S.F.	11.65	4.06	15.71
0560	1000 S.F. (42)(43)	5.95	2.53	8.48
0580	2000 S.F.	3.50	1.81	5.31

For expanded coverage of these items see *Means Mechanical Cost Data 1988*

283

Figure 8.11*a*

FIRE PROTECTION	B8.2-150	Firecycle Sprinkler Systems		
8.2-150	**Firecycle Sprinkler Systems**	COST PER S.F.		
		MAT.	INST.	TOTAL
0600	5000 S.F.	1.57	1.15	2.72
0620	10,000 S.F.	.94	.90	1.84
0640	50,000 S.F.	.48	.76	1.24
0660	Each additional floor of 500 S.F.	.62	1.12	1.74
0680	1000 S.F.	.45	1	1.45
0700	2000 S.F.	.36	.89	1.25
0720	5000 S.F.	.34	.78	1.12
0740	10,000 S.F.	.33	.72	1.05
0760	50,000 S.F.	.33	.68	1.01
1000	Ordinary hazard, one floor, 500 S.F.	11.75	4.20	15.95
1020	1000 S.F.	5.95	2.48	8.43
1040	2000 S.F.	3.55	1.90	5.45
1060	5000 S.F.	1.61	1.21	2.82
1080	10,000 S.F.	1.02	1.15	2.17
1100	50,000 S.F.	.74	1.18	1.92
1140	Each additional floor, 500 S.F.	.73	1.26	1.99
1160	1000 S.F.	.44	.99	1.43
1180	2000 S.F.	.45	.90	1.35
1200	5000 S.F.	.40	.85	1.25
1220	10,000 S.F.	.37	.84	1.21
1240	50,000 S.F.	.36	.80	1.16
1500	Extra hazard, one floor, 500 S.F.	12.60	4.77	17.37
1520	1000 S.F.	6.45	3.06	9.51
1540	2000 S.F.	3.61	2.31	5.92
1560	5000 S.F.	1.70	1.62	3.32
1580	10,000 S.F.	1.27	1.56	2.83
1600	50,000 S.F.	1.03	1.70	2.73
1660	Each additional floor, 500 S.F.	.78	1.50	2.28
1680	1000 S.F.	.63	1.41	2.04
1700	2000 S.F.	.57	1.41	1.98
1720	5000 S.F.	.49	1.26	1.75
1740	10,000 S.F.	.52	1.12	1.64
1760	50,000 S.F.	.52	1.05	1.57
2020	Grooved steel, black, sch. 40 pipe, light hazard, one floor			
2030	2000 S.F.	3.61	1.63	5.24
2060	10,000 S.F.	1.01	.80	1.81
2100	Each additional floor, 2000 S.F.	.53	.72	1.25
2150	10,000 S.F.	.39	.61	1
2200	Ordinary hazard, one floor, 2000 S.F.	3.63	1.70	5.33
2250	10,000 S.F.	1.20	1.05	2.25
2300	Each additional floor, 2000 S.F.	.56	.80	1.36
2350	10,000 S.F.	.49	.81	1.30
2400	Extra hazard, one floor, 2000 S.F.	3.76	2.04	5.80
2450	10,000 S.F.	1.30	1.24	2.54
2500	Each additional floor, 2000 S.F.	.74	1.15	1.89
2550	10,000 S.F.	.64	1.02	1.66
3050	Grooved steel, black, sch. 10 pipe light hazard, one floor,			
3060	2000 S.F.	3.57	1.62	5.19
3100	10,000 S.F.	.97	.78	1.75
3150	Each additional floor, 2000 S.F.	.50	.71	1.21
3200	10,000 S.F.	.37	.60	.97
3250	Ordinary hazard, one floor, 2000 S.F.	3.60	1.69	5.29
3300	10,000 S.F.	1.05	.97	2.02
3350	Each additional floor, 2000 S.F.	.53	.79	1.32
3400	10,000 S.F.	.44	.79	1.23
3450	Extra hazard, one floor, 2000 S.F.	3.74	2.04	5.78
3500	10,000 S.F.	1.22	1.22	2.44
3550	Each additional floor, 2000 S.F.	.71	1.14	1.85

284 For expanded coverage of these items see *Means Mechanical Cost Data 1988*

Figure 8.11b

FIRE PROTECTION	B8.2-310	Wet Standpipe Risers

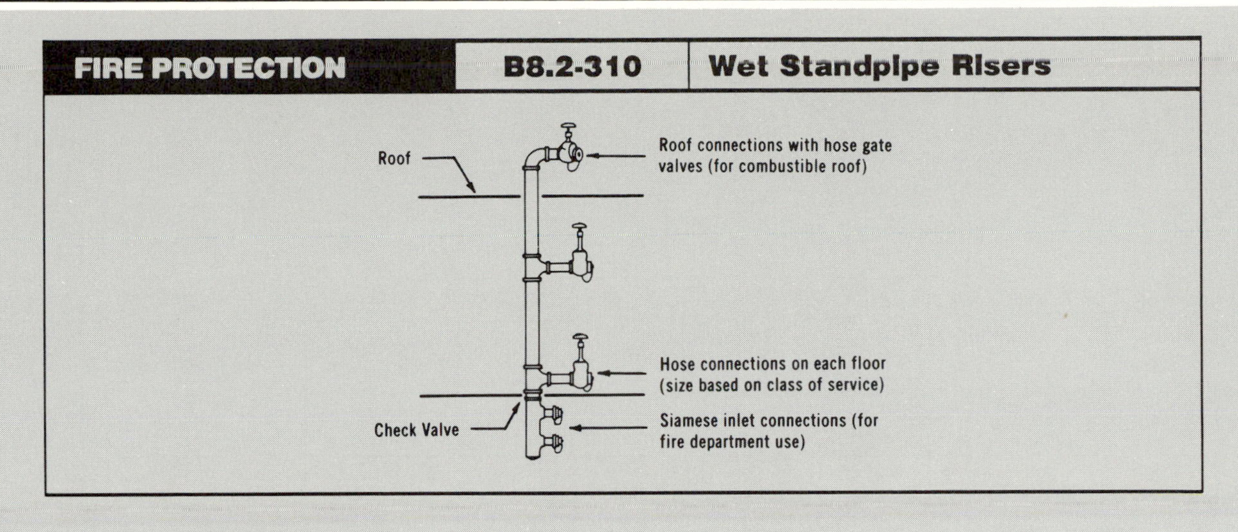

System Components	QUANTITY	UNIT	COST PER FLOOR MAT.	INST.	TOTAL
SYSTEM 082-310-0560					
WET STANDPIPE RISER, CLASS I, STEEL, BLACK, SCH. 40 PIPE, 10' HEIGHT					
4" DIAMETER PIPE, ONE FLOOR					
Pipe, steel, black, schedule 40, threaded, 4" diam	20.000	L.F.	210.60	289.40	500
Pipe, Tee, malleable iron, black, 150 lb threaded, 4" pipe size	2.000	Ea.	70.40	249.60	320
Pipe, 90° elbow, malleable iron, black, 150 lb threaded 4" pipe size	1.000	Ea.	24.20	85.80	110
Pipe, nipple, steel, black, schedule 40, 2-½" pipe size x 3" long	2.000	Ea.	13.42	62.58	76
Fire valve, gate, 300 lb, brass w/handwheel, 2-½" pipe size	1.000	Ea.	79.75	40.25	120
Fire valve, pressure restricting, adj, rgh brs, 2-½" pipe size	1.000	Ea.	196.36	83.64	280
Valve, swing check, w/ball drip, CI w/brs ftngs, 4" pipe size	1.000	Ea.	113.30	171.70	285
Standpipe conn wall dble flush brs w/plugs & chains 2-½"x2-½"x4"	1.000	Ea.	251.90	103.10	355
Valve, swing check, bronze, 125 lb, regrinding disc, 2-½" pipe size	1.000	Ea.	81.40	33.60	115
Roof manifold, fire, w/valves & caps, horiz/vert brs 2-½"x2-½"x4"	1.000	Ea.	266.20	108.80	375
Fire, hydrolator, vent & drain, 2-½" pipe size	1.000	Ea.	88	42	130
Valve, gate, iron body 125 lb, OS&Y, threaded, 4" pipe size	1.000	Ea.	192.50	52.50	245
TOTAL			1,588.03	1,322.97	2,911

8.2-310	Wet Standpipe Risers, Class I	COST PER FLOOR MAT.	INST.	TOTAL
0550	Wet standpipe risers, Class I, steel black sch. 40, 10' height			
0560	4" diameter pipe, one floor	1,600	1,325	2,925
0580	Additional floors	420	455	875
0600	6" diameter pipe, one floor	2,500	2,125	4,625
0620	Additional floors (43)	670	630	1,300
0640	8" diameter pipe, one floor	3,575	2,550	6,125
0660	Additional floors	855	760	1,615

8.2-310	Wet Standpipe Risers, Class II	COST PER FLOOR MAT.	INST.	TOTAL
1030	Wet standpipe risers, Class II, steel black sch. 40, 10' height			
1040	2" diameter pipe, one floor	535	475	1,010
1060	Additional floors	225	195	420
1080	2-½" diameter pipe, one floor	740	680	1,420
1100	Additional floors	245	230	475

286

For expanded coverage of these items see *Means Mechanical Cost Data 1988*

Figure 8.12*a*

FIRE PROTECTION	B8.2-310	Wet Standpipe Risers		
8.2-310	**Wet Standpipe Risers, Class III**	COST PER FLOOR		
		MAT.	INST.	TOTAL
1530	Wet standpipe risers, Class III, steel black sch. 40, 10′ height			
1540	4″ diameter pipe, one floor	1,625	1,325	2,950
1560	Additional floors	355	385	740
1580	6″ diameter pipe, one floor	2,525	2,125	4,650
1600	Additional floors	690	630	1,320
1620	8″ diameter pipe, one floor	3,625	2,550	6,175
1640	Additional floors	875	760	1,635

For expanded coverage of these items see *Means Mechanical Cost Data 1988*

287

Figure 8.12*b*

FIRE PROTECTION	B8.2-390	Standpipe Equipment		
8.2-390	**Standpipe Equipment**	COST EACH		
		MAT.	INST.	TOTAL
0100	Adapters, reducing, 1 piece, FxM, hexagon, cast brass, 2-½" x 1-½"	19.25		19.25
0200	Pin lug, 1-½" x 1"	8.95		8.95
0250	3" x 2-½"	30		30
0300	For polished chrome, add 75% mat.			
0400	Cabinets, D.S. glass in door, recessed, steel box, not equipped			
0500	Single extinguisher, steel door & frame	72	64	136
0550	Stainless steel door & frame	140	64	204
0600	Valve, 2-½" angle, steel door & frame	73	42	115
0650	Aluminum door & frame	79	46	125
0700	Stainless steel door & frame	145	46	191
0750	Hose rack assy, 2-½" x 1-½" valve & 100' hose, steel door & frame	115	86	201
0800	Aluminum door & frame	115	85	200
0850	Stainless steel door & frame	255	87	342
0900	Hose rack assy,& extinguisher,2-½"x1-½" valve & hose,steel door & frame	125	105	230
0950	Aluminum	125	100	225
1000	Stainless steel	270	105	375
1550	Compressor, air, dry pipe system, automatic, 200 gal., ⅓ H.P.	490	220	710
1600	520 gal., 1 H.P.	565	225	790
1650	Alarm, electric pressure switch (circuit closer)	58	10.95	68.95
2500	Couplings, hose, rocker lug, cast brass, 1-½"	15.40		15.40
2550	2-½"	33		33
3000	Escutcheon plate, for angle valves, polished brass, 1-½"	8.25		8.25
3050	2-½"	15.40		15.40
3500	Fire pump, electric, w/controller, fittings, relief valve			
3550	4" pump, 30 H.P., 500 G.P.M.	10,500	1,650	12,150
3600	5" pump, 40 H.P., 1000 G.P.M.	12,200	1,900	14,100
3650	5" pump, 100 H.P., 1000 G.P.M.	15,100	2,075	17,175
3700	For jockey pump system, add	1,375	250	1,625
5000	Hose, per linear foot, synthetic jacket, lined,			
5100	300 lb. test, 1-½" diameter	1.31		1.31
5150	2-½" diameter	1.84		1.84
5200	500 lb. test, 1-½" diameter	1.49		1.49
5250	2-½" diameter	2.34		2.34
5500	Nozzle, plain stream, polished brass, 1-½" x 10"	18.70		18.70
5550	2-½" x 15" x 13/16" or 1-½"	61		61
5600	Heavy duty combination adjustable fog and straight stream w/handle 1-½"	215		215
5650	2-½" direct connection	275		275
6000	Rack, for 1-½" diameter hose 100 ft. long, steel	22	26	48
6050	Brass	38	26	64
6500	Reel, steel, for 50 ft. long 1-½" diameter hose	50	38	88
6550	For 75 ft. long 2-½" diameter hose	61	37	98
7050	Siamese, w/plugs & chains, polished brass, sidewalk, 4" x 2-½" x 2-½"	305	210	515
7100	6" x 2-½" x 2-½"	400	260	660
7200	Wall type, flush, 4" x 2-½" x 2-½"	250	105	355
7250	6" x 2-½" x 2-½"	310	115	425
7300	Projecting, 4" x 2-½" x 2-½"	225	105	330
7350	6" x 2-½" x 2-½"	265	110	375
7400	For chrome plate, add 15% mat.			
8000	Valves, angle, wheel handle, 300 Lb., rough brass, 1-½"	24	24	48
8050	2-½"	56	41	97
8100	Combination pressure restricting, 1-½"	35	24	59
8150	2-½"	78	42	120
8200	Pressure restricting, adjustable, satin brass, 1-½"	58	24	82
8250	2-½"	98	42	140
8300	Hydrolator, vent and drain, rough brass, 1-½"	38	24	62
8350	2-½"	88	42	130
8400	Cabinet assy, incls. 2-½" valve, adapter, rack, hose, nozzle & hydrolator	455	195	650

290

For expanded coverage of these items see *Means Mechanical Cost Data 1988*

Figure 8.13

PLUMBING	B8.1-010	Sample Estimate

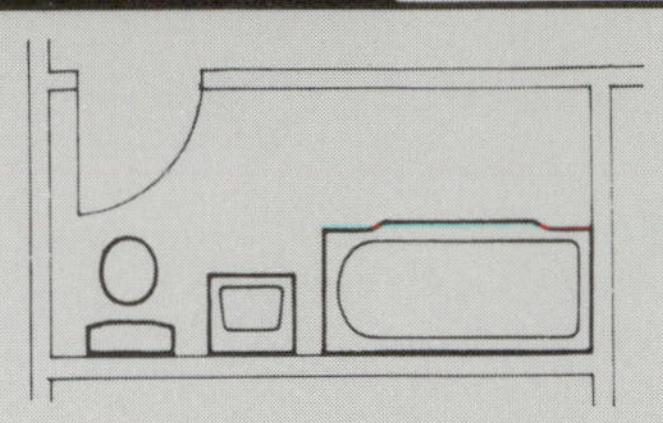

Example of Plumbing Cost Calculations: The bathroom system includes the individual fixtures such as bathtub, lavatory, shower and water closet. These fixtures are listed below as separate items merely as a checklist.

8.1-010 Plumbing Systems 20 Unit, 2 Story Apartment Building

	FIXTURE	SYSTEM	LINE	QUANTITY	UNIT	COST EACH		
						MAT.	INST.	TOTAL
0440	Bathroom	B8.1-630	3640	20	Ea.	17,100	22,900	40,000
0480	Bathtub							
0520	Booster pump [1]	not req'd.						
0560	Drinking fountain							
0600	Garbage disposal [1]	not incl.						
0660								
0680	Grease interceptor							
0720	Water heater	B8.1-170	2140	1	Ea.	3,325	1,575	4,900
0760	Kitchen sink	B8.1-431	1960	20	Ea.	7,050	9,975	17,025
0800	Laundry sink	B8.1-432	1840	4	Ea.	2,000	1,625	3,625
0840	Lavatory							
0900								
0920	Roof drain, 1 floor	B8.1-310	4200	2	Ea.	505	725	1,230
0960	Roof drain, add'l floor	B8.1-310	4240	20	L.F.	97	185	282
1000	Service sink	B8.1-434	4300	1	Ea.	455	615	1,070
1040	Sewage ejector [1]	not req'd.						
1080	Shower							
1100								
1160	Sump pump							
1200	Urinal							
1240	Water closet							
1320								
1360	**SUB TOTAL**					30,600	37,600	68,200
1480	Water controls	C8.1-031		10% [2]		3,050	3,775	6,825
1520	Pipe & fittings [3]	C8.1-031		30% [2]		9,175	11,300	20,475
1560	Other							
1600	Quality/complexity	C8.1-031		15% [2]		4,575	5,650	10,225
1680								
1720	**TOTAL**					47,400	58,500	105,900

[1] **Note:** Cost for items such as booster pumps, backflow preventers, sewage ejectors, water meters, etc., may be obtained from R.S. Means' "Mechanical Cost Data" book, or be added as an increase in the Water Controls percentage. Water controls, Pipe and Fittings, and the Quality/Complexity factors come from Table C8.1-031.

[2] Percentage of subtotal.

[3] Long easily discernable runs of pipe would be more accurately priced from Table 8.1-040. If this is done, reduce the miscellaneous percentage in proportion.

For expanded coverage of these items see *Means Mechanical Cost Data 1988*

Figure 8.14

The fire sprinkler price includes all necessary piping. However, it must be adjusted for quality and complexity along with the cost of the other parts of the plumbing system.

The Plumbing Estimate System

The combined assemblies/unit price method for estimating plumbing costs can be summarized as follows:

1. Estimate the cost of roof drains, fixture assemblies, and stand pipe systems using the assemblies method.
2. Estimate the cost of pumps and floor drains using the unit price method.
3. Add the costs of the roof drains, fixture assemblies, stand pipe systems, pumps, and floor drains to arrive at Subtotal A.
4. Calculate and add percentages of Subtotal A for water control, and piping and fittings.
5. Estimate the cost of the sprinkler system using the assemblies method.
6. Add the subtotal, the percentages, and the sprinkler system costs to arrive at Subtotal B.
7. Calculate a quality/complexity modifier using Subtotal B.
8. Add Subtotal B and the quality/complexity qualifier to complete the estimate.

Table 8.1-031 Plumbing Approximations for Quick Estimating

Water Control

Water Meter; Backflow Preventer;
Shock Absorbers; Vacuum Breakers; .... 10 to 15% of Fixtures
Mixer.

Pipe And Fittings: .... 30 to 60% of Fixtures

Note: Lower percentage for compact buildings or larger buildings with plumbing in one area.
Larger percentage for large buildings with plumbing spread out.
In extreme cases pipe may be more than 100% of fixtures.
Percentages **do not** include special purpose or process piping.

Plumbing Labor:

1 & 2 Story Residential .... Rough-in Labor = 80% of Materials
Apartment Buildings .... Rough-in Labor = 90 to 100% of Materials
Labor for handling and placing fixtures is approximately 25 to 30% of fixtures.

Quality/Complexity Multiplier (For all installations)

Economy installation, add .... 0 to 5%
Good quality, medium complexity, add .... 5 to 15%
Above average quality and complexity, add .... 15 to 25%

Figure 8.15

151 | Pipe and Fittings

	151 100 Miscellaneous Fittings		Crew	Daily Output	Man-Hours	Unit	Bare Costs Mat.	Bare Costs Labor	Bare Costs Equip.	Bare Costs Total	Total Incl O&P
105	5600	Flanged, iron									
	5660	2-½" pipe size	Q-1	5	3.200	Ea.	1,150	69		1,219	1,375
	5680	3" pipe size		4.50	3.560		1,310	76		1,386	1,550
	5700	4" pipe size		3	5.330		1,840	115		1,955	2,200
	5720	6" pipe size	Q-2	3	8		2,810	180		2,990	3,350
	5740	8" pipe size		2	12		6,140	265		6,405	7,150
	5760	10" pipe size		1	24		8,140	535		8,675	9,750
110	0010	**CLEANOUTS**									
	0060	Floor type									
	0080	Round or square, scoriated nickel bronze top									
	0100	2" pipe size	1 Plum	10	.800	Ea.	38	19.10		57.10	70
	0120	3" pipe size		8	1		44.50	24		68.50	84
	0140	4" pipe size		6	1.330		54.75	32		86.75	105
	0980	Round top, recessed for terrazzo									
	1000	2" pipe size	1 Plum	9	.889	Ea.	69	21		90	105
	1080	3" pipe size		6	1.330		69	32		101	125
	1100	4" pipe size		4	2		81	48		129	160
	1120	5" pipe size	Q-1	6	2.670		130	57		187	225
115	0010	**CLEANOUT TEE** Cast iron with countersunk plug									
	0200	2" pipe size	1 Plum	4	2	Ea.	10.75	48		58.75	82
	0220	3" pipe size		3.60	2.220		20.50	53		73.50	100
	0240	4" pipe size		3.30	2.420		24.75	58		82.75	110
	0280	6" pipe size	Q-1	5	3.200		46	69		115	150
	0500	For round smooth access cover, add					20%				
120	0010	**CONNECTORS** Flexible, corrugated, ⅞" O.D., ½" I.D.									
	0050	Gas, seamless brass, steel fittings									
	0200	12" long	1 Plum	36	.222	Ea.	5	5.30		10.30	13.30
	0220	18" long		36	.222		6.20	5.30		11.50	14.60
	0240	24" long		34	.235		7.30	5.60		12.90	16.25
	0280	36" long		32	.250		8.75	5.95		14.70	18.40
	0340	60" long		30	.267		13.20	6.35		19.55	24
	2000	Water, copper tubing, dielectric separators									
	2100	12" long	1 Plum	36	.222	Ea.	4.55	5.30		9.85	12.80
	2260	24" long	"	34	.235	"	6.75	5.60		12.35	15.65
125	0010	**DRAINS**									
	0140	Cornice, C.I., 45° or 90° outlet									
	0180	1-½" & 2" pipe size	Q-1	14	1.140	Ea.	48.50	25		73.50	89
	0200	3" and 4" pipe size	"	12	1.330		66	29		95	115
	0260	For galvanized body, add					9			9	9.90
	0280	For polished bronze dome, add					7.70			7.70	8.45
	0400	Deck, auto park, C.I., 13" top									
	0440	3", 4", 5", and 6" pipe size	Q-1	8	2	Ea.	160	43		203	240
	0480	For galvanized body, add				"	73			73	80
	2000	Floor, medium duty, C.I., deep flange, 7" top									
	2040	2" and 3" pipe size	Q-1	12	1.330	Ea.	28.25	29		57.25	73
	2080	For galvanized body, add					12			12	13.20
	2120	For polished bronze top, add					15.50			15.50	17.05
	2400	Heavy duty, with sediment bucket, C.I., 12" loose grate									
	2420	3", 4", 5", and 6" pipe size	Q-1	9	1.780	Ea.	96.50	38		134.50	160
	2460	For polished bronze top, add				"	40			40	44
	2500	Heavy duty, cleanout & trap w/bucket, C.I., 15" top									
	2540	2", 3", and 4" pipe size	Q-1	6	2.670	Ea.	700	57		757	855
	2560	For galvanized body, add					205			205	225
	2580	For polished bronze top, add					215			215	235
	2780	Shower, with strainer, uniform diam. trap, bronze top									
	2800	1-½", 2" and 3" pipe size	Q-1	8	2	Ea.	59	43		102	130

For expanded coverage of these items see *Means Mechanical Cost Data 1988* 283

Figure 8.16

152 | Plumbing Fixtures

		152 400 Pumps	CREW	DAILY OUTPUT	MAN-HOURS	UNIT	BARE COSTS MAT.	BARE COSTS LABOR	BARE COSTS EQUIP.	BARE COSTS TOTAL	TOTAL INCL O&P
470	0010	**PUMPS, SPRINKLER** With check valve, steel base, 15' lift									
	0100	37 GPM, ¾ HP	1 Plum	4	2	Ea.	320	48		368	420
	0140	56 GPM, 1-½ HP	"	2	4		610	95		705	810
	0180	68 GPM, 2 H.P.	Q-1	2	8	↓	625	170		795	940
480	0010	**PUMPS, SUBMERSIBLE** Dewatering									
	0020	Sand & sludge, 20' head, starter & level control									
	0050	Cast iron									
	0100	2" discharge, 10 GPM	1 Plum	4	2	Ea.	345	48		393	450
	0160	60 GPM		3	2.670		480	64		544	620
	0200	120 GPM	↓	2	4		890	95		985	1,125
	1000	160 GPM	Q-1	1.50	10.670		1,860	230		2,090	2,375
	1100	3" discharge, 220 GPM	"	1.20	13.330	↓	2,280	285		2,565	2,925
	7000	Sump pump, 10' head, automatic									
	7100	Bronze, 22 GPM., ¼ HP, 1-¼" discharge	1 Plum	6	1.330	Ea.	205	32		237	270
	7140	68 GPM, ½ HP, 1-¼" or 1-½" discharge		5	1.600		335	38		373	425
	7160	94 GPM, ½ HP, 1-¼" or 1-½" discharge		5	1.600		455	38		493	555
	7180	105 GPM, ½ HP, 2" or 3" discharge		4	2		435	48		483	550
	7500	Cast iron, 23 GPM, ¼ HP, 1-¼" discharge		6	1.330		97	32		129	155
	7540	35 GPM, ⅓ HP, 1-¼" discharge		6	1.330		108	32		140	165
	7560	68 GPM, ½ HP, 1-¼" or 1-½" discharge	↓	5	1.600	↓	225	38		263	305
490	0010	**PUMPS, WELL** Water system, with pressure control									
	0020										
	1000	Deep well, multi-stage jet, 42 gal. tank									
	1040	110' lift, 40 lb. discharge, 5 GPM, ¾ HP	1 Plum	.80	10	Ea.	455	240		695	850
	2000	Shallow well, reciprocating, 25 gal. tank									
	2040	25' lift, 5 GPM, ⅓ HP	1 Plum	2	4	Ea.	350	95		445	525
	3000	Shallow well, single stage jet, 42 gal. tank,									
	3040	15' lift, 40 lb. discharge, 16 GPM, ¾ HP	1 Plum	2	4	Ea.	345	95		440	520

153 | Plumbing Appliances

		153 100 Water Appliances	CREW	DAILY OUTPUT	MAN-HOURS	UNIT	BARE COSTS MAT.	BARE COSTS LABOR	BARE COSTS EQUIP.	BARE COSTS TOTAL	TOTAL INCL O&P
105	0010	**WATER COOLER**									
	0030	See line 153-105-9800 for rough-in, waste & vent									
	0040	for all water coolers									
	0100	Wall mounted, non-recessed									
	0140	4 GPH	Q-1	4	4	Ea.	285	86		371	440
	0180	8.2 GPH	"	4	4		335	86		421	495
	0600	For hot and cold water, add					160			160	175
	0640	For stainless steel cabinet, add					38			38	42
	1000	Dual height, 8.2 GPH	Q-1	3.80	4.210		455	90		545	635
	1040	14.3 GPH		3.80	4.210		480	90		570	660
	1080	16.1 GPH	↓	3.80	4.210		525	90		615	710
	1240	For stainless steel cabinet, add					72			72	79
	2600	Wheelchair type, 8 GPH	Q-1	4	4	↓	865	86		951	1,075
	2610										
	3300	Semi-recessed, 8.1 GPH	Q-1	4	4	Ea.	445	86		531	615
	3320	12 GPH	"	4	4	"	465	86		551	640
	4600	Floor mounted, flush-to-wall									
	4640	4 GPH	1 Plum	3	2.670	Ea.	305	64		369	430
	4680	8.2 GPH		3	2.670		345	64		409	475
	4720	14.3 GPH	↓	3	2.670	↓	355	64		419	485

302

For expanded coverage of these items see *Means Mechanical Cost Data 1988*

Figure 8.17

The Chambers Estimate

To illustrate the combined assemblies/unit price method, we will perform an estimate of the plumbing costs for our sample office building project in Chambers, Georgia. Plans and drawings are shown in Figures 8.18 through 8.20. The quantities shown in the sample estimate represent realistic conditions. Assumptions have also been made for items not shown that would normally be included in the plans and specifications of a project of this type.

Figure 8.18, the Parking Garage Plan, shows six floor drains, a 6" stand pipe, an electric water heater, and a fire hose cabinet. Figure 8.19 shows the details of a partial plan of the penthouse floor and main roof, and plumbing details at the service core. The partial plan indicates that there are 12 roof drains, and a fire hose cabinet at the penthouse. The core plumbing details show that each tenant floor has a fire hose cabinet and stand pipe, a service sink, six water closets, two urinals, four wall-hung lavatories, and three vanity-mounted lavatories. Figure 8.20, the Mechanical Riser Diagram, shows the supply and waste systems that are connected to the fixtures, and connections for a sprinkler system.

In preparing the estimate shown in Figure 8.21, the roof drains, fixture assemblies, and stand pipe systems are priced using cost information from the 1988 edition of *Means Assemblies Cost Data*; the floor drains are priced using cost information from the 1988 *Means Building Construction Cost Data*. These costs are added to arrive at Subtotal A.

Next, percentages of Subtotal A are added for control devices, and pipe and fittings. Figure 8.15 shows a range of 10% to 15% for water control. Based on experience, we know that the higher value, 15%, should be used. Figure 8.15 also shows a range of 30% to 60% for pipe and fittings. In extreme cases, this percentage can actually range from 30% for very small projects with closely grouped fixtures, to 110% for a chemistry laboratory building. Getting this percentage right is critical to the accuracy of the estimate. The choice must be made based on experience with other projects of a similar type. We will assume that other projects similar to our office building required a 50% factor. Therefore, 50% will be used here.

Next, the sprinkler system is estimated by the following calculations. First, multiply the square foot area of the basement by the price from line 8.2-150-0620 (Figure 8.11b) for the first floor of the system. Then, multiply the area of floors 1 through 3 and the penthouse by the price in line 8.2-150-0740 for the remainder of the system. Subtotal A, the water control adjustment, the piping and fittings adjustment, and the sprinkler systems costs are all added together to arrive at Subtotal B.

Subtotal B is then multiplied by a percentage in order to obtain a quality/complexity cost. It is judged that the Chambers project falls in the *good quality, medium complexity* range. Again, experience has shown that the high end of this range should be used. Therefore, 15% is added for quality/complexity. Subtotal B and the quality/complexity cost are added to complete the estimate.

Just as with the other estimates, the price must be adjusted for sales tax, General Conditions, overhead and profit, bid conditions, location, and inflation. If this price is part of a total project estimate, the adjustments are made on the summary sheet as shown in the cost control estimate in Chapter 7. If, however, this estimate is used to check the cost of the plumbing system alone, then the adjustments must be made to this price in order to calculate the total cost of the system. This price represents

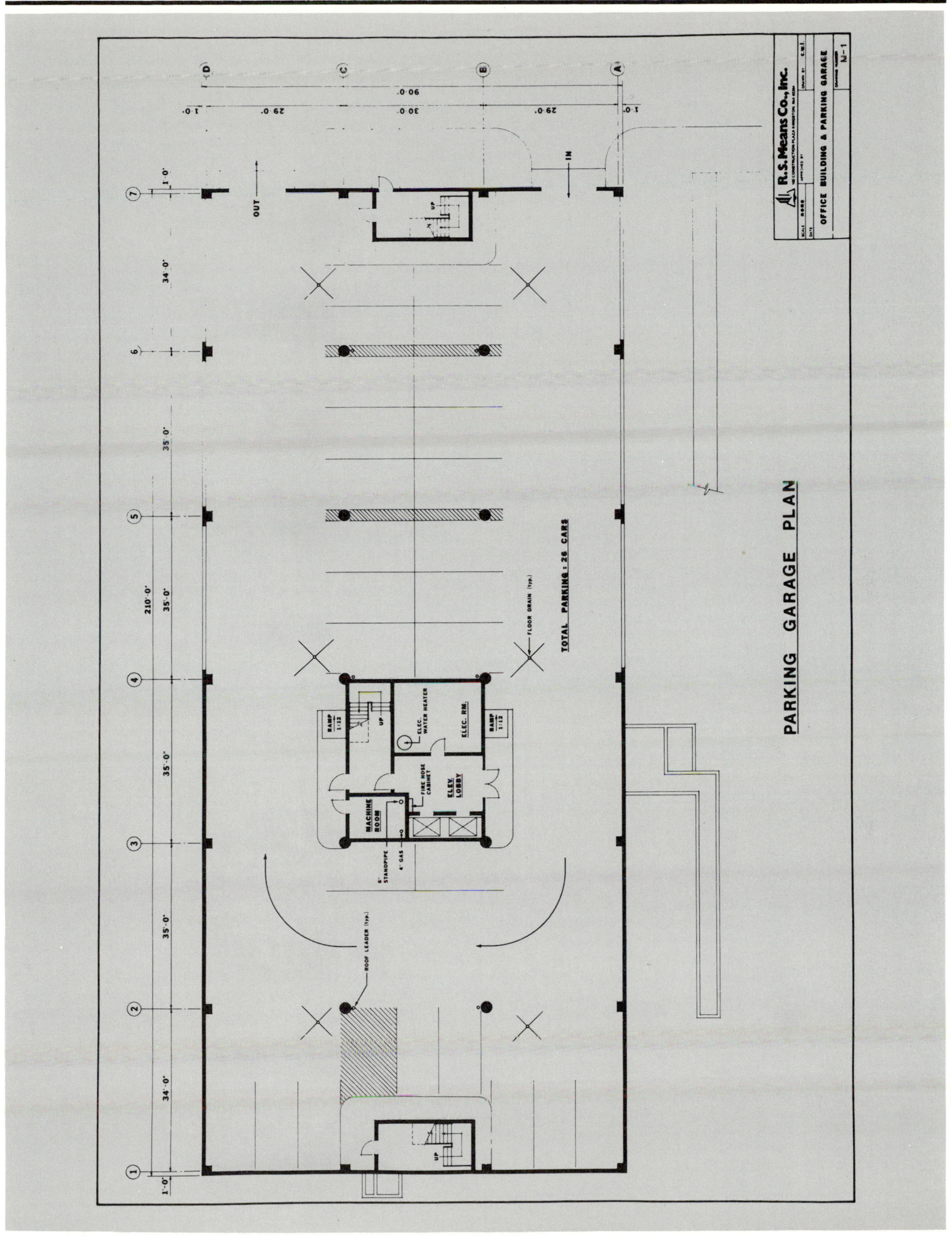
PARKING GARAGE PLAN
TOTAL PARKING: 26 CARS
FLOOR DRAIN (typ.)
ROOF LEADER (typ.)
MACHINE ROOM
ELEV. LOBBY
ELEC. RM.
ELEC. WATER HEATER
FIRE HOSE CABINET
RAMP 1:12
6" STANDPIPE
4" GAS
UP
OUT
IN
210'-0"
90'-0"
R.S. Means Co., Inc.
OFFICE BUILDING & PARKING GARAGE
M-1

Figure 8.18

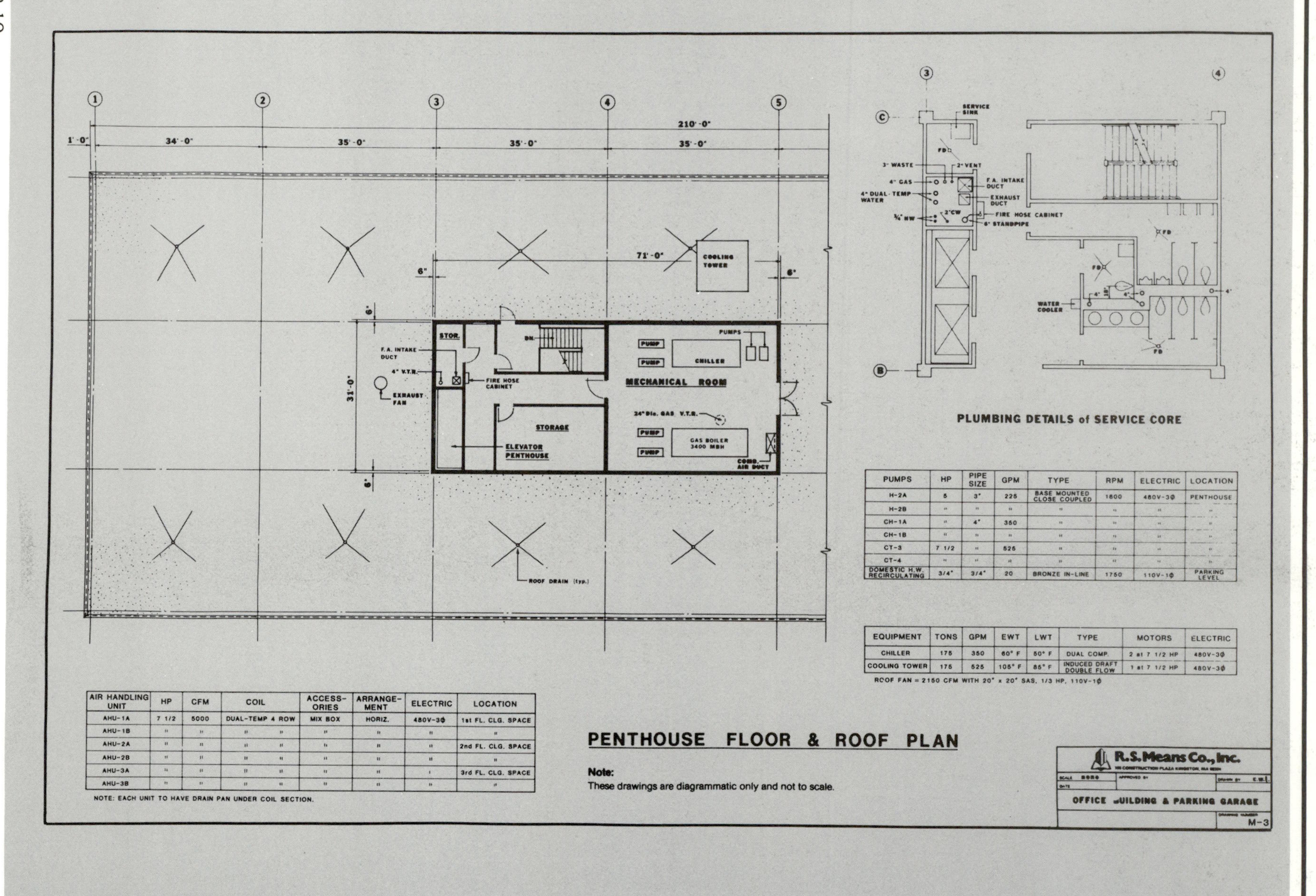

PUMPS	HP	PIPE SIZE	GPM	TYPE	RPM	ELECTRIC	LOCATION
H-2A	5	3"	225	BASE MOUNTED CLOSE COUPLED	1800	480V-3Ø	PENTHOUSE
H-2B	"	"	"	"	"	"	"
CH-1A	"	4"	350	"	"	"	"
CH-1B	"	"	"	"	"	"	"
CT-3	7 1/2	"	525	"	"	"	"
CT-4	"	"	"	"	"	"	"
DOMESTIC H.W. RECIRCULATING	3/4"	3/4"	20	BRONZE IN-LINE	1750	110V-1Ø	PARKING LEVEL

EQUIPMENT	TONS	GPM	EWT	LWT	TYPE	MOTORS	ELECTRIC
CHILLER	175	350	60° F	50° F	DUAL COMP.	2 at 7 1/2 HP	480V-3Ø
COOLING TOWER	175	525	105° F	85° F	INDUCED DRAFT DOUBLE FLOW	1 at 7 1/2 HP	480V-3Ø

ROOF FAN = 2150 CFM WITH 20" x 20" SAS, 1/3 HP, 110V-1Ø

AIR HANDLING UNIT	HP	CFM	COIL	ACCESS-ORIES	ARRANGE-MENT	ELECTRIC	LOCATION
AHU-1A	7 1/2	5000	DUAL-TEMP 4 ROW	MIX BOX	HORIZ.	480V-3Ø	1st FL. CLG. SPACE
AHU-1B	"	"	" "	"	"	"	"
AHU-2A	"	"	" "	"	"	"	2nd FL. CLG. SPACE
AHU-2B	"	"	" "	"	"	"	"
AHU-3A	"	"	" "	"	"	"	3rd FL. CLG. SPACE
AHU-3B	"	"	" "	"	"	"	"

NOTE: EACH UNIT TO HAVE DRAIN PAN UNDER COIL SECTION.

Figure 8.19

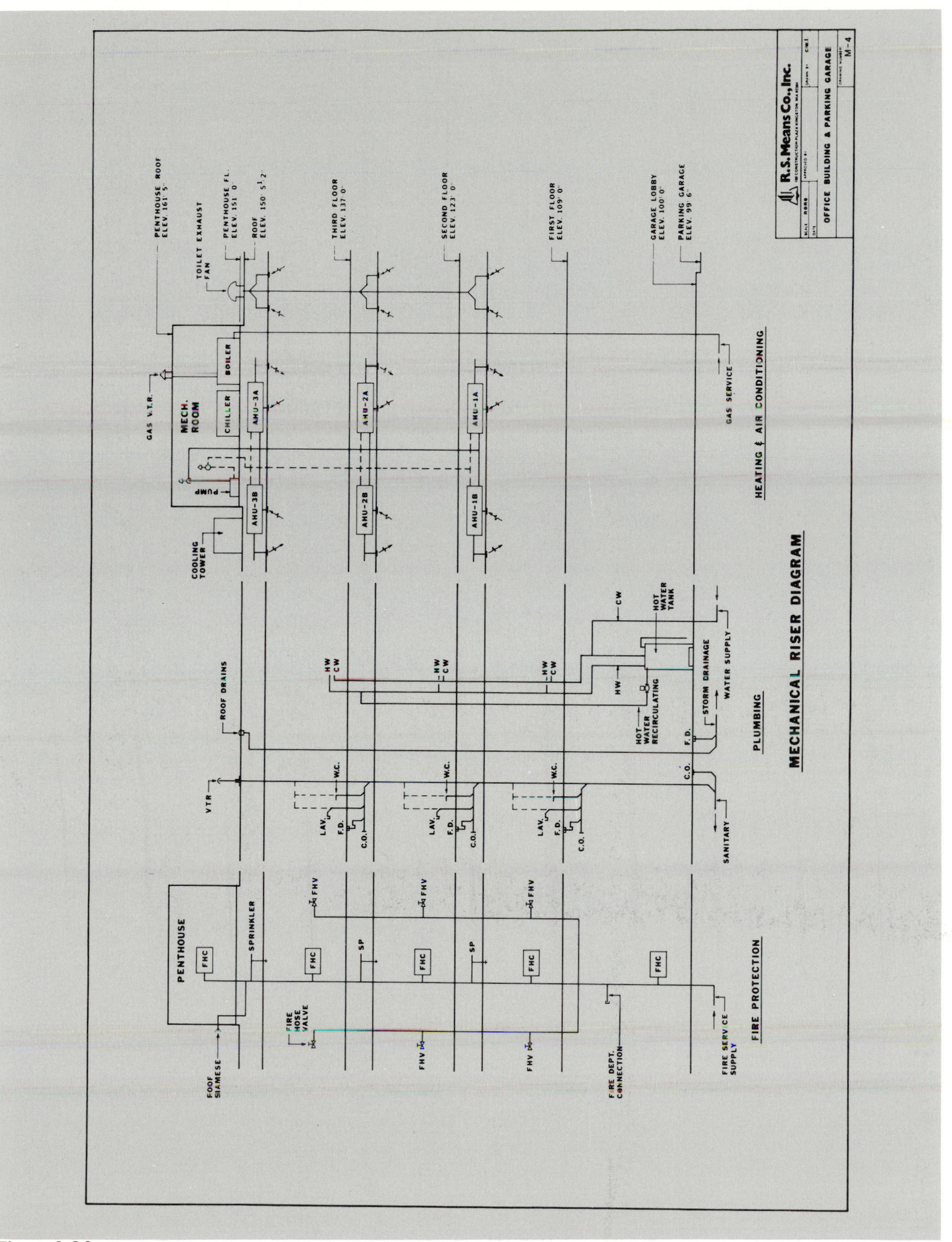
MECHANICAL RISER DIAGRAM
FIRE PROTECTION
PLUMBING
HEATING & AIR CONDITIONING
PENTHOUSE
FHC
SPRINKLER
SP
FHV
FIRE HOSE VALVE
ROOF SIAMESE
FIRE DEPT. CONNECTION
FIRE SERVICE SUPPLY
VTR
ROOF DRAINS
W.C.
LAV.
F.D.
C.O.
SANITARY
STORM DRAINAGE
WATER SUPPLY
HOT WATER RECIRCULATING
HOT WATER TANK
HW
CW
COOLING TOWER
PUMP
MECH. ROOM
CHILLER
BOILER
AHU-3A
AHU-3B
AHU-2A
AHU-2B
AHU-1A
AHU-1B
GAS V.T.R.
GAS SERVICE
TOILET EXHAUST FAN
PENTHOUSE ROOF ELEV. 161' 5"
PENTHOUSE FL. ELEV. 151' 0"
ROOF ELEV. 150' 5 1/2"
THIRD FLOOR ELEV. 137' 0"
SECOND FLOOR ELEV. 123' 0"
FIRST FLOOR ELEV. 109' 0"
GARAGE LOBBY ELEV. 100' 0"
PARKING GARAGE ELEV. 99' 6"
R.S. Means Co., Inc.
OFFICE BUILDING & PARKING GARAGE
M-4

Figure 8.20

ASSEMBLY NUMBER	DESCRIPTION	QTY	UNIT	TOTAL COST UNIT	TOTAL COST TOTAL	COST PER S.F.
8.0	**Mechanical System**					
8.1-160-1860	WH - 80gal	1	ea	2435	2435	
8.1-310-4120+	3" CI Roof Drain	12	ea	581	6972	
8.1-310-4160+	✓ addl floors	12	ea	456	5472	
8.1-433-2040	Lavatory - Wall hung PE on CI 18"x15"	12	ea	735	8820	
8.1-433-1880	✓ - @ van 19" ϕ	9	ea	480	4320	
8.1-434-4340	Service Sink - Wall hung PE on CI 24"x30"	3	ea	1095	3285	
8.1-450-2000	Urinals - wall hung	6	ea	755	4530	
8.1-460-1880	EWC - Wall hung - dual ht	3	ea	840	2520	
8.1-510-3000	WC - 1st pr	3	ea	1315	3945	
8.1-510-3100	- addl pr	3	ea	1290	3870	
8.1-510-1800	- addl sgl	6	ea	795	4770	
8.2-310-1080	Std Pipe - 2½" 1st flr	1	ea	1420	1420	
8.2-310-1100	✓ ✓ addl flr	4	ea	475	1900	
8.2-390-0800	FHC	6	ea	200	1200	
151-125-0440	Auto Deck Drain	6	ea	240	1440	
	Subtotal "A"				56899	
	Water Control @ 15%				8535	
	Pipe & Fittings @ 50%				28450	
8.2-150-0620	Firecycle Sprinkler - 1st flr	18,900	sf	1.84	34776	
8.2-150-0740	✓ ✓ - addl flr	58,901	sf	1.05	61846	
	Subtotal "B"				190506	
	Quality / Complexity @ 15%				28576	
	Total				219082	
9.0	**Electrical**					

Page 4 of 6

Figure 8.21

the plumbing contractor's quote to the general contractor; therefore, the general contractor's General Conditions costs and overhead and profit have not yet been added. Only sales tax, bid conditions, location, and inflation adjustments are needed. The adjustment is performed as follows.

Total Plumbing Costs	$219,082
Sales Tax @ 4% × .5	4,382
Subtotal	$223,464
Bid Conditions @ 8%	17,877
Subtotal	$241,341
Location Factor @ 89.2	× .892
Subtotal	$215,276
Inflation @ 4%/yr.	× 1.061
Adjusted Total Plumbing Costs	**$228,408**

Many buildings contain other equipment (e.g., food service and laboratory equipment) that requires supply and waste piping. Although the cost of this equipment is not included in the plumbing estimate, a percentage should be added for pipe and fittings. For example, assume that the Chambers project has $10,000 of food service equipment that requires supply and waste hook-ups. $5,000 ($10,000 x the chosen pipe and fittings adjustment, 50%) should be added to the plumbing estimate as follows:

Subtotal "A"	$56,899
Water Control @ 15%	8,535
Pipe & Fittings @ 50%	28,450
Food Service Pipe & Fittings $10,000 × .5	5,000
Sprinkler 1st Floor	34,776
Sprinkler Addl. Floors	61,846
Subtotal "B"	195,506
Quality/Complexity @ 15%	29,326
Total	**$224,832**

Summary

Estimating plumbing costs can be something of a challenge for the layman. However, a combined assemblies and unit cost method is within the capabilities of those who are not plumbing experts, and can produce an accurate check estimate. The basic steps involved in this process are:

- pricing fixture assemblies
- adding percentages for water control, pipe and fittings, and other equipment pipe and fittings
- pricing the sprinkler system
- adding a quality/complexity modifier
- adjusting the total for sales tax, bid conditions, location factor, and inflation.

The critical part of the procedure is choosing the proper adjustment percentage to use for pipe and fittings. Guidance for this choice comes primarily from experience with other projects of the same type.

9

ESTIMATING HVAC COSTS

9

ESTIMATING HVAC COSTS

The challenges involved in obtaining accurate estimates of HVAC costs are similar to those for plumbing estimates. Sufficient information for accurate estimating is not available until near the end of the design development phase, and the estimator often lacks the special training and experience needed to accurately estimate HVAC costs. Like plumbing, early HVAC check estimates must be made using the square foot method. Also like plumbing, when sufficient information becomes available (near the end of the detailed design phase), a simplified combination of assemblies and unit price methods can be used to provide an estimate sufficiently accurate for design cost control.

This chapter presents the simplified HVAC estimating system. First, we will outline typical components of an HVAC system. This section is followed by a discussion of HVAC assemblies in *Means Assemblies Cost Data*, showing which components can be estimated using the assemblies method, and which require the unit cost method. Finally, we will illustrate the HVAC estimating system, using the HVAC costs for the Chambers office building.

The HVAC System

Heating, Ventilating, and Air Conditioning (HVAC) systems cool, heat, clean, humidify, dehumidify, and replace air within a building by mechanical means. The methods used to accomplish these functions are diverse. The energy used may be supplied by electricity, natural gas, coal, and/or oil. The heating/cooling medium may be water, air, steam, or chemical refrigerants. The system design, the choice of energy source, and the choice of heating/cooling medium are based on climate, building use, and local energy costs. Regardless of the choices and design, each system consists of the same basic components: heating equipment, cooling equipment, air handling equipment, exhaust systems, and piping.

Heating Equipment

Heating equipment generates heat to warm the building. Included in this category are electrical resistance heaters, heat pumps, gas- and oil-fired unit heaters, boiler systems, and solar heating systems. Boiler systems may use either water or steam as a medium, and electricity, gas, or oil as an energy source. Boiler systems include pumps, piping, expansion tanks, terminal units, vents, and controls. Solar heating systems include collector panels, tanks, piping, and controls. Figure 9.1 is an example of a boiler system; Figure 9.2 shows a solar heating system.

Cooling Equipment

Cooling equipment cools a building by removing excess heat. Examples of such equipment are water- and air-cooled chillers, water- and air-cooled condensers, cooling towers, coils, piping, pumps, and controls. Cooling systems use water, air, and/or chemical refrigerants as a medium, and electricity as an energy source. Figure 9.3 shows some examples of chillers.

Air Handling Equipment

Air handling equipment moves conditioned air to the areas where it is needed. This category includes duct work, fan coil units, air handling units, VAV boxes, fans, dampers, duct heaters, and coils. See Figure 9.4 for examples of air handling units, and Figure 9.5 for duct work.

Exhaust Systems

Exhaust systems remove air to help control fumes, odors, and smoke. This equipment includes duct work, exhaust fans, and ventilators. Examples of exhaust systems are shown in Figure 9.6.

Piping

Piping is the means by which water, steam, and chemical refrigerants are distributed throughout the building. Piping systems include pipe, couplings, fittings, valves, and hangers.

More detailed information about HVAC systems and the pricing of specific items may be found in *Means Mechanical Estimating*.

Assemblies and Unit Cost Applications

Like plumbing, the cost of the HVAC equipment is estimated using assemblies costs where possible, and unit costs where necessary. *Means Assemblies Cost Data* provides prices for complete heating and cooling assemblies. Typical electric and fossil fuel boiler assemblies, solar heating assemblies, and cooling assemblies are included. Some examples of a boiler assembly, chiller assembly, and computer cooling are shown in Figures 9.7 through 9.9. Costs (in the form of unit prices) for chimneys and vents are shown in Figures 9.10a and b.

Of the equipment categories listed above in "The HVAC System" discussion, only the exhaust system is not covered in the assemblies prices in *Means Assemblies Cost Data*. The "System Components" listing in Figure 9.7 shows the boiler heating assembly including the boiler, piping system, terminal units, expansion tank, circulation pump, and controls. The System Components listing in Figure 9.8 shows that the typical cooling assembly includes air handling equipment, refrigeration equipment, cooling tower, piping, and pumps. Figures 9.10a and b show unit price cost listings for vents, vent fittings, and complete computer cooling packages. A cost per square foot of floor area is provided for both heating and cooling assemblies. The cost of computer room cooling units is listed for each installation. Vertical linear foot prices are used for vents and chimneys, and unit prices for vent fittings.

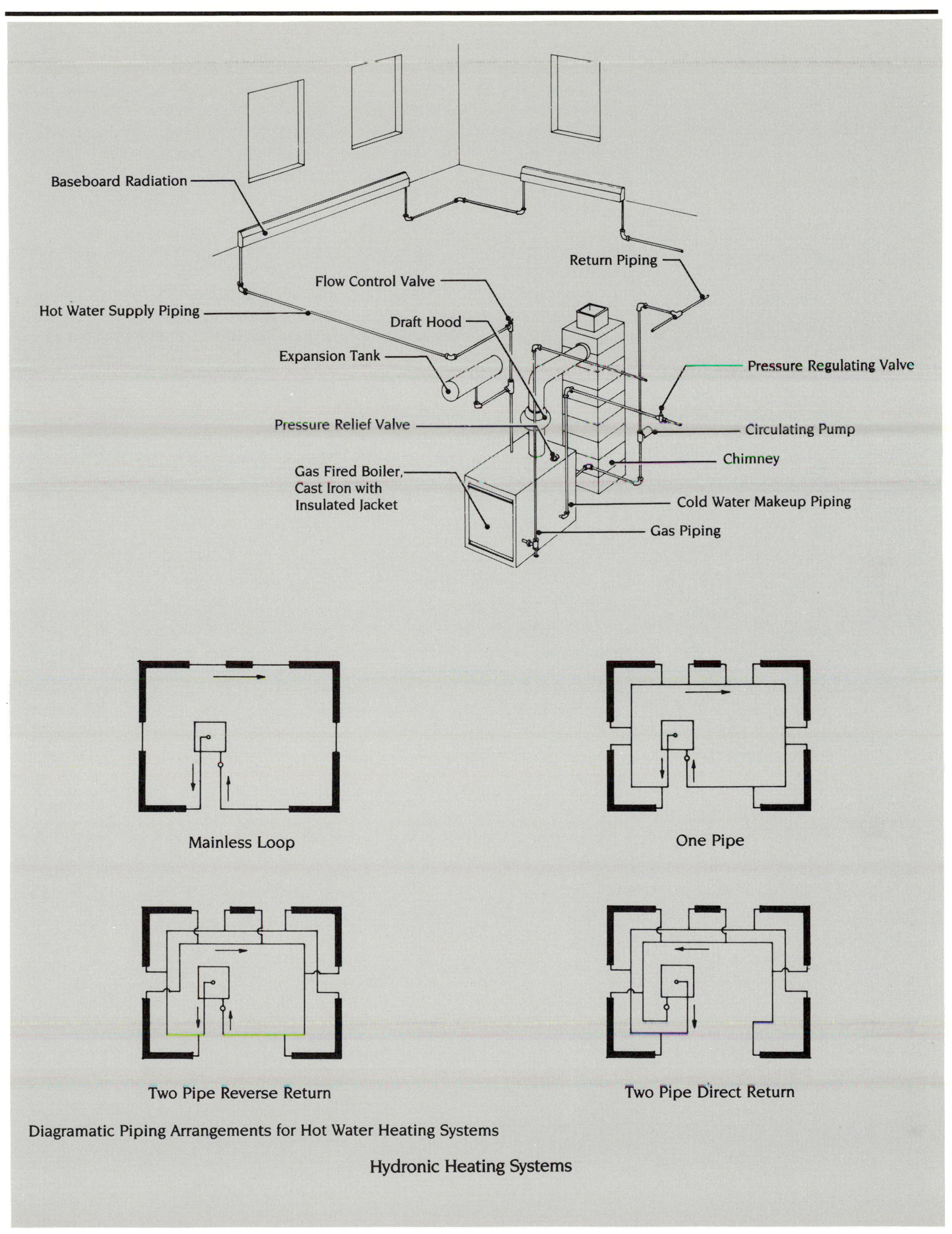

Hydronic Heating Systems

Figure 9.1

The cost of the exhaust system equipment is estimated using the unit price method. Figures 9.11a and b show some examples of unit prices for ductwork and exhaust fans from the 1988 edition of *Means Building Construction Cost Data*. Exhaust fans are priced as "each", and ductwork is priced by pound or per linear foot. Occasionally, some miscellaneous equipment is specified that is not included in the assemblies costs. For example, the Chambers office building project is to have electric heaters in the stairs (Figure 9.12) and the duct heater shown at the building entrance (Figure 9.13). This equipment must also be estimated by the unit cost method.

Just as with plumbing, the Means HVAC assemblies costs must be adjusted using a quality/complexity modifier. The ranges are the same as for plumbing systems: 0% to 5% for economy installations, 5% to 15% for good quality and moderate complexity installations, and 15% to 25% for above average quality and complexity installations.

The Estimating System

Very simply, the cost of the heating and cooling systems is estimated using the assemblies method, and the costs of the exhaust system and any miscellaneous equipment are estimated using the unit price method. A percentage of the total of these systems is then added as a quality/complexity modifier.

The Chambers Estimate

This procedure can be demonstrated by preparing an HVAC estimate for the Chambers project. The drawings are shown in Figures 9.12 through 9.14. The quantities shown in the sample estimate represent realistic conditions. Assumptions have also been made for items not shown that would normally be included in the plans and specifications for a project of this type. Figure 9.13, the First Floor Plan, shows a typical floor HVAC layout including ductwork, air handling units, and an electric duct heater.

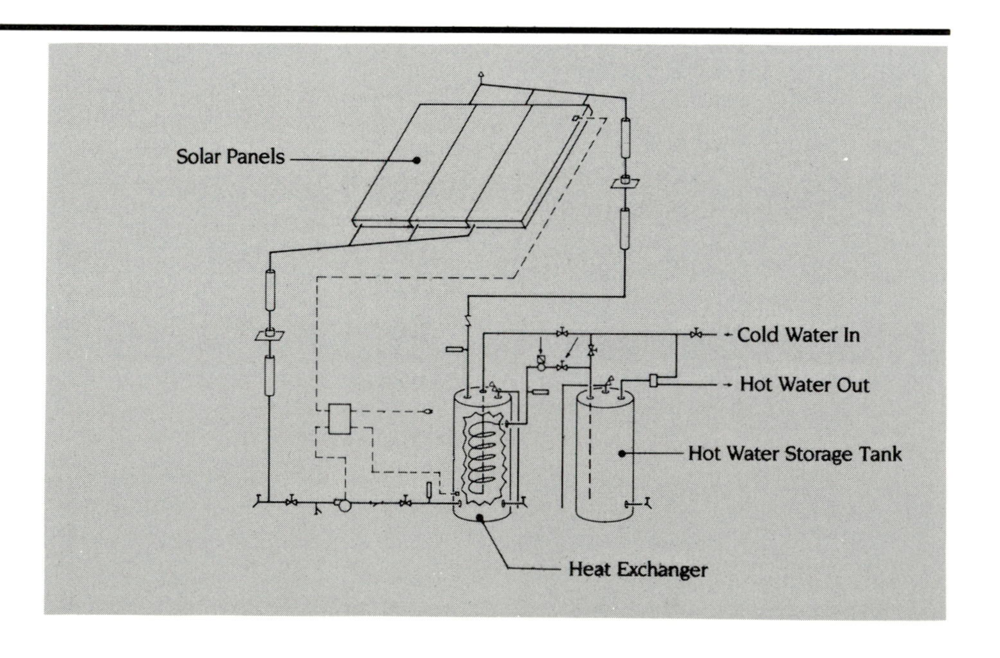

Figure 9.2

Also included are an exhaust system for the rest rooms and janitor's closets. Figure 9.12, the Parking Garage Plan, shows electric heaters in the stairs. Figure 9.14 contains a Penthouse Floor and Roof Plan, Plumbing Details at Service Core, and a Heating and Cooling Piping Flow Diagram. These drawings show a gas-fired boiler, a chiller/cooling tower system, and a four-pipe hot and chilled water distribution system. The estimate is shown in Figure 9.15.

The assemblies shown in Figures 9.7 and 9.8 are a close match to the ones shown for the Chambers office building. The chiller/cooling tower, chilled water distribution piping, and fan coil air handling units are included in the chilled water/cooling tower assembly. The building system has large fan coil units with a duct distribution system in lieu of many small fan coil units shown in the assembly. A note in Figure 9.8 states that the cost of the two pipe systems is approximately the same, and this assembly can therefore be used for the Chambers system. The price given for office buildings of approximately 60,000 S.F. in size (from line 8.4-120-4040) is multiplied by the gross area of heated space to estimate the cost of the cooling system.

The boiler system and hot water distribution piping are included in the boiler assembly in Figure 9.7. Of the components listed, all are required for the building except unit heaters. The fan coil units in the cooling system make the unit heaters unnecessary. The cost of the unit heaters can be removed by reducing the appropriate line item price by the percentage of the assemblies cost that is attributable to unit heaters. Using the price for buildings of approximately 67,000 S.F. in size from line 8.3-141-1480, the reduction is performed as follows:

$$\frac{(\$8{,}542.51 - \$790)}{\$8{,}542.51} \times \$2.96/\text{S.F.} = \$2.69/\text{S.F.}$$

The gross area of heated space is then multiplied by $2.69 to estimate the cost of the heating system.

The heating and cooling assemblies do not include the stair heaters, the duct heater, and the exhaust system. The costs of these items are estimated using unit prices from *Means Building Construction Cost Data* (1988 edition), as shown in Figure 9.15.

In our example, a minor modification of the boiler assembly price was necessary to match the actual building system. Major modifications can be made as well. For example, assume that the Chambers office building is to be built in an industrial park that supplies heating and chilled water to all buildings from a central utility plant. In such a case, unit heaters, the boiler, and breeching would not be needed. The heating assembly price from line 8.3-141-1480 in Figure 9.7 would then be modified as follows:

$$\frac{[\$8{,}542.51 - (\$1{,}650 + \$790 + \$82.51)]}{8{,}542.51} \times \$2.96\ \text{S.F.} = \$2.09\ \text{S.F.}$$

In the cooling system, the chiller and cooling tower would not be needed. The cooling tower pumps and piping would be retained for the circulation system. The cooling assembly price from line 8.4-120-4040 in Figure 9.8 would be modified as follows:

$$\frac{[\$28{,}707.73 - (\$7{,}877.60 + \$505.78)]}{\$28{,}707.73} \times \$6.76\ \text{S.F.} = \$4.79\ \text{S.F.}$$

As with the plumbing estimate, the total HVAC system cost can be transposed to the assemblies summary sheet to be included in a total project estimate; or, it can be adjusted for sales tax, bid conditions, location, and inflation to be used for an individual check estimate.

Summary

Heating, Ventilating, and Air Conditioning costs can be estimated easily and accurately using a combination of assemblies and unit price methods. The heating and cooling systems are estimated using the assemblies method; the exhaust system and miscellaneous equipment require the unit price method. Using assemblies costs from *Means Assemblies Cost Data* provides some flexibility since these costs can be modified to match the actual HVAC design.

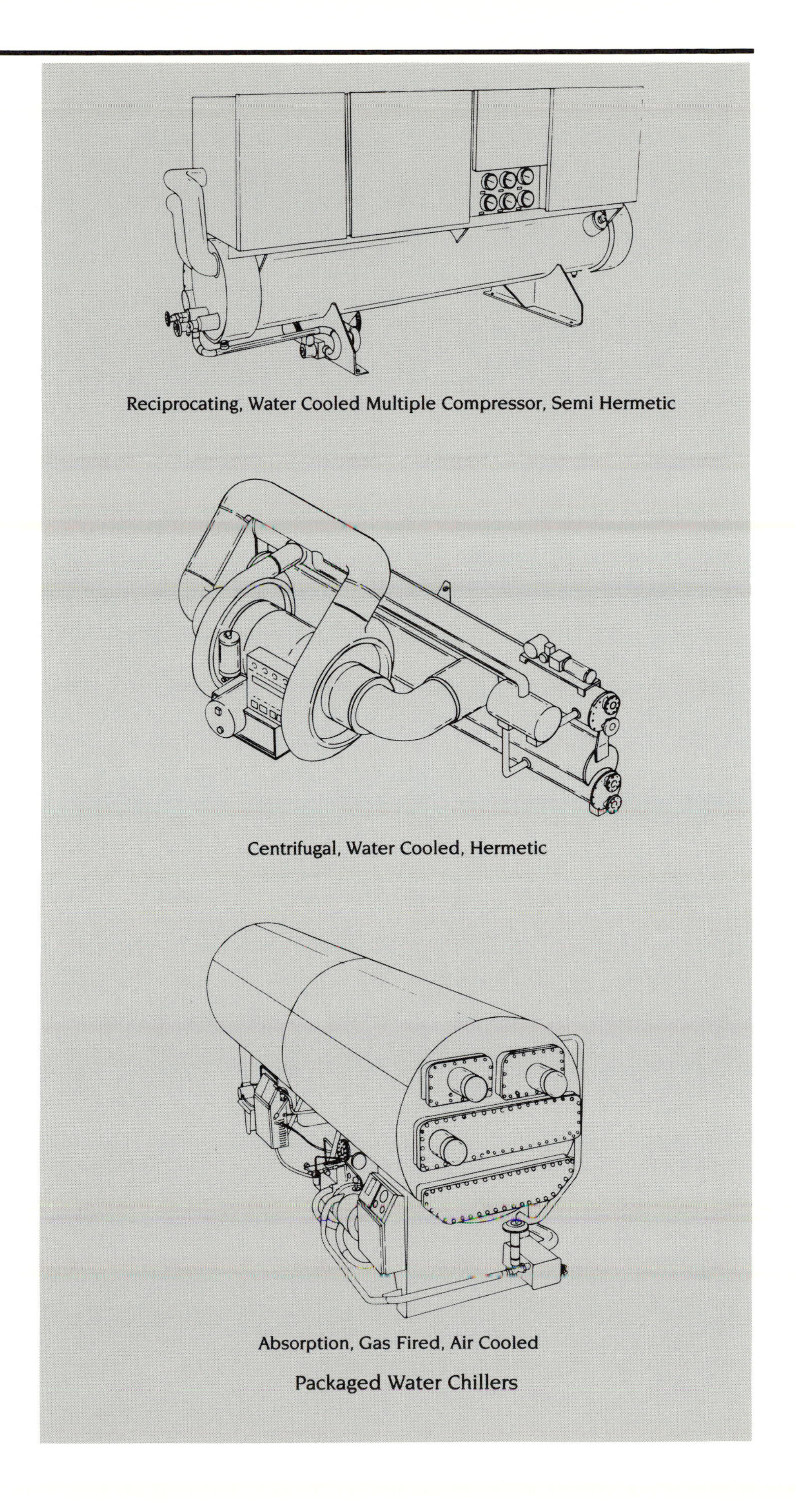

Packaged Water Chillers

Figure 9.3

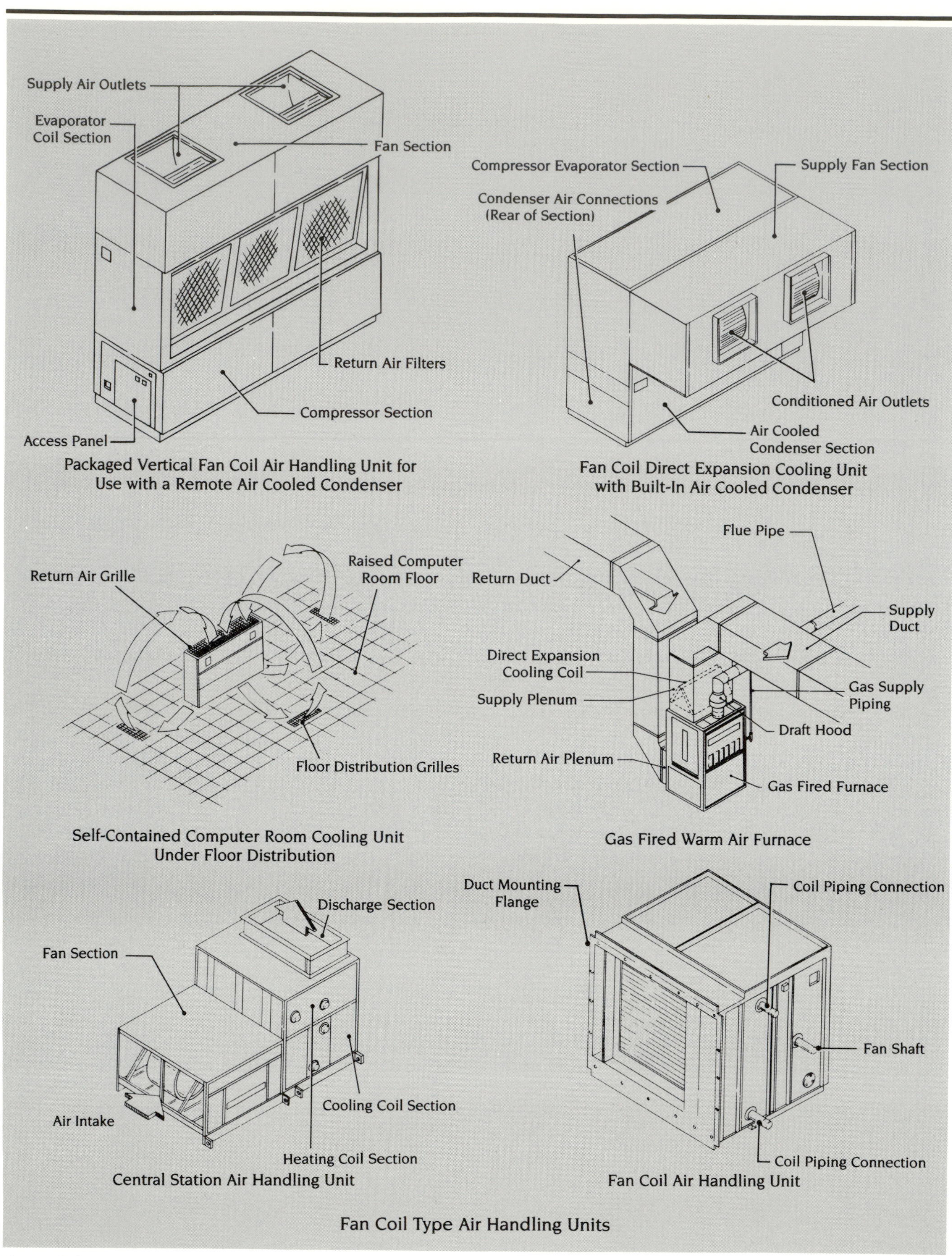
Supply Air Outlets
Evaporator Coil Section
Fan Section
Return Air Filters
Compressor Section
Access Panel
Packaged Vertical Fan Coil Air Handling Unit for Use with a Remote Air Cooled Condenser
Compressor Evaporator Section
Supply Fan Section
Condenser Air Connections (Rear of Section)
Conditioned Air Outlets
Air Cooled Condenser Section
Fan Coil Direct Expansion Cooling Unit with Built-In Air Cooled Condenser
Return Air Grille
Raised Computer Room Floor
Floor Distribution Grilles
Self-Contained Computer Room Cooling Unit Under Floor Distribution
Flue Pipe
Return Duct
Supply Duct
Direct Expansion Cooling Coil
Supply Plenum
Gas Supply Piping
Draft Hood
Return Air Plenum
Gas Fired Furnace
Gas Fired Warm Air Furnace
Discharge Section
Fan Section
Cooling Coil Section
Air Intake
Heating Coil Section
Central Station Air Handling Unit
Duct Mounting Flange
Coil Piping Connection
Fan Shaft
Coil Piping Connection
Fan Coil Air Handling Unit
Fan Coil Type Air Handling Units

Figure 9.4

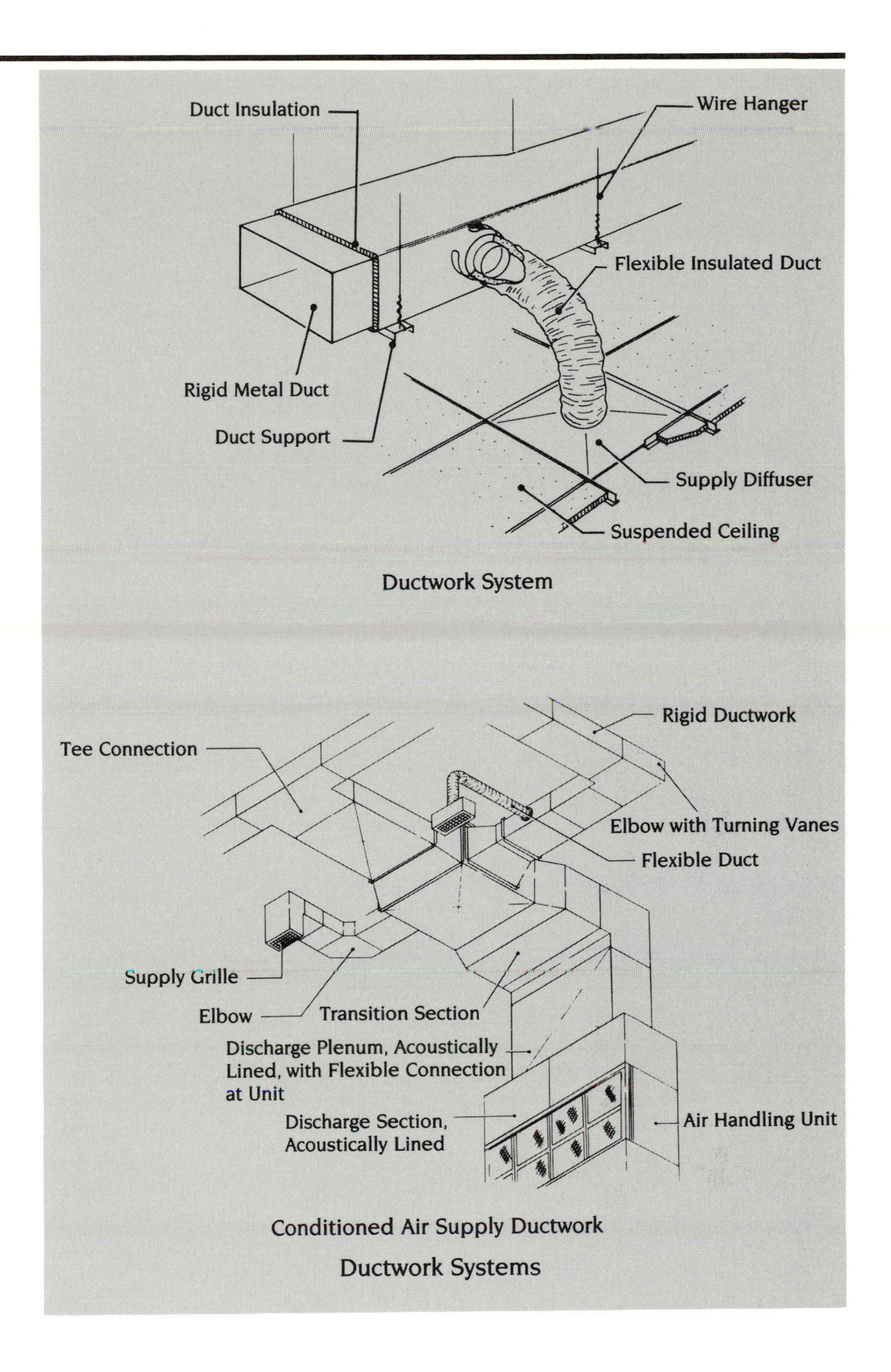

Ductwork Systems

Figure 9.5

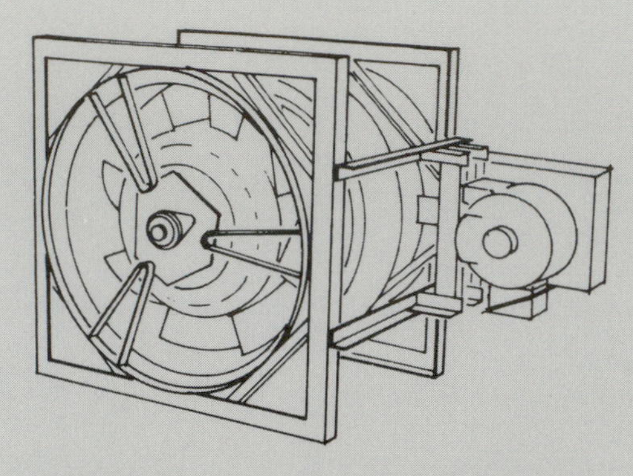

Axial Flow, Belt Drive, Centrifugal Fan

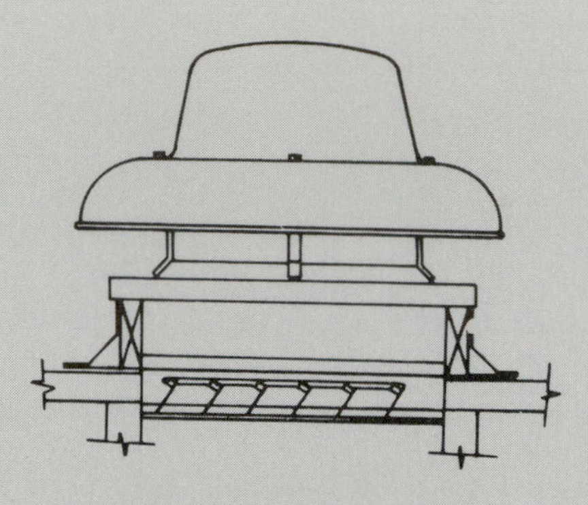

Centrifugal Roof Exhaust Fan

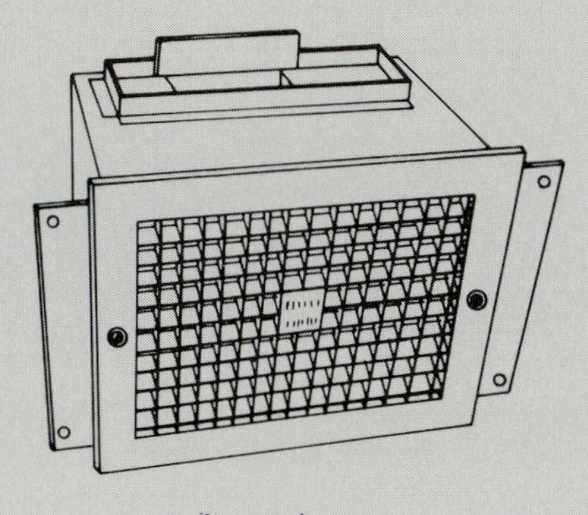

Ceiling Exhaust Fan

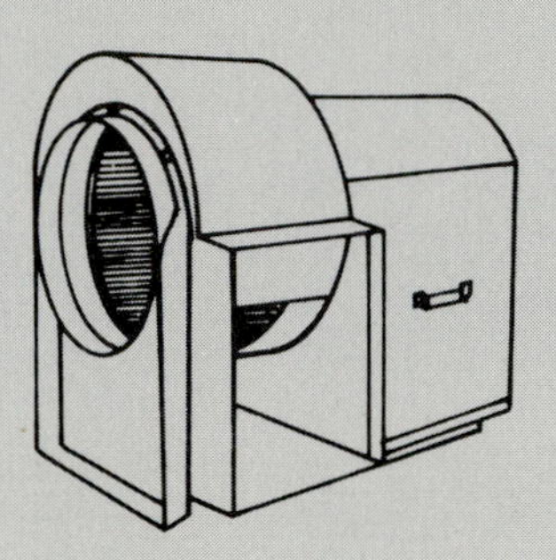

Belt Drive, Utility Set

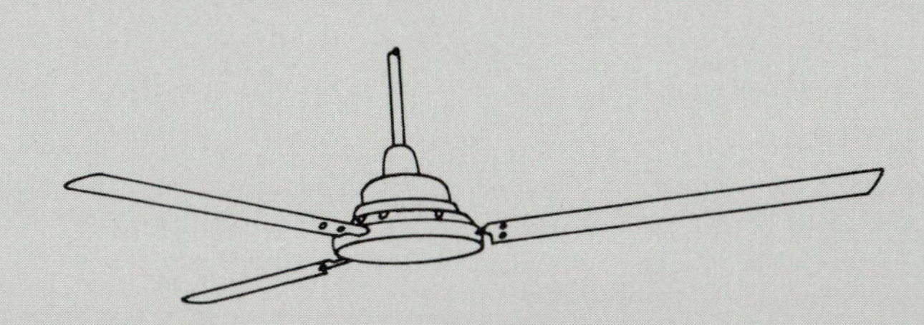

Paddle Blade Air Circulator

Belt Drive Propeller Fan with Shutter

Fans and Ventilators

Figure 9.6

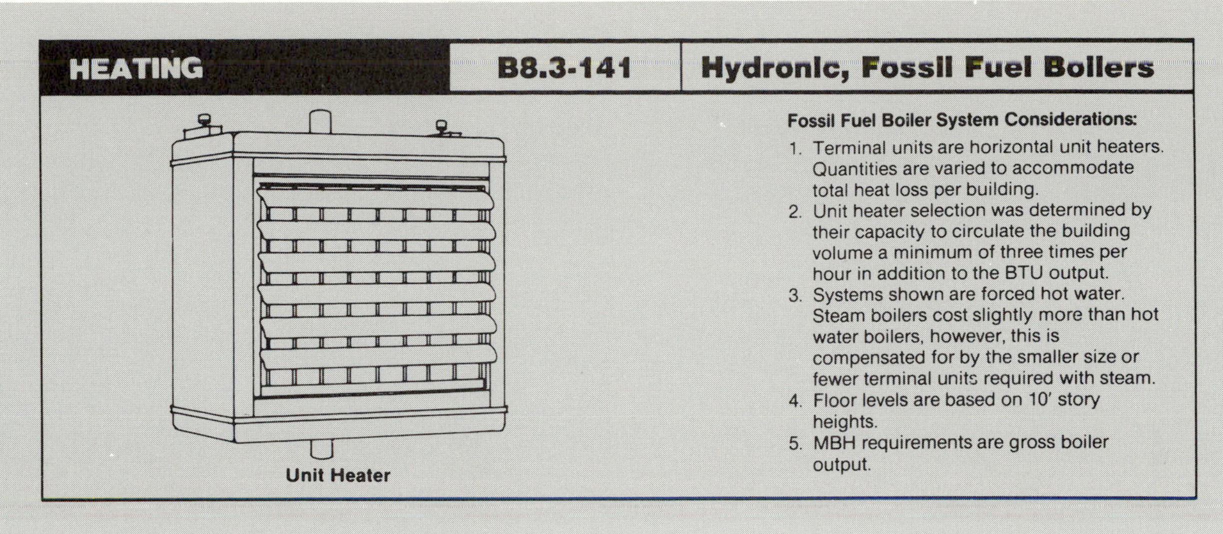

HEATING | B8.3-141 | Hydronic, Fossil Fuel Boilers

Fossil Fuel Boiler System Considerations:

1. Terminal units are horizontal unit heaters. Quantities are varied to accommodate total heat loss per building.
2. Unit heater selection was determined by their capacity to circulate the building volume a minimum of three times per hour in addition to the BTU output.
3. Systems shown are forced hot water. Steam boilers cost slightly more than hot water boilers, however, this is compensated for by the smaller size or fewer terminal units required with steam.
4. Floor levels are based on 10′ story heights.
5. MBH requirements are gross boiler output.

System Components	QUANTITY	UNIT	COST EACH MAT.	COST EACH INST.	COST EACH TOTAL
SYSTEM 08.3-141-1280					
HEATING SYSTEM, HYDRONIC, FOSSIL FUEL, TERMINAL UNIT HEATERS					
CAST IRON BOILER, GAS, 80 MBH, 1,070 S.F. BUILDING					
Boiler, gas, hot water, CI, burner, controls & insulation 80 MBH	1.000	Ea.	918.50	731.50	1,650
Pipe, steel, black, schedule 40, threaded, cplg & hngr 10′OC, 2″ diam	200.000	L.F.	638	1,572	2,210
Unit heater, 1 speed propeller, horizontal, 200° EWT, 72.7 MBH	2.000	Ea.	649	141	790
Unit heater piping hookup with controls	2.000	Set	457.62	1,202.38	1,660
Boiler breeching	1.000	System	45.93	36.58	82.51
Expansion tank, painted steel, ASME, 18 Gal capacity	1.000	Ea.	748	42	790
Circulating pump, CI, flange connection, 1/12 HP	1.000	Ea.	137.50	82.50	220
Pipe covering, calcium silicate w/cover, 1″ wall, 2″ diam	200.000	L.F.	498	642	1,140
TOTAL			4,092.55	4,449.96	8,542.51
COST PER S.F.			3.82	4.16	7.98

8.3-141	Heating Systems, Hydronic, Fossil Fuel, Unit Heaters	COST PER S.F. MAT.	COST PER S.F. INST.	COST PER S.F. TOTAL
1260	Heating systems, hydronic, fossil fuel, terminal unit heaters,			
1280	Cast iron boiler, gas, 80 M.B.H., 1,070 S.F. bldg.	3.82	4.16	7.98
1320	163 M.B.H., 2,140 S.F. bldg. (44)	2.61	2.83	5.44
1360	544 M.B.H., 7,250 S.F. bldg.	1.92	1.93	3.85
1400	1,088 M.B.H., 14,500 S.F. bldg. (45)	1.71	1.84	3.55
1440	3,264 M.B.H., 43,500 S.F. bldg.	1.40	1.36	2.76
1480	5,032 M.B.H., 67,100 S.F. bldg.	1.52	1.44	2.96
1520	Oil, 109 M.B.H., 1,420 S.F. bldg.	4.31	3.67	7.98
1560	235 M.B.H., 3,150 S.F. bldg.	2.99	2.65	5.64
1600	940 M.B.H., 12,500 S.F. bldg.	2.13	1.75	3.88
1640	1,600 M.B.H., 21,300 S.F. bldg.	2.06	1.68	3.74
1680	2,480 M.B.H., 33,100 S.F. bldg.	2.06	1.54	3.60
1720	3,350 M.B.H., 44,500 S.F. bldg.	1.74	1.54	3.28
1760	Coal, 148 M.B.H., 1,975 S.F. bldg.	3.04	2.43	5.47
1800	300 M.B.H., 4,000 S.F. bldg.	2.36	1.89	4.25
1840	2,360 M.B.H., 31,500 S.F. bldg.	1.68	1.52	3.20
1880	Steel boiler, gas, 72 M.B.H., 1,020 S.F. bldg.	3.76	3.09	6.85
1920	240 M.B.H., 3,200 S.F. bldg.	2.65	2.40	5.05
1960	480 M.B.H., 6,400 S.F. bldg.	2.17	1.78	3.06
2000	800 M.B.H., 10,700 S.F. bldg.	1.88	1.59	3.47
2040	1,960 M.B.H., 26,100 S.F. bldg.	1.69	1.43	3.12
2080	3,000 M.B.H., 40,000 S.F. bldg.	1.65	1.46	3.11

296

For expanded coverage of these items see *Means Mechanical Cost Data 1988*

Figure 9.7

AIR CONDITIONING	B8.4-120	Chilled Water, Water Cooled

Reciprocating Package Chiller
Condenser Water
Chilled Water Supply & Return Piping
Cooling Tower
Cooling Tower Water Makeup
Roof Structure
Insulate
Return
Supply
Fan Coil Unit
Finish Ceiling

General: Water cooled chillers are available in the same sizes as air cooled units. They are also available in larger capacities.

Design Assumptions: The chilled water systems with water cooled condenser, include reciprocating hermetic compressors, water cooling tower, pumps, piping and expansion tanks and are based on a two pipe system. Chilled water piping is insulated. No ducts are included and fan-coil units are cooling only. Area distribution is through use of multiple fan coil units. Fewer but larger fan coil units with duct distribution would be approximately the same S.F. cost. Water treatment and balancing are not included.

System Components	QUANTITY	UNIT	COST EACH MAT.	COST EACH INST.	COST EACH TOTAL
SYSTEM 08.4-120-1320					
PACKAGED CHILLER, WATER COOLED, WITH FAN COIL UNIT					
APARTMENT CORRIDORS, 4,000 S.F., 7.33 TON					
Fan coil air conditioner unit, cabinet mounted & filters, chilled water	2.000	Ea.	2,566.37	304.15	2,870.52
Water chiller, reciprocating, water cooled, 1 compressor semihermetic	1.000	Ea.	6,146.36	1,731.24	7,877.60
Cooling tower, draw thru single flow, belt drive	1.000	Ea.	443.47	62.31	505.78
Cooling tower pumps & piping	1.000	System	221.73	152.10	373.83
Chilled water unit coil connections	2.000	Ea.	741.10	1,258.90	2,000
Chilled water distribution piping	520.000	L.F.	4,763.20	10,316.80	15,080
TOTAL			14,882.23	13,825.50	28,707.73
COST PER S.F.			3.72	3.46	7.18

8.4-120	Chilled Water, Cooling Tower Systems	COST PER S.F. MAT.	COST PER S.F. INST.	COST PER S.F. TOTAL
1300	Packaged chiller, water cooled, with fan coil unit			
1320	Apartment corridors, 4,000 S.F., 7.33 ton	3.72	3.46	7.18
1600	Banks and libraries, 4,000 S.F., 16.66 ton (46)	5.10	3.87	8.97
1800	60,000 S.F., 250.00 ton	4.32	3.27	7.59
1880	Bars and taverns, 4,000 S.F., 44.33 ton (49)	8.65	4.94	13.59
2000	20,000 S.F., 221.66 ton	8.85	4.69	13.54
2160	Bowling alleys, 4,000 S.F., 22.66 ton	6.05	4.19	10.24
2320	40,000 S.F., 226.66 ton	5.10	3.22	8.32
2440	Department stores, 4,000 S.F., 11.66 ton	4.44	3.69	8.13
2640	60,000 S.F., 175.00 ton	3.61	2.93	6.54
2720	Drug stores, 4,000 S.F., 26.66 ton	6.45	4.33	10.78
2880	40,000 S.F., 266.67 ton	5.85	3.68	9.53
3000	Factories, 4,000 S.F., 13.33 ton	4.37	3.67	8.04
3200	60,000 S.F., 200.00 ton	3.86	3.07	6.93
3280	Food supermarkets, 4,000 S.F., 11.33 ton	4.36	3.67	8.03
3480	60,000 S.F., 170.00 ton	3.57	2.92	6.49
3560	Medical centers, 4.000 S.F., 9.33 ton	3.76	3.37	7.13
3760	60,000 S.F., 140.00 ton	3.29	2.87	6.16
3840	Offices, 4,000 S.F., 12.66 ton	4.23	3.63	7.86
4040	60,000 S.F., 190.00 ton	3.74	3.02	6.76
4120	Restaurants, 4,000 S.F., 20.00 ton	5.45	3.89	9.34
4320	60,000 S.F., 300.00 ton	4.83	3.44	8.27
4400	Schools and colleges, 4,000 S.F., 15.33 ton	4.81	3.79	8.60
4600	60,000 S.F., 230.00 ton	4.02	3.14	7.16

For expanded coverage of these items see *Means Mechanical Cost Data 1988*

Figure 9.8

AIR CONDITIONING	B8.4-810	Computer Cooling

Computer rooms impose special requirements on air conditioning systems. A prime requirement is reliability, due to the potential monetary loss that could be incurred by a system failure. A second basic requirement is the tolerance of control with which temperature and humidity are regulated, and dust eliminated. As the air conditioning system reliability is so vital, the additional cost of reserve capacity and redundant components is often justified.

System Descriptions: Computer areas may be environmentally controlled by one of three methods as follows:

1. **Self-contained Units**
   These are units built to higher standards of performance and reliability. They usually contain alarms and controls to indicate component operation failure, filter change, etc. It should be remembered that these units in the room will occupy space that is relatively expensive to build and that all alterations and service of the equipment will also have to be accomplished within the computer area.
2. **Decentralized Air Handling Units**
   In operation these are similar to the self-contained units except that their cooling capability comes from remotely located refrigeration equipment as refrigerant or chilled water. As no compressors or refrigerating equipment are required in the air units, they are smaller and require less service than self-contained units. An added plus for this type of system occurs if some of the computer components themselves also require chilled water for cooling.
3. **Central System Supply**
   Cooling is obtained from a central source which, since it is not located within the computer room, may have excess capacity and permit greater flexibility without interfering with the computer components. System performance criteria must still be met.

Note: The costs shown below do not include an allowance for ductwork or piping.

8.4-810	Computer Room Cooling Units	COST EACH		
		MAT.	INST.	TOTAL
0560	Computer room unit, air cooled, includes remote condenser			
0580	3 ton	7,125	1,000	8,125
0600	5 ton (49)	7,950	1,125	9,075
0620	8 ton	14,100	1,875	15,975
0640	10 ton	15,500	2,050	17,550
0660	15 ton	18,000	2,350	20,350
0680	20 ton	20,600	2,725	23,325
0700	23 ton	21,900	2,800	24,700
0800	Chilled water, for connection to existing chiller system			
0820	5 ton	5,875	690	6,565
0840	8 ton	8,625	1,000	9,625
0860	10 ton	9,000	1,000	10,000
0880	15 ton	9,775	1,025	10,800
0900	20 ton	10,400	1,100	11,500
0920	23 ton	10,900	1,175	12,075
1000	Glycol system, complete except for interconnecting tubing			
1020	3 ton	10,500	1,275	11,775
1040	5 ton	11,300	1,325	12,625
1060	8 ton	18,000	2,250	20,250
1080	10 ton	19,300	2,400	21,700
1100	15 ton	23,200	3,050	26,250
1120	20 ton	25,700	3,325	29,025
1140	23 ton	27,000	3,600	30,600
1240	Water cooled, not including condenser water supply or cooling tower			
1260	3 ton	7,425	815	8,240
1280	5 ton	8,575	930	9,505
1300	8 ton	13,500	1,575	15,075
1320	15 ton	16,700	1,875	18,575
1340	20 ton	19,300	2,100	21,400
1360	23 ton	20,600	2,325	22,925

For expanded coverage of these items see *Means Mechanical Cost Data 1988*

Figure 9.9

SPECIAL	B8.5-410	Chimney & Vent		

8.5-410	Gas Vent, Galvanized Steel, Double Wall	COST PER V.L.F.		
		MAT.	INST.	TOTAL
0030	Gas vent, galvanized steel, double wall, round pipe diameter, 3"	2.42	7.15	9.57
0040	4"	2.95	7.55	10.50
0050	5"	3.50	8	11.50
0060	6"	4.09	8.55	12.64
0070	7"	5.55	9.20	14.75
0080	8"	6.20	9.85	16.05
0090	10"	13.10	10.90	24
0100	12"	17.50	11.50	29
0110	14"	29	11.95	40.95
0120	16	39	12.60	51.60
0130	18"	49	13.15	62.15
0140	20"	58	22	80
0150	22"	73	23	96
0160	24"	90	25	115
0180	30"	120	29	149
0220	38"	175	34	209
0300	48"	275	40	315
0400				

8.5-410	Chimney, Stainless Steel, Insulated, Wood & Oil	COST PER V.L.F.		
		MAT.	INST.	TOTAL
0530	Chimney, stainless steel, insulated, diameter, 6"	16.80	8.20	25
0540	7"	21	9.20	30.20
0550	8"	25	9.95	34.95
0560	10"	35	11.05	46.05
0570	12"	46	11.60	57.60
0580	14"	61	12.05	73.05

8.5-410	Gas Vent Fittings, Galvanized Steel, Double Wall	COST EACH		
		MAT.	INST.	TOTAL
1020	Gas vent, fittings, 45° Elbow, adjustable thru 8" diameter			
1030	Pipe diameter, 3"	5.95	14.05	20
1040	4"	6.95	15.05	22
1050	5"	8.20	15.80	24
1060	6"	9.80	17.20	27
1070	7"	13	18	31
1080	8"	14.95	20	34.95
1090	10"	37	22	59
1100	12"	48	23	71
1110	14"	66	25	91
1120	16"	83	27	110
1130	18"	105	29	134
1140	20"	120	45	165
1150	22"	180	45	225
1170	24"	230	49	279
1200	30"	490	56	546
1250	38"	640	67	707
1300	48"	840	79	919
1410	90° Elbow, adjustable, pipe diameter, 3"	10.75	14.25	25
1420	4"	12.10	14.90	27
1430	5"	13.95	16.05	30
1440	6"	16.90	17.10	34
1450	7"	24	18.60	42.60
1460	8"	26	20	46
1510	Tees, standard, pipe diameter, 3"	14.60	19.40	34
1520	4"	15.25	19.75	35
1530	5"	16.10	21	37.10
1540	6"	17.90	21	38.90
1550	7"	22	22	44
1560	8"	24	24	48

316

For expanded coverage of these items see *Means Mechanical Cost Data 1988*

Figure 9.10*a*

SPECIAL	B8.5-410	Chimney & Vent		

8.5-410	Gas Vent Fittings, Galvanized Steel, Double Wall	COST EACH		
		MAT.	INST.	TOTAL
1570	10"	68	24	92
1580	12"	81	24	105
1590	14"	130	28	158
1600	16"	185	33	218
1610	18"	220	35	255
1620	20"	305	48	353
1630	22"	375	61	436
1640	24"	440	65	505
1660	30"	760	81	841
1700	38"	1,025	110	1,135
1750	48"	1,550	160	1,710
1810	Tee, access caps, pipe diameter, 3"	1.12	11.40	12.52
1820	4"	1.39	12.20	13.59
1830	5"	1.51	12.85	14.36
1840	6"	1.73	13.85	15.58
1850	7"	2.62	14.70	17.32
1860	8"	3.08	15.05	18.13
1870	10"	6.05	15.95	22
1880	12"	8.05	16.95	25
1890	14"	16.65	18.35	35
1900	16"	19.35	21	40.35
1910	18"	21	21	42
1920	20"	22	30	52
1930	22"	25	37	62
1940	24"	29	38	67
1950	30"	67	43	110
1970	38"	100	58	158
1990	48"	155	72	227
2110	Tops, bird proof, pipe diameter, 3"	5.45	11.20	16.65
2120	4"	5.80	11.65	17.45
2130	5"	7.85	12.15	20
2140	6"	10.10	12.90	23
2150	7"	14.05	13.95	28
2160	8"	17.70	14.30	32
2170	10"	42	15.40	57.40
2180	12"	64	16.30	80.30
2190	14"	72	16.85	88.85
2200	16"	130	17.40	147.40
2210	18"	210	17.70	227.70
2220	20"	240	27	267
2230	22"	320	38	358
2240	24"	405	38	443
2410	Roof flashing, tall cone/adjustable, pipe diameter, 3"	4.19	14.25	18.44
2420	4"	5.70	15.30	21
2430	5"	6.35	15.65	22
2440	6"	8.90	17.10	26
2450	7"	10.10	17.90	28
2460	8"	11	20	31
2470	10"	31	21	52
2480	12"	39	23	62
2490	14"	47	25	72
2500	16"	59	28	87
2510	18"	75	30	105
2520	20"	98	42	140
2530	22"	125	57	182
2540	24"	150	67	217

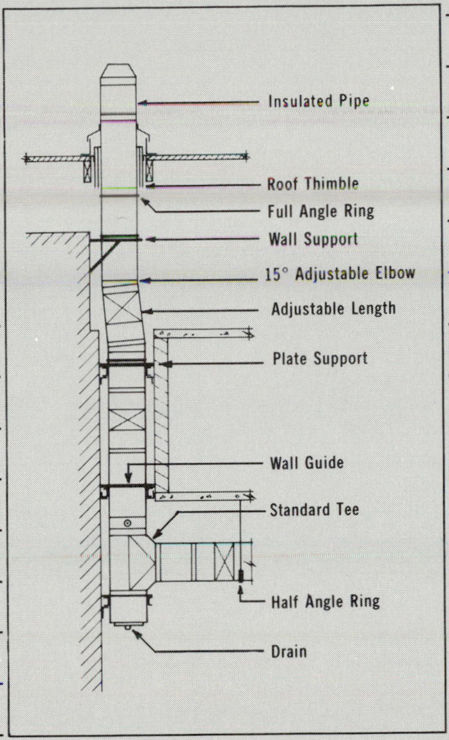

Insulated Pipe
Roof Thimble
Full Angle Ring
Wall Support
15° Adjustable Elbow
Adjustable Length
Plate Support
Wall Guide
Standard Tee
Half Angle Ring
Drain

For expanded coverage of these items see *Means Mechanical Cost Data 1988* 317

Figure 9.10b

157 | Air Conditioning/Ventilating

		157 100 A.C. & Vent. Units	CREW	DAILY OUTPUT	MAN-HOURS	UNIT	BARE COSTS MAT.	BARE COSTS LABOR	BARE COSTS EQUIP.	BARE COSTS TOTAL	TOTAL INCL O&P	
190	1620	200 ton cooling	Q-8	.05	640	Ea.	65,500	14,700	385	80,585	94,000	190
	1660	250 ton cooling	"	.04	800	"	74,500	18,400	485	93,385	109,500	
		157 200 System Components										
201	0010	COILS, FLANGED										201
	0500	Chilled water cooling, 6 rows, 24" x 48"	Q-5	2	8	Ea.	1,970	175		2,145	2,425	
	1000	Direct expansion cooling, 6 rows, 24" x 48"		2	8		2,140	175		2,315	2,600	
	1500	Hot water heating, 1 row, 24" x 48"		3	5.330		780	115		895	1,025	
	2000	Steam heating, 1 row, 24" x 48"		3	5.330		1,110	115		1,225	1,400	
225	0010	CONDENSERS Ratings are for 30°F TD, R-22										225
	0080	Air cooled, belt drive, propeller fan										
	0100	20 ton	Q-5	1	16	Ea.	3,650	345		3,995	4,525	
	0180	30 ton	"	.60	26.670		3,860	575		4,435	5,100	
	0240	48 ton	Q-6	.48	50		5,630	1,125		6,755	7,850	
	0280	60 ton		.32	75		6,380	1,675		8,055	9,500	
	0320	72 ton		.29	82.760		7,510	1,850		9,360	11,000	
	0360	82 ton		.27	88.890		8,870	2,000		10,870	12,700	
	0380	90 ton		.26	92.310		9,610	2,075		11,685	13,600	
	1550	Air cooled, direct drive, propeller fan										
	1600	1-½ ton	Q-5	3.60	4.440	Ea.	395	96		491	575	
	1620	2 ton		3.20	5		435	110		545	635	
	1640	5 ton		2	8		1,010	175		1,185	1,375	
	1660	10 ton		1.40	11.430		1,650	245		1,895	2,175	
	1690	16 ton		1.10	14.550		2,140	315		2,455	2,825	
	1720	26 ton		.74	21.620		3,000	470		3,470	3,975	
	1760	41 ton	Q-6	.54	44.440		4,960	1,000		5,960	6,925	
	1800	63 ton		.32	75		6,720	1,675		8,395	9,875	
240	0010	COOLING TOWERS Packaged units										240
	0080	Draw thru, single flow										
	0100	Belt drive, 60 tons	Q-6	90	.267	Ton	55	6		61	69	
	0150	90 tons		100	.240		49	5.40		54.40	62	
	0200	100 tons		109	.220		47	4.94		51.94	59	
	1500	Induced air, double flow										
	1800	Gear drive, 125 ton	Q-6	120	.200	Ton	48.60	4.49		53.09	60	
	1900	150 ton		126	.190		47.85	4.28		52.13	59	
	2000	300 ton		129	.186		43.30	4.18		47.48	54	
	2100	600 ton		132	.182		34.20	4.08		38.28	44	
	2150	840 ton		142	.169		31	3.79		34.79	40	
	2200	Up to 1000 tons		150	.160		27	3.59		30.59	35	
	3000	For higher capacities, use multiples										
	3500	For pumps and piping, add	Q-6	38	.632	Ton	27.50	14.20		41.70	51	
	4000	For absorption systems, add				"	75%	75%				
	4500	For rigging, see division 016-460										
250	0010	DUCTWORK										250
	0020	Fabricated rectangular, includes fittings, joints, supports,										
	0030	allowance for flexible connections, no insulation										
	0031	NOTE: Fabrication and installation are combined										
	0040	as LABOR cost.										
	0100	Aluminum, alloy 3003-H14, under 300 lb.	Q-10	75	.320	Lb.	2.05	7.15		9.20	12.90	
	0160	Over 10,000 lb.		145	.166		1.30	3.70		5	6.95	
	0500	Galvanized steel, under 400 lb.		235	.102		1.35	2.28		3.63	4.89	
	0520	400 to 1000 lb.		255	.094		.65	2.10		2.75	3.85	
	0540	1000 to 2000 lb.		265	.091		.47	2.02		2.49	3.53	
	0560	2000 to 5000 lb.		275	.087		.37	1.95		2.32	3.31	
	0580	Over 10,000 lb.		300	.080		.34	1.79		2.13	3.03	
	1000	Stainless steel, type 304, under 300 lb.		165	.145		2.69	3.25		5.94	7.80	
	1060	Over 10,000 lb.		235	.102		1.19	2.28		3.47	4.71	

322

For expanded coverage of these items see *Means Mechanical Cost Data 1988*

Figure 9.11a

	157 200 System Components		CREW	DAILY OUTPUT	MAN-HOURS	UNIT	BARE COSTS MAT.	LABOR	EQUIP.	TOTAL	TOTAL INCL O&P	
250	1100	For medium pressure ductwork, add				Lb.		15%				250
	1200	For high pressure ductwork, add				"		40%				
	1300	Flexible, coated fiberglass fabric on corr. resist. metal helix										
	1400	pressure to 12" (WG) UL-181										
	1500	Non-insulated, 3" diameter	Q-9	400	.040	L.F.	.63	.86		1.49	1.97	
	1540	5" diameter		320	.050		.73	1.08		1.81	2.40	
	1560	6" diameter		280	.057		.86	1.23		2.09	2.78	
	1580	7" diameter		240	.067		1.01	1.44		2.45	3.25	
	1600	8" diameter		200	.080		1.12	1.72		2.84	3.80	
	1640	10" diameter		160	.100		1.41	2.16		3.57	4.76	
	1660	12" diameter		120	.133		1.68	2.87		4.55	6.15	
	1900	Insulated, 1" thick with ¾ lb., PE jacket		340	.047		1.02	1.01		2.03	2.63	
	1920	5" diameter		300	.053		1.22	1.15		2.37	3.05	
	1940	6" diameter		260	.062		1.39	1.33		2.72	3.50	
	1960	7" diameter		220	.073		1.59	1.57		3.16	4.08	
	1980	8" diameter		180	.089		1.70	1.92		3.62	4.72	
	2020	10" diameter		140	.114		2.10	2.46		4.56	6	
	2040	12" diameter	↓	100	.160	↓	2.47	3.45		5.92	7.85	
	3490	Rigid fiberglass duct board, foil reinf. kraft facing										
	3500	Rectangular, 1" thick, alum. faced, (FRK), std. weight	Q-10	350	.069	SF Surf	.45	1.53		1.98	2.78	
290	0010	FANS (135)										290
	0020	Air conditioning and process air handling										
	0030	Axial flow, compact, low sound, 2.5" S.P.										
	0050	3800 CFM, 5 HP	Q-20	3.40	5.880	Ea.	2,275	130		2,405	2,700	
	0080	6400 CFM, 5 HP		2.80	7.140		2,550	155		2,705	3,025	
	0100	10,500 CFM, 7-½ HP		2.40	8.330		3,170	185		3,355	3,750	
	0120	15,600 CFM, 10 HP	↓	1.60	12.500	↓	3,990	275		4,265	4,800	
	0150											
	0200	In-line centrifugal, supply/exhaust booster,										
	0220	aluminum wheel/hub, disconnect switch, ¼" S.P.										
	0240	500 CFM, 10" diameter connection	Q-20	3	6.670	Ea.	435	145		580	695	
	0260	1380 CFM, 12" diameter connection		2	10		610	220		830	995	
	0280	1520 CFM, 16" diameter connection		2	10		655	220		875	1,050	
	0300	2560 CFM, 18" diameter connection		1	20		875	440		1,315	1,625	
	0320	3480 CFM, 20" diameter connection		.80	25		1,050	550		1,600	1,975	
	1500	Vaneaxial, low pressure, 2000 CFM, ½ HP		3.60	5.560		1,300	120		1,420	1,600	
	1520	4000 CFM, 1 HP		3.20	6.250		1,400	135		1,535	1,750	
	1540	8000 CFM, 2 HP	↓	2.80	7.140	↓	1,700	155		1,855	2,100	
	2500	Ceiling fan, right angle, extra quiet, 0.10" S.P.										
	2520	95 CFM	Q-20	20	1	Ea.	120	22		142	165	
	2540	210 CFM		19	1.050		130	23		153	175	
	2560	385 CFM		18	1.110		165	24		189	220	
	2580	885 CFM		16	1.250		315	27		342	385	
	2600	1650 CFM		13	1.540		415	34		449	505	
	2620	2960 CFM	↓	11	1.820		560	40		600	675	
	2640	For wall or roof cap, add	1 Shee	16	.500		73	12		85	98	
	2660	For straight thru fan, add					10%					
	2680	For speed control switch, add	1 Elec	16	.500	↓	40	11.80		51.80	61	
	3000	Paddle blade air circulator, 3 speed switch										
	3020	36", 4000 CFM high, 3000 CFM low	Q-20	6	3.330	Ea.	160	73		233	285	
	3040	52", 7000 CFM high, 4000 CFM low	"	4	5	"	185	110		295	365	
	3100	For antique white motor, same cost										
	3200	For brass plated motor, add				Ea.	22.50			22.50	25	
	3300	For light adaptor kit, add				"	15			15	16.50	
	3500	Centrifugal, airfoil, motor and drive, complete										
	3520	1000 CFM, ½ HP	Q-20	2.50	8	Ea.	1,000	175		1,175	1,350	
	3540	2000 CFM, 1 HP		2	10		1,100	220		1,320	1,525	
	3560	4000 CFM, 3 HP	↓	1.80	11.110	↓	1,500	245		1,745	2,000	

For expanded coverage of these items see *Means Mechanical Cost Data 1988*

Figure 9.11b

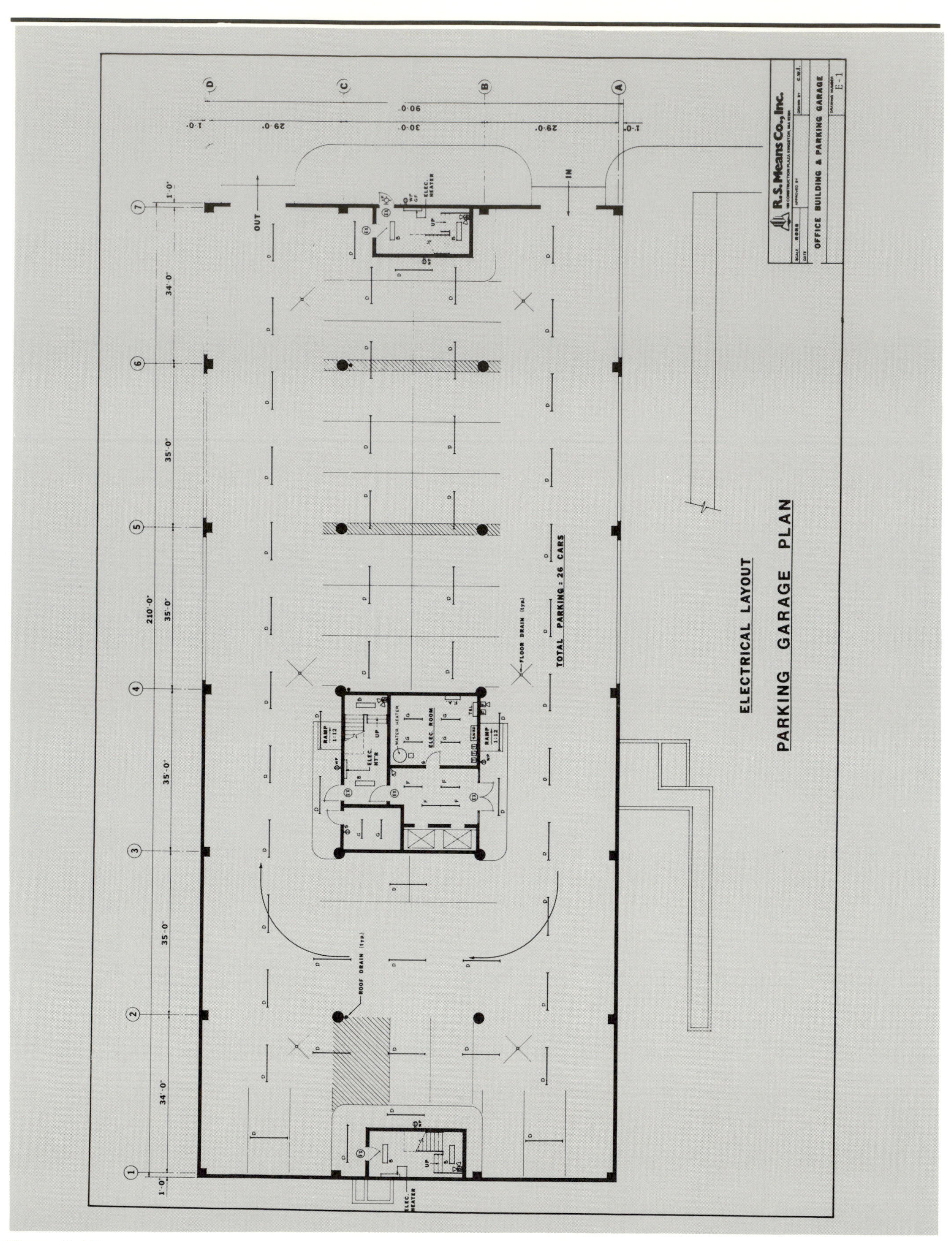
ELECTRICAL LAYOUT
PARKING GARAGE PLAN
TOTAL PARKING: 26 CARS
R.S. Means Co., Inc.
OFFICE BUILDING & PARKING GARAGE
E-1
210'-0"
90'-0"
IN
OUT
ELEC. ROOM
WATER HEATER
ROOF DRAIN (typ)
FLOOR DRAIN (typ)
ELEC. HEATER
RAMP 1:12

Figure 9.12

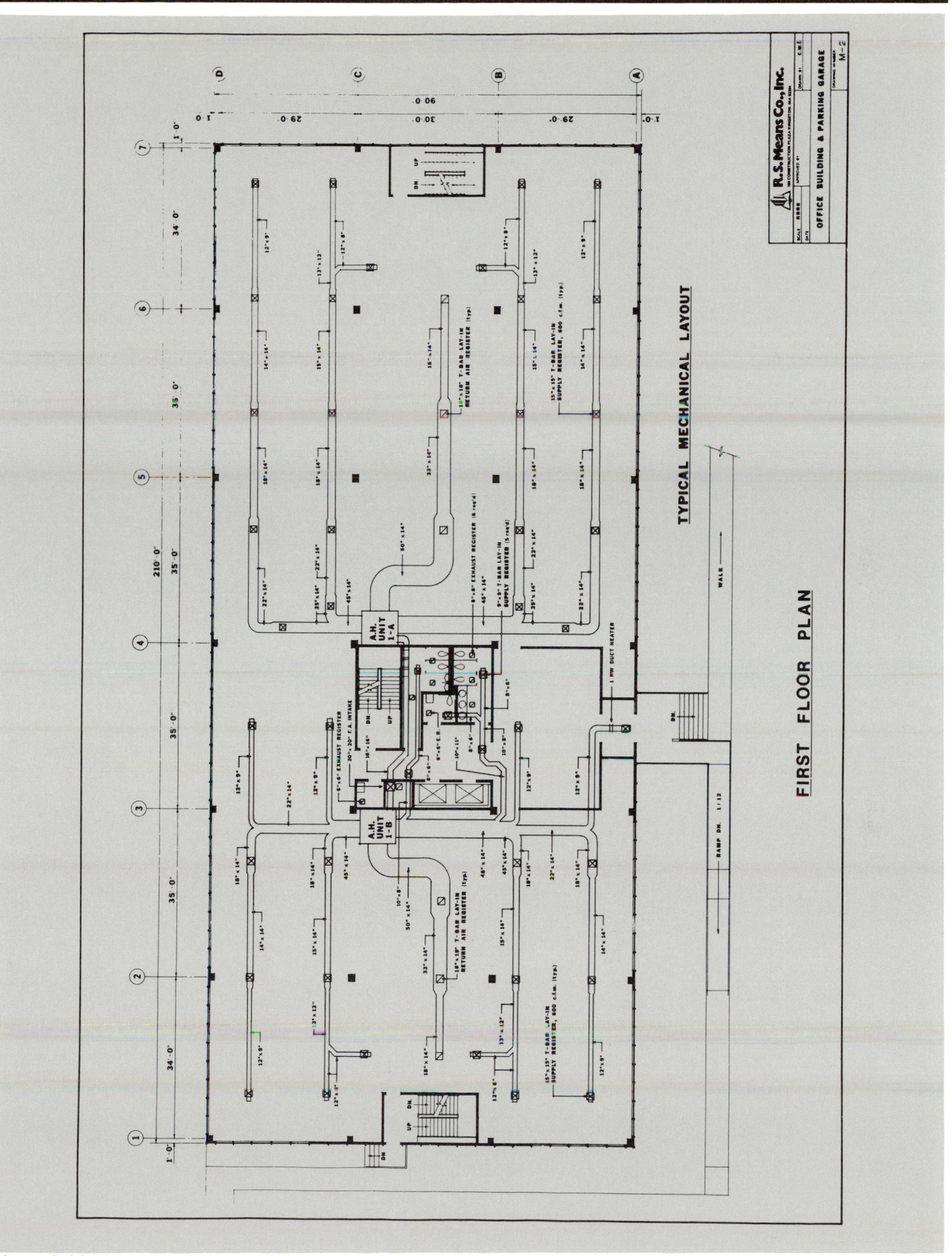
FIRST FLOOR PLAN
TYPICAL MECHANICAL LAYOUT
R.S. Means Co., Inc.
OFFICE BUILDING & PARKING GARAGE
M-2
A.H. UNIT 1-A
A.H. UNIT 1-B
210'-0"
90'-0"
WALK
RAMP DN. 1:12
1 KW DUCT HEATER

Figure 9.13

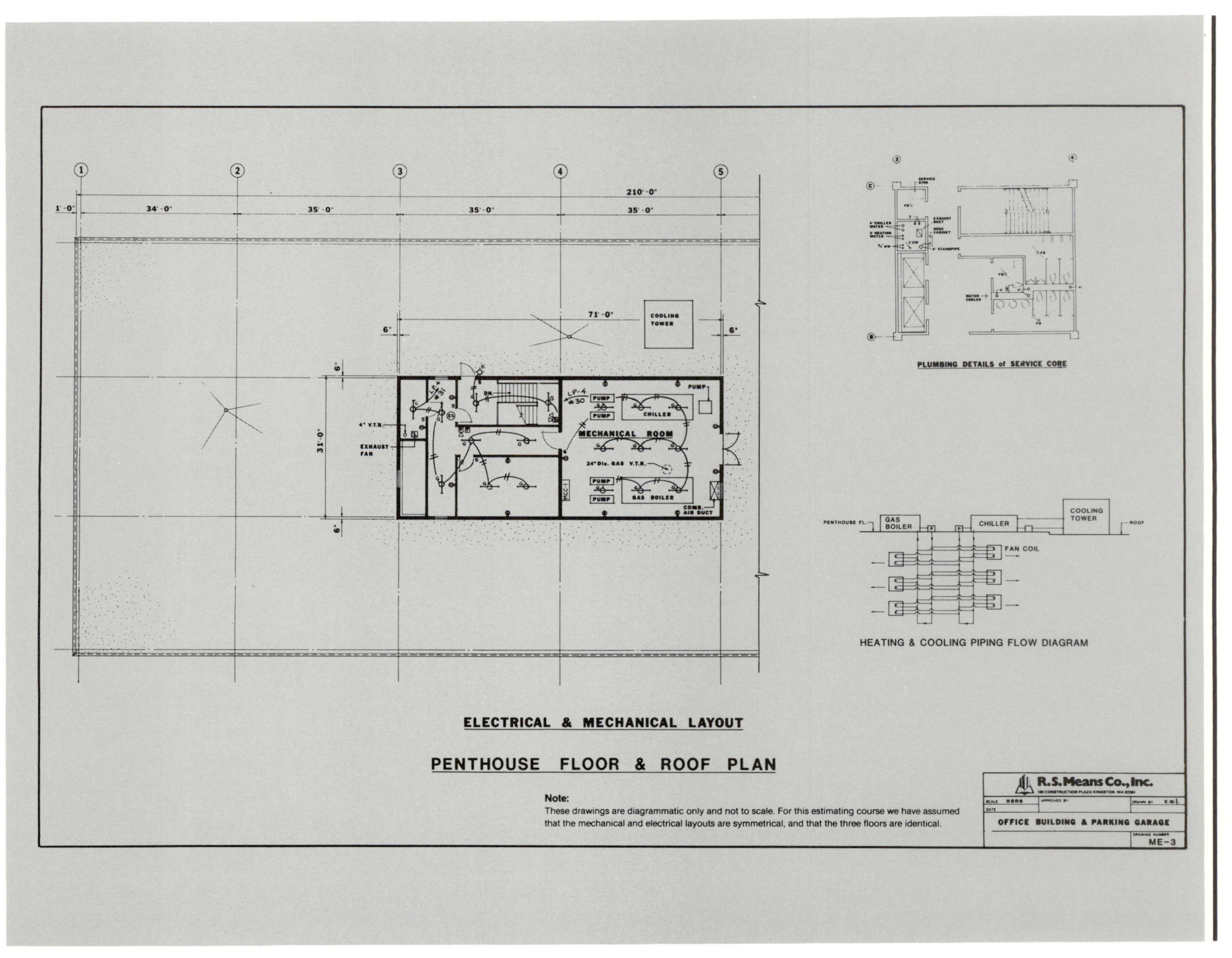
ELECTRICAL & MECHANICAL LAYOUT
PENTHOUSE FLOOR & ROOF PLAN
Note:
These drawings are diagrammatic only and not to scale. For this estimating course we have assumed that the mechanical and electrical layouts are symmetrical, and that the three floors are identical.
210'-0"
1'-0"
34'-0"
35'-0"
35'-0"
35'-0"
71'-0"
6"
6'
31'-0"
COOLING TOWER
MECHANICAL ROOM
CHILLER
GAS BOILER
PUMP
24" Dia. GAS V.T.R.
COMB. AIR DUCT
MCC-1
LP-4 #30
4" V.T.R.
EXHAUST FAN
PLUMBING DETAILS of SERVICE CORE
HEATING & COOLING PIPING FLOW DIAGRAM
PENTHOUSE FL.
ROOF
FAN COIL
R.S. Means Co., Inc.
OFFICE BUILDING & PARKING GARAGE
ME-3

Figure 9.14

ASSEMBLY NUMBER	DESCRIPTION	QTY	UNIT	TOTAL COST UNIT	TOTAL COST TOTAL	COST PER S.F.
8.0	**Mechanical System**					
8.3-141-1480 +	Boiler Assembly - CI, gas	56,700	sf	2.11	119637	
8.4-120-4040	Chilled Water Cooling Tower System	✓	sf	6.76	383292	
8.5-410-0060	Boiler Vent - 6" galv	15	lf	12.64	190	
8.5-410-1440	✓ ✓ - 90° elbow	1	ea	34	34	
8.5-410-2440	✓ ✓ - Roof flashing	1	ea	26	26	
157-250-0540	Exhaust Duct	2000	lbs	3.53	7060	
157-290-7140	Roof Exhaust Fan	1	ea	960	960	
157-460-1020	Registers	23	ea	20	460	
155-408-0100	Duct Heater	1	ea	300	300	
155-420-1020	Stair Heaters	3	ea	445	1335	
	Subtotal				513294	
	Quality / Complexity @ 15%				76994	
					590288	
9.0	**Electrical**					

Page 4 of 6

Figure 9.15

10

ESTIMATING ELECTRICAL COSTS

10

ESTIMATING ELECTRICAL COSTS

The same factors that present obstacles to accurate cost control estimates for plumbing and HVAC also apply to electrical estimates. Again, the problem is a lack of detailed information until near the end of the detailed design phase, and a lack of specific training and experience on the part of the estimator. Just as with plumbing and HVAC, early electrical cost control estimates require the square foot method. When sufficient information is available near the end of the design development phase, a combination of the assemblies and square foot methods can be used.

In this chapter we will briefly describe the components of an electrical system; explore the cost information available in *Means Assemblies Cost Data* to determine which components are estimated using the assemblies method, and which are estimated using the square foot method; outline the estimating procedure; and then demonstrate that procedure by estimating the electrical costs for the Chambers office building.

The Electrical System

The electrical system supplies and controls electrical power, lighting, and specialty systems for a building. The components of the electrical system are service and distribution, lighting, wiring devices, equipment connections, basic materials, and special systems. An abbreviated example of a complete electrical system is shown in Figure 10.1.

Service and Distribution

The service and distribution system provides electrical power to the building, and controls and distributes that power to the building's major electrical systems. This equipment includes the primary feeder from the electrical company, transformers, meters, switchgear, power and lighting panels, and associated wiring. A basic service and distribution system is shown in Figure 10.1.

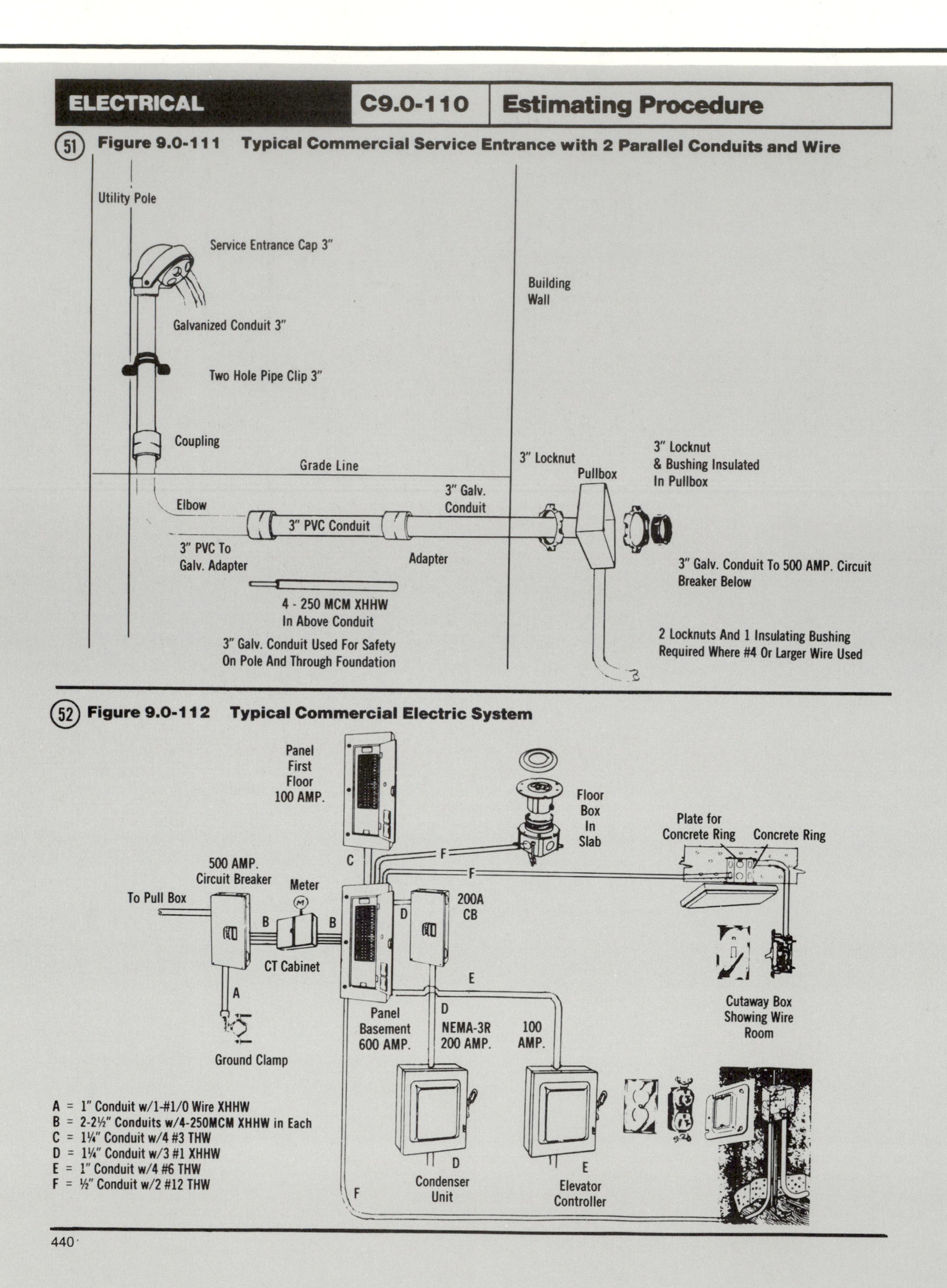
ELECTRICAL
C9.0-110
Estimating Procedure
51 Figure 9.0-111 Typical Commercial Service Entrance with 2 Parallel Conduits and Wire
Utility Pole
Service Entrance Cap 3"
Building Wall
Galvanized Conduit 3"
Two Hole Pipe Clip 3"
Coupling
Grade Line
3" Locknut
Pullbox
3" Locknut & Bushing Insulated In Pullbox
3" Galv. Conduit
Elbow
3" PVC Conduit
3" PVC To Galv. Adapter
Adapter
3" Galv. Conduit To 500 AMP. Circuit Breaker Below
4 - 250 MCM XHHW In Above Conduit
3" Galv. Conduit Used For Safety On Pole And Through Foundation
2 Locknuts And 1 Insulating Bushing Required Where #4 Or Larger Wire Used
52 Figure 9.0-112 Typical Commercial Electric System
Panel First Floor 100 AMP.
Floor Box In Slab
Plate for Concrete Ring
Concrete Ring
500 AMP. Circuit Breaker
Meter
To Pull Box
200A CB
CT Cabinet
Cutaway Box Showing Wire Room
Ground Clamp
Panel Basement 600 AMP.
NEMA-3R 200 AMP.
100 AMP.
A = 1" Conduit w/1-#1/0 Wire XHHW
B = 2-2½" Conduits w/4-250MCM XHHW in Each
C = 1¼" Conduit w/4 #3 THW
D = 1¼" Conduit w/3 #1 XHHW
E = 1" Conduit w/4 #6 THW
F = ½" Conduit w/2 #12 THW
Condenser Unit
Elevator Controller
440

Figure 10.1

Lighting

Lighting includes all building light fixtures and associated wiring. The types of lighting include decorative, illumination, and emergency lighting. The fixtures on the exterior of the building are included in this system for estimating purposes. The cost of parking area lighting and other site lighting is estimated with site work (see Chapter 7). Figure 10.2 shows some examples of lighting fixtures.

Wiring Devices

Wiring devices include all outlet boxes, receptacles, lighting switches, dimmers, cover plates, and associated wiring. Examples are shown in Figure 10.3.

Equipment Connections

Equipment connections include all materials and equipment for providing electrical power for HVAC, elevator, food service, and other equipment. Figure 10.4 shows some examples.

Basic Materials

Included in this category are all separate disconnects, wiring raceways, pull boxes, junction boxes, supports, fittings, grounding devices, wireways, buses, and cable systems. Some examples are shown in Figure 10.5.

Special Systems

Special systems include the equipment required for fire alarm and detection, lightning protection, intercom systems, closed circuit TV, computer circuits, and emergency generators. See Figure 10.6 for some illustrations.

More detailed information about a wide range of electrical systems and estimating techniques may be found in *Means Electrical Estimating*.

Cost Data

Means Assemblies Cost Data provides both square foot and assemblies costs for electrical systems. The electrical component square foot costs shown in Figures 10.7a and b were used to prepare the need analysis estimate in Chapter 6 and the cost control estimate in Chapter 7. This information represents the average cost of each component for the different building types. While these prices are appropriately used when information is limited, they do not provide the kind of flexibility that is needed for accurate cost control estimates. Published costs can, however, be used as a supplement in electrical estimating for design cost control, and are a useful source of information for those who are not particularly knowledgeable in electrical estimating. Just as with plumbing and HVAC estimates, the assemblies method is used where possible to tailor the estimate to the job. Where assemblies costs are not available, the square foot method is used.

Means Assemblies Cost Data provides assemblies costs for high voltage cables, electrical service, feeder assemblies, switchgear, light fixtures, receptacles, light switches, equipment connection assemblies, electric baseboard heaters, and some special systems. Linear foot prices are given for cables and feeders. A *per each* price is provided for electrical service systems, switchgear assemblies, and baseboard heaters. Square foot prices are listed for lighting, receptacles, light switches, and specialty systems. Square foot and *per each* assembly prices are provided for equipment connection devices. Each assembly includes the basic material components needed to complete the installation. Example costs for feeders, service assemblies, light fixtures, receptacles, and equipment connections are shown in Figures 10.8 through 10.12.

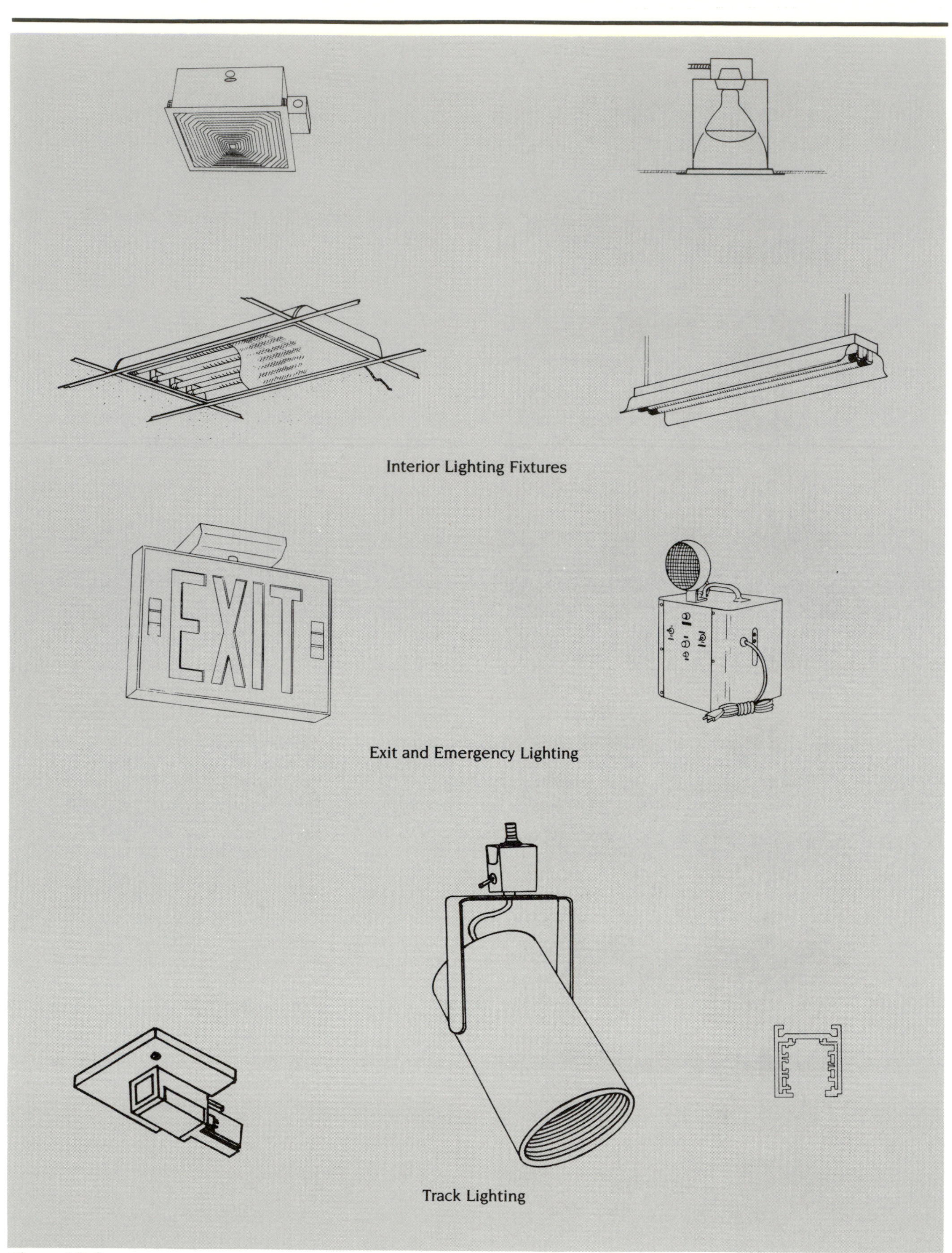

Interior Lighting Fixtures

Exit and Emergency Lighting

Track Lighting

Figure 10.2

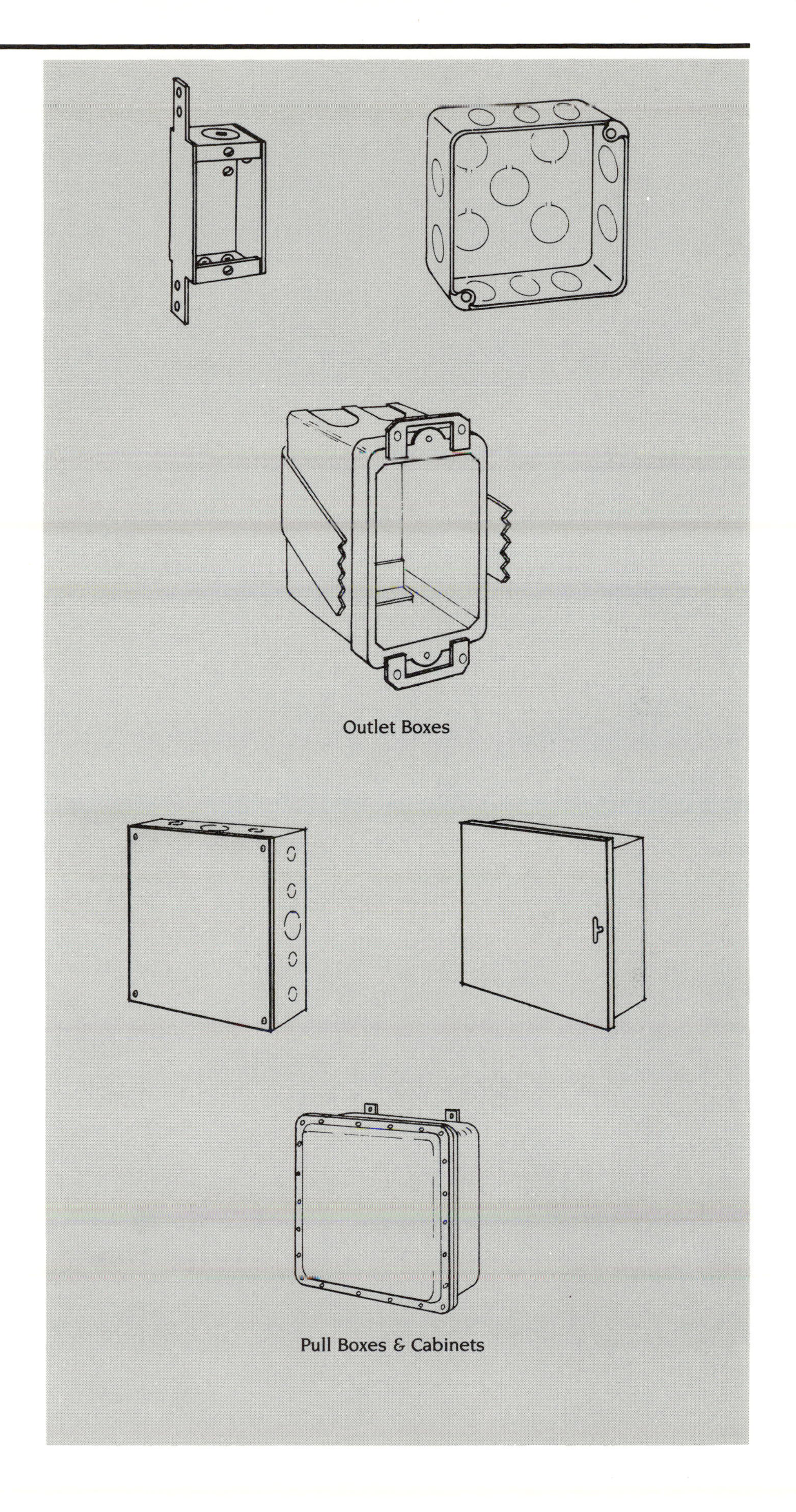

Figure 10.3

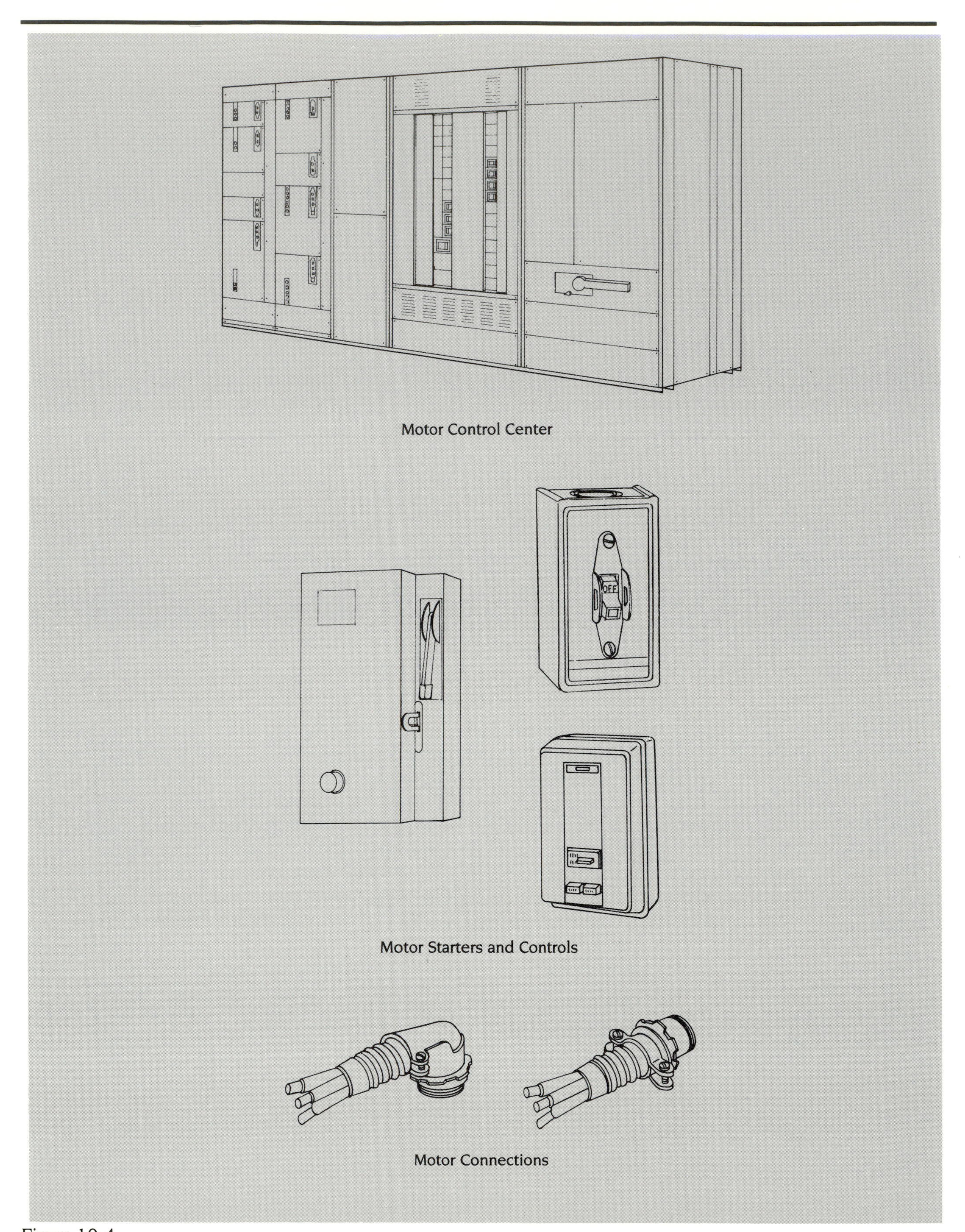
Motor Control Center
Motor Starters and Controls
Motor Connections

Figure 10.4

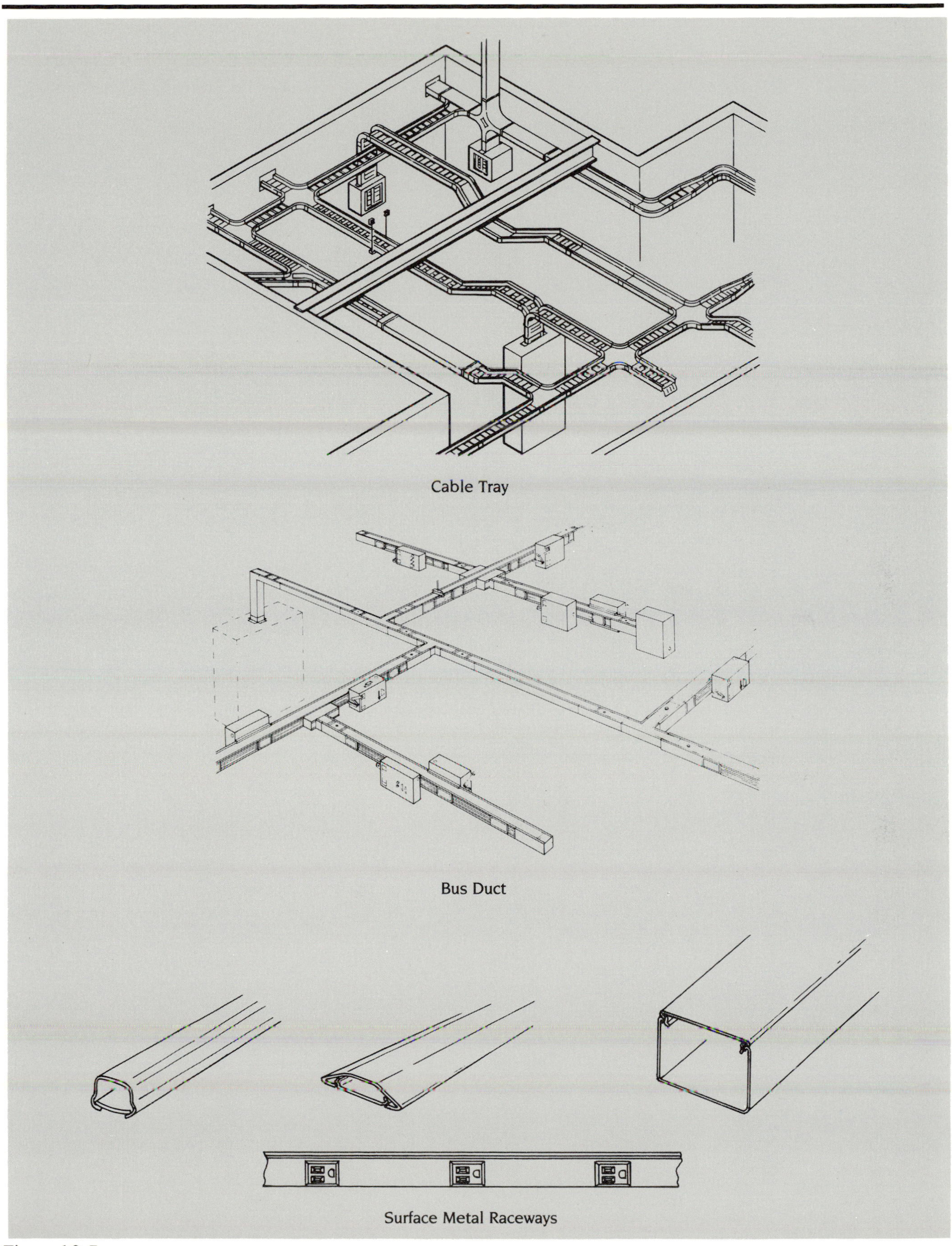

Figure 10.5

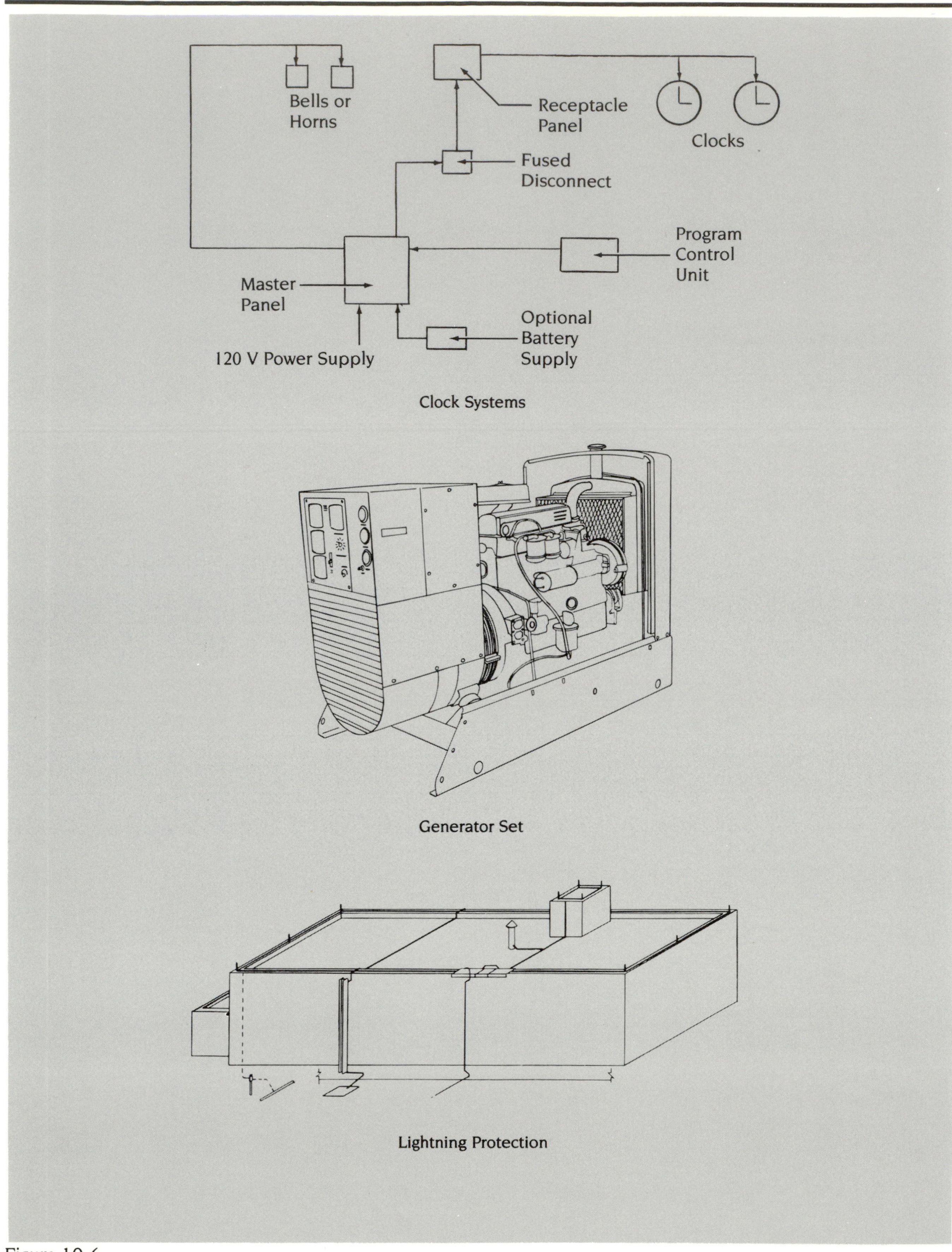
Bells or Horns
Receptacle Panel
Clocks
Fused Disconnect
Program Control Unit
Master Panel
Optional Battery Supply
120 V Power Supply
Clock Systems
Generator Set
Lightning Protection

Figure 10.6

ELECTRICAL	C9.0-110	Estimating Procedure

A Conceptual Estimate of the costs for a building, when final drawings are not available, can be quickly figured by using **Table B9.0-117 Cost Per S.F. For Electrical Systems For Various Building Types.** The following definitions apply to this table.

1. **Service And Distribution:** This system includes the incoming primary feeder from the power company, the main building transformer, metering arrangement, switchboards, distribution panel boards, stepdown transformers, power and lighting panels. Items marked (*) include the cost of the primary feeder and transformer. In all other projects the cost of the primary feeder and transformer is paid for by the local power company.
2. **Lighting:** Includes all interior fixtures for decor, illumination, exit and emergency lighting. Fixtures for exterior building lighting are included but parking area lighting is not included unless mentioned. See also Section B9.2 for detailed analysis of lighting requirements and costs.
3. **Devices:** Includes all outlet boxes, receptacles, switches for lighting control, dimmers and cover plates.
4. **Equipment Connections:** Includes all materials and equipment for making connections for Heating, Ventilating and Air Conditioning, Food Service and other motorized items requiring connections.
5. **Basic Materials:** This category includes all disconnect power switches not part of service equipment, raceways for wires, pull boxes, junction boxes, supports, fittings, grounding materials, wireways, busways and cable systems.
6. **Special Systems:** Includes installed equipment only for the particular system such as fire detection and alarm, sound, emergency generator and others as listed in the table.

(56) Table 9.0-117 Cost Per S.F. for Electric Systems for Various Building Types

Type Construction	1. Service & Distrib.	2. Lighting	3. Devices	4. Equipment Connections	5. Basic Materials	6. Special Systems: Fire Alarm & Detection	Lighting Protection	Master TV Antenna
Apartment, luxury high rise	$ .95	$ .66	$ .47	$ .61	$1.62	$ .28		$ .18
Apartment, low rise	.54	.56	.41	.50	.94	.24		
Auditorium	1.23	3.41	.36	.90	1.97	.39		
Bank, branch office	1.44	3.72	.61	.90	1.84	1.09		
Bank, main office	1.10	2.03	.18	.40	1.99	.57		
Church	.70	2.00	.24	.19	.90	.57		
* College, science building	1.26	2.46	.75	.64	1.97	.48		
* College library	1.01	1.49	.15	.43	1.17	.57		
* College, physical education center	1.59	2.11	.25	.34	.90	.31		
Department store	.53	1.47	.15	.60	1.60	.24		
* Dormitory, college	.71	1.87	.16	.40	1.56	.42		.25
Drive-in donut shop	2.03	5.63	.88	.91	2.55	—		
Garage, commercial	.27	.68	.10	.27	.53	—		
* Hospital, general	3.92	2.89	1.03	.73	3.19	.35	$ .07	
* Hospital, pediatric	3.43	4.33	.86	2.64	5.92	.41		.31
* Hotel, airport	1.54	2.38	.16	.37	2.33	.31	.16	.28
Housing for the elderly	.43	.56	.25	.69	1.97	.41		.24
Manufacturing, food processing	.96	2.93	.13	1.33	2.15	.24		
Manufacturing apparel	.63	1.52	.19	.50	1.15	.20		
Manufacturing, tools	1.44	3.59	.17	.60	1.93	.25		
Medical clinic	.51	1.13	.29	.87	1.41	.40		
Nursing home	1.02	2.38	.31	.26	1.97	.54		.18
Office building	1.39	3.20	.13	.51	2.06	.27	.13	
Radio-TV studio	.92	3.11	.45	.91	2.29	.37		
Restaurant	3.68	3.13	.58	1.48	2.93	.20		
Retail store	.77	1.65	.16	.37	.90	—		
School, elementary	1.30	2.98	.36	.37	2.48	.34		.11
School, junior high	.78	2.49	.16	.66	1.97	.41		
* School, senior high	.87	1.96	.33	.86	2.19	.35		
Supermarket	.89	1.69	.22	1.42	1.89	.14		
* Telephone exchange	2.03	.66	.10	.57	1.21	.64		
Theater	1.68	2.26	.36	1.22	1.89	.48		
Town Hall	1.02	1.79	.36	.46	2.50	.31		
* U.S. Post Office	3.05	2.34	.37	.68	1.81	.31		
Warehouse, grocery	.55	1.00	.10	.37	1.34	.18		

* Includes cost of primary feeder and transformer. Cont'd on next page.

444

Figure 10.7a

ELECTRICAL | **C9.0-110** | **Estimating Procedure**

COST ASSUMPTIONS:

Each of the projects analyzed in Table C9.0-117 were bid within the last 10 years in the Northeastern part of the United States. Bid prices have been adjusted to Jan. 1, 1988 levels. The list of projects is by no means all-inclusive, yet by carefully examining the various systems for a particular building type, certain cost relationships will emerge. The use of Section C14 with the S.F. and C.F. electrical costs should produce a budget S.F. cost for the electrical portion of a job that is consistent with the amount of design information normally available at the conceptual estimate stage.

(56) Table 9.0-117 (Cont.) Cost Per S.F. for Electric Systems for Various Building Types

Type Construction	6. Special Systems, Cont'd.						
	Intercom Systems	Sound Systems	Closed Circuit TV	Snow Melting	Emergency Generator	Security	Master Clock Sys.
Apartment, luxury high rise	$ .40						
Apartment, low rise	.28						
Auditorium		$1.05	$ .49		$ .81		
Bank, branch office	.54		1.13			$ .95	
Bank, main office	.31		.23		.65	.52	$ .20
Church	.39						
* College, science building	.39				.81		.24
* College, library					.43		
* College, physical education center		.53					
Department store					.15		
* Dormitory, college	.52						
Drive-in donut shop							.07
Garage, commercial							.05
* Hospital, general	.41		.14		1.13		
* Hospital, pediatric	2.80	.28	.31		.71		
* Hotel, airport	.41				.42		
Housing for the elderly	.49						
Manufacturing food processing		.16			1.46		
Manufacturing, apparel		.24					
Manufacturing, tools		.31		$ .18			
Medical clinic							
Nursing home	.93				.36		
Office building		.12			.36	.14	.05
Radio-TV studio	.53				.92		.38
Restaurant		.24					
Retail store							
School, elementary		.14					.14
School, junior high		.45			.30		.31
* School, senior high	.37		.24		.43	.20	.21
Supermarket		.17			.38	.24	
* Telephone exchange					3.71	.10	
Theater		.36					
Town Hall							.14
* U.S. Post Office	.36			.05	.42		
Warehouse, grocery	.22						

*Includes cost of primary feeder and transformer. Cont'd on next page.

445

Figure 10.7*b*

SERVICE	B9.1-310	Electrical Feeder

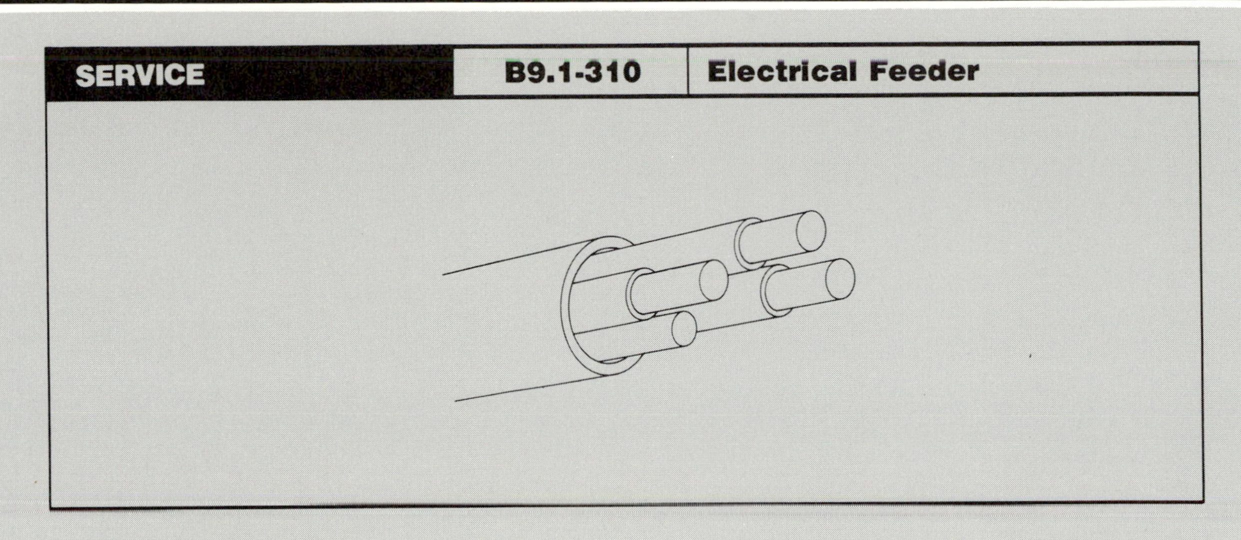

System Components	QUANTITY	UNIT	COST PER L.F. MAT.	INST.	TOTAL
SYSTEM 09.1-310-0200					
FEEDERS, INCLUDING STEEL CONDUIT & WIRE, 60 AMPERES					
Rigid galvanized steel conduit, ¾", including fittings	1.000	L.F.	1.49	3.43	4.92
Wire 600 volt, type XHHW copper stranded #6	.040	C.L.F.	.92	1.68	2.60
TOTAL			2.41	5.11	7.52

9.1-310	Feeder Installation	COST PER L.F. MAT.	INST.	TOTAL
0200	Feeder installation, including conduit and wire, 60 amperes	2.41	5.10	7.51
0240	100 amperes	4.35	6.75	11.10
0280	200 amperes	8.95	10.60	19.55
0320	400 amperes (59)	17.95	21	38.95
0360	600 amperes	37	34	71
0400	800 amperes	50	41	91
0440	1000 amperes	63	53	116
0480	1200 amperes	74	54	128
0520	1600 amperes	100	82	182
0560	2000 amperes	125	105	230

320 For expanded coverage of these items see *Means Electrical Cost Data 1988*

Figure 10.8

SERVICE	B9.1-210	Electric Service

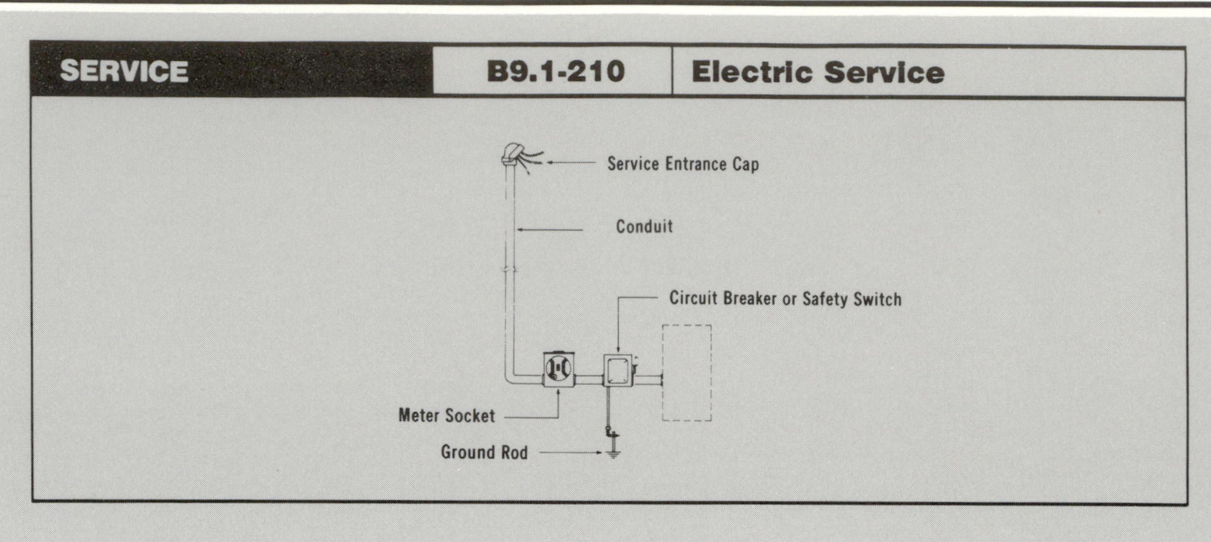

System Components	QUANTITY	UNIT	COST EACH MAT.	COST EACH INST.	COST EACH TOTAL
SYSTEM 09.1-210-0200					
SERVICE INSTALLATION, INCLUDES BREAKERS, METERING, 20′ CONDUIT & WIRE					
3 PHASE, 4 WIRE, 60 AMPS					
Circuit breaker, enclosed (NEMA 1), 600 volt, 3 pole, 60 amp	1.000	Ea.	220	100	320
Meter socket, single position, 4 terminal, 100 amp	1.000	Ea.	22	88	110
Rigid galvanized steel conduit, ¾″, including fittings	20.000	L.F.	29.80	68.60	98.40
Wire, 600V type XHHW, copper stranded #6	.900	C.L.F.	20.79	37.71	58.50
Service entrance cap ¾″ diameter	1.000	Ea.	3.85	21.15	25
Conduit LB fitting with cover, ¾″ diameter	1.000	Ea.	5.83	21.17	27
Ground rod, copper clad, 8′ long, ¾″ diameter	1.000	Ea.	18.26	51.74	70
Ground rod clamp, bronze, ¾″ diameter	1.000	Ea.	3.30	8.60	11.90
Ground wire, bare armored, #6-1 conductor	.200	C.L.F.	16.28	30.72	47
TOTAL			340.11	427.69	767.80

9.1-210	Electric Service, 3 Phase - 4 Wire	COST EACH MAT.	COST EACH INST.	COST EACH TOTAL
0200	Service installation, includes breakers, metering, 20′ conduit & wire			
0220	3 phase, 4 wire, 120/208 volts, 60 amp	340	430	770
0240	100 amps	450	515	965
0280	200 amps	620	795	1,415
0320	400 amps (51)	1,325	1,450	2,775
0360	600 amps	2,550	1,950	4,500
0400	800 amps	3,725	2,350	6,075
0440	1000 amps	4,725	2,700	7,425
0480	1200 amps	5,975	2,750	8,725
0520	1600 amps	10,300	3,975	14,275
0560	2000 amps	11,400	4,525	15,925
0570	Add 25% for 277/480 volt			
0580				
0610	1 phase, 3 wire, 120/240 volts, 100 amps	220	470	690
0620	200 amps	455	755	1,210

For expanded coverage of these items see *Means Electrical Cost Data 1988*

319

Figure 10.9

LIGHTING & POWER	B9.2-212	Fluorescent Fixture (by Type)

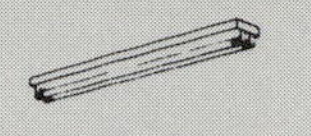

A. Strip Fixture

B. Surface Mounted

C. Recessed

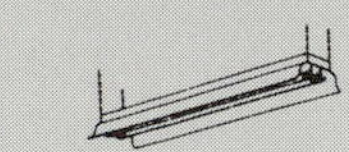

D. Pendent Mounted

Design Assumptions:

1. A 100 footcandle average maintained level of illumination.
2. Ceiling heights range from 9' to 11'.
3. Average reflectance values are assumed for ceilings, walls and floors.
4. Cool white (CW) fluorescent lamps with 3150 lumens for 40 watt lamps and 6300 lumens for 8' slimline lamps.
5. Four 40 watt lamps per 4' fixture and two 8' lamps per 8' fixture.
6. Average fixture efficiency values and spacing to mounting height ratios.
7. Installation labor is average U.S. rate as of January 1, 1988.

System Components	QUANTITY	UNIT	COST PER S.F. MAT.	COST PER S.F. INST.	COST PER S.F. TOTAL
SYSTEM 09.2-212-0520					
FLUORESCENT FIXTURES MOUNTED 9'-11" ABOVE FLOOR, 100 FC					
TYPE A, 8 FIXTURES PER 400 S.F.					
Steel intermediate conduit, (IMC) ½" diam	.404	L.F.	.40	1.11	1.51
Wire, 600V, type THWN-THHN, copper, solid, #12	.008	C.L.F.	.04	.20	.24
Fluorescent strip fixture 8' long, surface mounted, two 75W SL	.020	Ea.	.92	.90	1.82
Steel outlet box 4" concrete	.020	Ea.	.09	.28	.37
Steel outlet box plate with stud, 4" concrete	.020	Ea.	.04	.07	.11
TOTAL			1.49	2.56	4.05

9.2-212	Fluorescent Fixtures (by Type)	COST PER S.F. MAT.	COST PER S.F. INST.	COST PER S.F. TOTAL
0520	Fluorescent fixtures, type A, 8 fixtures per 400 S.F.	1.49	2.56	4.05
0560	11 fixtures per 600 S.F.	1.40	2.48	3.88
0600	17 fixtures per 1000 S.F. (65)	1.37	2.43	3.80
0640	23 fixtures per 1600 S.F.	1.23	2.30	3.53
0680	28 fixtures per 2000 S.F.	1.23	2.30	3.53
0720	41 fixtures per 3000 S.F.	1.20	2.29	3.49
0800	53 fixtures per 4000 S.F.	1.18	2.23	3.41
0840	64 fixtures per 5000 S.F.	1.18	2.23	3.41
0880	Type B, 11 fixtures per 400 S.F.	3.04	3.66	6.70
0920	15 fixtures per 600 S.F.	2.80	3.51	6.31
0960	24 fixtures per 1000 S.F.	2.73	3.50	6.23
1000	35 fixtures per 1600 S.F.	2.54	3.34	5.88
1040	42 fixtures per 2000 S.F.	2.49	3.35	5.84
1080	61 fixtures per 3000 S.F.	2.50	3.23	5.73
1160	80 fixtures per 4000 S.F.	2.41	3.31	5.72
1200	98 fixtures per 5000 S.F.	2.40	3.29	5.69
1240	Type C, 11 fixtures per 400 S.F.	2.49	3.96	6.45
1280	14 fixtures per 600 S.F.	2.18	3.69	5.87
1320	23 fixtures per 1000 S.F.	2.17	3.67	5.84
1360	34 fixtures per 1600 S.F.	2.09	3.62	5.71
1400	43 fixtures per 2000 S.F.	2.11	3.59	5.70
1440	63 fixtures per 3000 S.F.	2.06	3.53	5.59
1520	81 fixtures per 4000 S.F.	2	3.49	5.49
1560	101 fixtures per 5000 S.F.	2	3.49	5.49
1600	Type D, 8 fixtures per 400 S.F.	2.42	3.09	5.51
1640	12 fixtures per 600 S.F.	2.42	3.08	5.50
1680	19 fixtures per 1000 S.F.	2.33	2.98	5.31
1720	27 fixtures per 1600 S.F.	2.16	2.90	5.06
1760	34 fixtures per 2000 S.F.	2.15	2.88	5.03
1800	48 fixtures per 3000 S.F.	2.05	2.80	4.85
1880	64 fixtures per 4000 S.F.	2.05	2.80	4.85
1920	79 fixtures per 5000 S.F.	2.05	2.80	4.85

322

For expanded coverage of these items see *Means Electrical Cost Data 1988*

Figure 10.10

LIGHTING & POWER	B9.2-522	Receptacle (by Wattage)

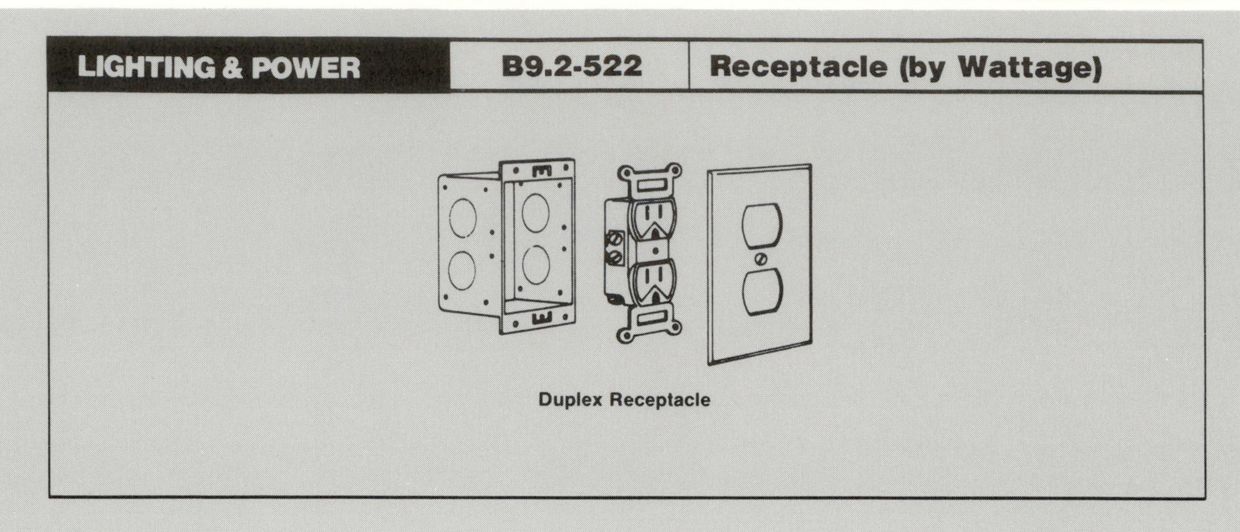
Duplex Receptacle

System Components	QUANTITY	UNIT	COST PER S.F. MAT.	INST.	TOTAL
SYSTEM 09.2-522-0200					
RECEPTACLES INCL. PLATE, BOX, CONDUIT, WIRE & TRANS. WHEN REQUIRED					
2.5 PER 1000 S.F., .3 WATTS PER S.F.					
Steel intermediate conduit, (IMC) ½" diam	167.000	L.F.	.17	.46	.63
Wire 600V type THWN-THHN, copper solid #12	3.382	C.L.F.	.02	.08	.10
Wiring device, receptacle, duplex, 120V grounded, 15 amp	2.500	Ea.	.01	.02	.03
Wall plate, 1 gang, brown plastic	2.500	Ea.		.01	.01
Steel outlet box 4" square	2.500	Ea.		.03	.03
Steel outlet box 4" plaster rings	2.500	Ea.		.01	.01
TOTAL			.20	.61	.81

9.2-522	Receptacle (by Wattage)	COST PER S.F. MAT.	INST.	TOTAL
0190	Receptacles include plate, box, conduit, wire & transformer when required			
0200	2.5 per 1000 S.F., .3 watts per S.F.	.20	.61	.81
0240	With transformer (54)	.22	.64	.86
0280	4 per 1000 S.F., .5 watts per S.F.	.22	.71	.93
0320	With transformer	.25	.76	1.01
0360	5 per 1000 S.F., .6 watts per S.F.	.26	.84	1.10
0400	With transformer	.30	.90	1.20
0440	8 per 1000 S.F., .9 watts per S.F.	.27	.93	1.20
0480	With transformer	.34	1.02	1.36
0520	10 per 1000 S.F., 1.2 watts per S.F.	.29	1.02	1.31
0560	With transformer	.39	1.15	1.54
0600	16.5 per 1000 S.F., 2.0 watts per S.F.	.34	1.27	1.61
0640	With transformer	.52	1.50	2.02
0680	20 per 1000 S.F.,2.4 watts per S.F.	.35	1.38	1.73
0720	With transformer	.56	1.65	2.21

For expanded coverage of these items see *Means Electrical Cost Data 1988*

339

Figure 10.11

LIGHTING & POWER	B9.2-610	Central A.C. Power

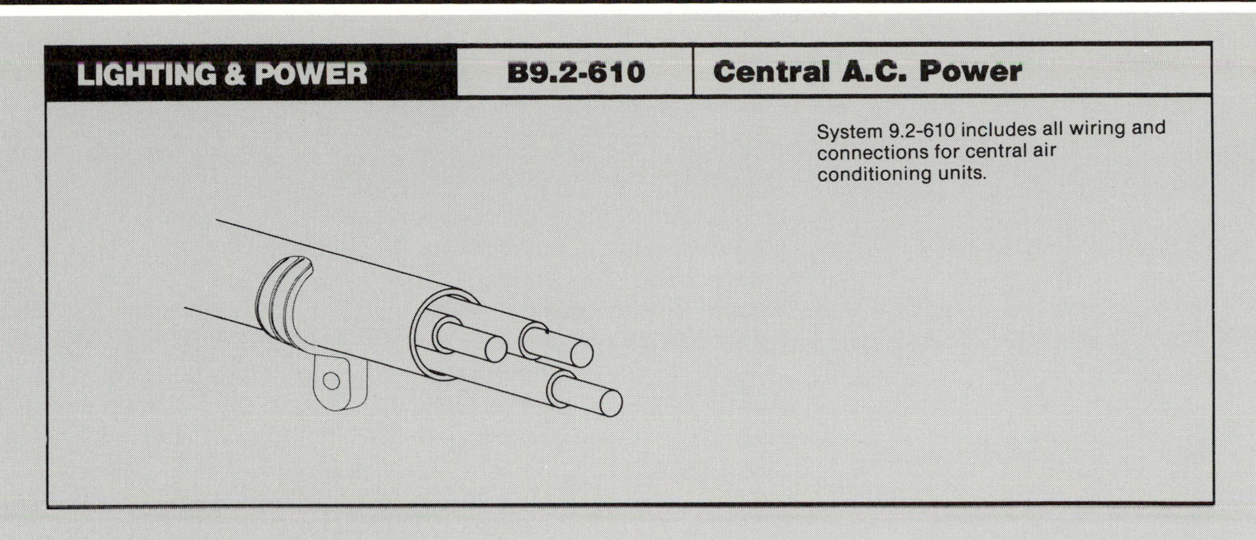

System 9.2-610 includes all wiring and connections for central air conditioning units.

System Components	QUANTITY	UNIT	COST PER S.F. MAT.	INST.	TOTAL
SYSTEM 09.2-610-0200					
CENTRAL AIR CONDITIONING POWER, 1 WATT					
Steel intermediate conduit, ½" diam.	.030	L.F.	.03	.08	.11
Wire 600V type THWN-THHN, copper solid #12	.001	C.L.F.	.01	.02	.03
TOTAL			.04	.10	.14

9.2-610	Central A. C. Power (by Wattage)	COST PER S.F. MAT.	INST.	TOTAL
0200	Central air conditioning power, 1 watt	.04	.10	.14
0220	2 watts	.05	.12	.17
0240	3 watts	.06	.13	.19
0280	4 watts	.08	.18	.26
0320	6 watts	.14	.26	.40
0360	8 watts	.17	.27	.44
0400	10 watts	.21	.31	.52

For expanded coverage of these items see *Means Electrical Cost Data 1988*

343

Figure 10.12

Of the components listed, those most easily estimated by the layman using the assemblies method are service system, lighting, and devices. The components of the service system, service and switchgear, are priced as *each*, based on the size of service. Lighting fixtures and devices, receptacles, and light switches, are priced by the number of fixtures or devices per 1000 square feet of floor area.

The more difficult components to estimate are equipment connections, basic materials, and special systems. To estimate equipment connections, the number of watts per square foot of floor area must be determined based on the building plans. If these figures are available, there should be no trouble using this method. If, however, that information must be gleaned from the panel schedules, there is a problem for the layman. Since assemblies costs are not available for every kind of special system, we will use the assemblies method for service, lighting, and devices, and square foot prices for equipment connections and special systems.

The Electrical Estimating Procedure

The basic steps involved in electrical estimating are as follows:

1. Estimate the cost of the electrical service and switchgear.
2. Determine the average density of light fixtures, receptacles, and light switches.
3. Price light fixtures, receptacles, and light switches.
4. Estimate the costs for equipment connections and special systems using the square foot prices from Figure 10.7.

The Chambers Estimate

The electrical drawings for the Chambers office building are shown in Figures 10.13 through 10.17. The quantities shown in the sample estimate represent realistic conditions. Assumptions have been made for items not shown that would normally be included in the plans and specifications for this type of project.

Figure 10.13, the Parking Garage Plan, shows the lighting fixtures and receptacles at the garage. Figure 10.14, the First Floor Plan, shows the lighting fixtures and receptacles for a typical floor. Figure 10.15, the Penthouse Floor & Roof Plan, shows the lighting fixtures and receptacles for the penthouse. Figure 10.16, the Electrical Riser Diagram, shows 1200 amp service and switchgear assemblies, and an equipment connection system. Figure 10.17, the Second & Third Floor Plan, shows an under-carpet power system for a typical floor and a poke-through telephone system requested by the owner as an alternate. A fire detection and alarm system is also shown throughout the building.

Figure 10.18 is the estimate. The cost of the 1200 amp electrical service is estimated using line 9.1-210-0480 from Figure 10.9. The cost of the switchgear is estimated in the same way using an appropriate assemblies cost from *Means Assemblies Cost Data* (1988 edition).

Next, the density of light fixtures and devices is calculated by dividing the number of fixtures by the floor area. Using the Parking Garage Plan in Figure 10.13 as an example, a count determines that there are 75 light fixtures in the parking area. The parking area is 210′ by 90′, therefore:

$$\frac{75 \text{ each}}{(210' \times 90')} = .0039 \text{ fixtures/S.F.}$$

Because fluorescent strip fixtures are by far the most common type in the garage and penthouse, the price for this type of fixture is used to estimate the cost of the fixtures in these areas.

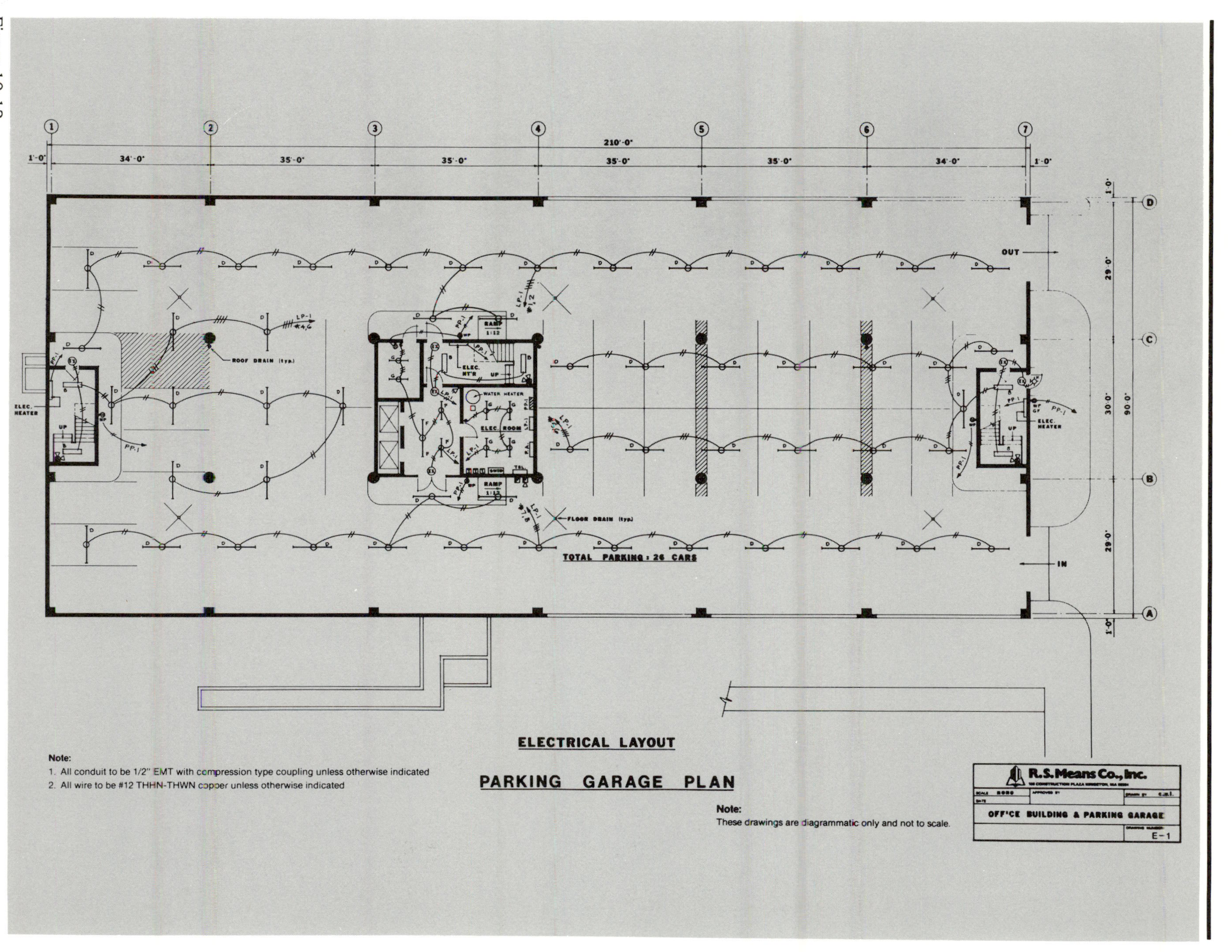

210'-0"
34'-0"
35'-0"
1'-0"
29'-0"
30'-0"
90'-0"
OUT
IN
ROOF DRAIN (typ.)
FLOOR DRAIN (typ.)
WATER HEATER
ELEC. ROOM
ELEC. HEATER
TOTAL PARKING : 26 CARS
ELECTRICAL LAYOUT
PARKING GARAGE PLAN
Note:
1. All conduit to be 1/2" EMT with compression type coupling unless otherwise indicated
2. All wire to be #12 THHN-THWN copper unless otherwise indicated
Note:
These drawings are diagrammatic only and not to scale.
R.S. Means Co., Inc.
OFFICE BUILDING & PARKING GARAGE
E-1

Figure 10.13

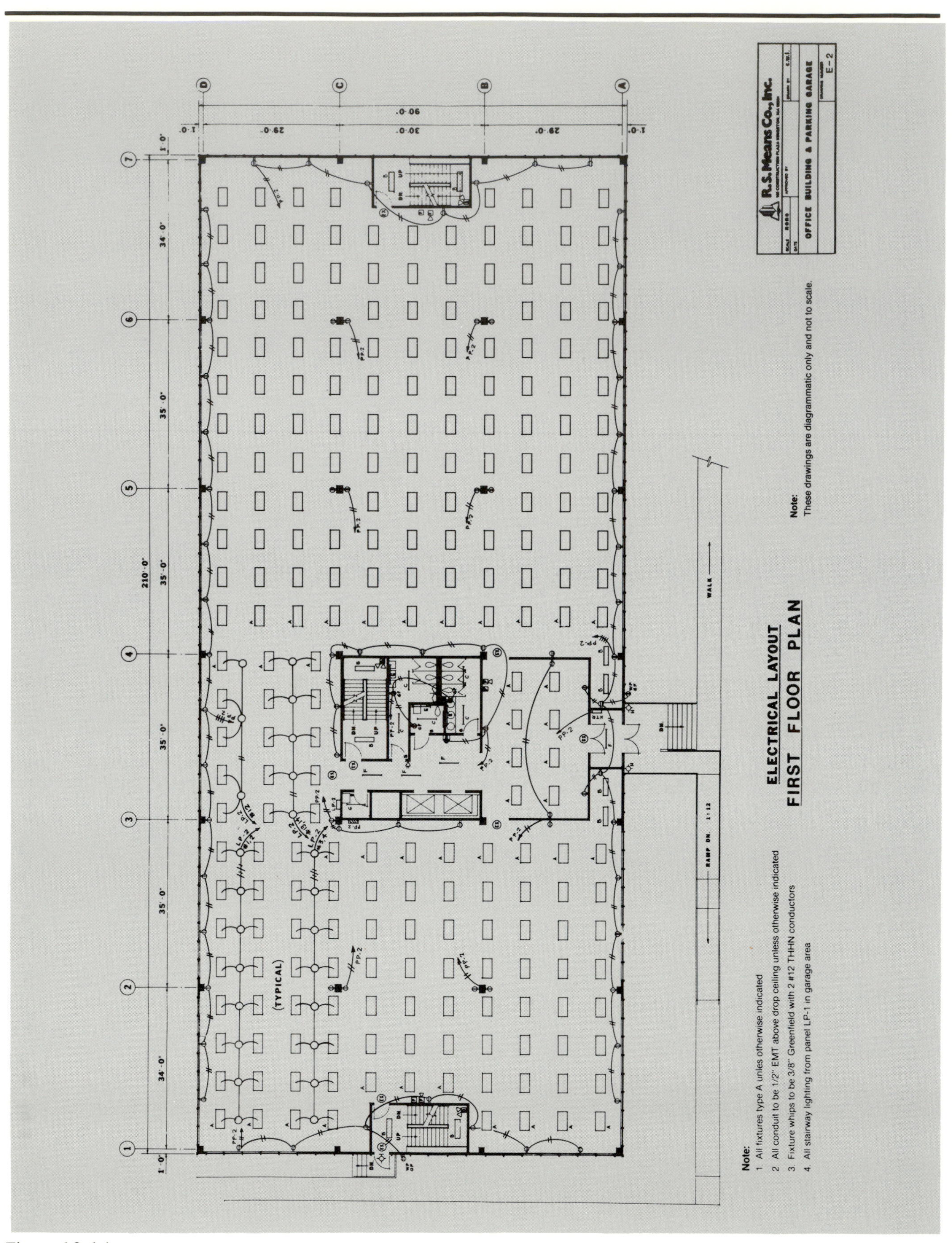
ELECTRICAL LAYOUT
FIRST FLOOR PLAN
Note:
1. All fixtures type A unles otherwise indicated
2. All conduit to be 1/2" EMT above drop ceiling unless otherwise indicated
3. Fixture whips to be 3/8" Greenfield with 2 #12 THHN conductors
4. All stairway lighting from panel LP-1 in garage area
Note:
These drawings are diagrammatic only and not to scale.
R.S. Means Co., Inc.
OFFICE BUILDING & PARKING GARAGE
E-2
210'-0"
34'-0"
35'-0"
90'-0"
29'-0"
30'-0"
1'-0"
(TYPICAL)
WALK
RAMP DN.
PP-2

Figure 10.14

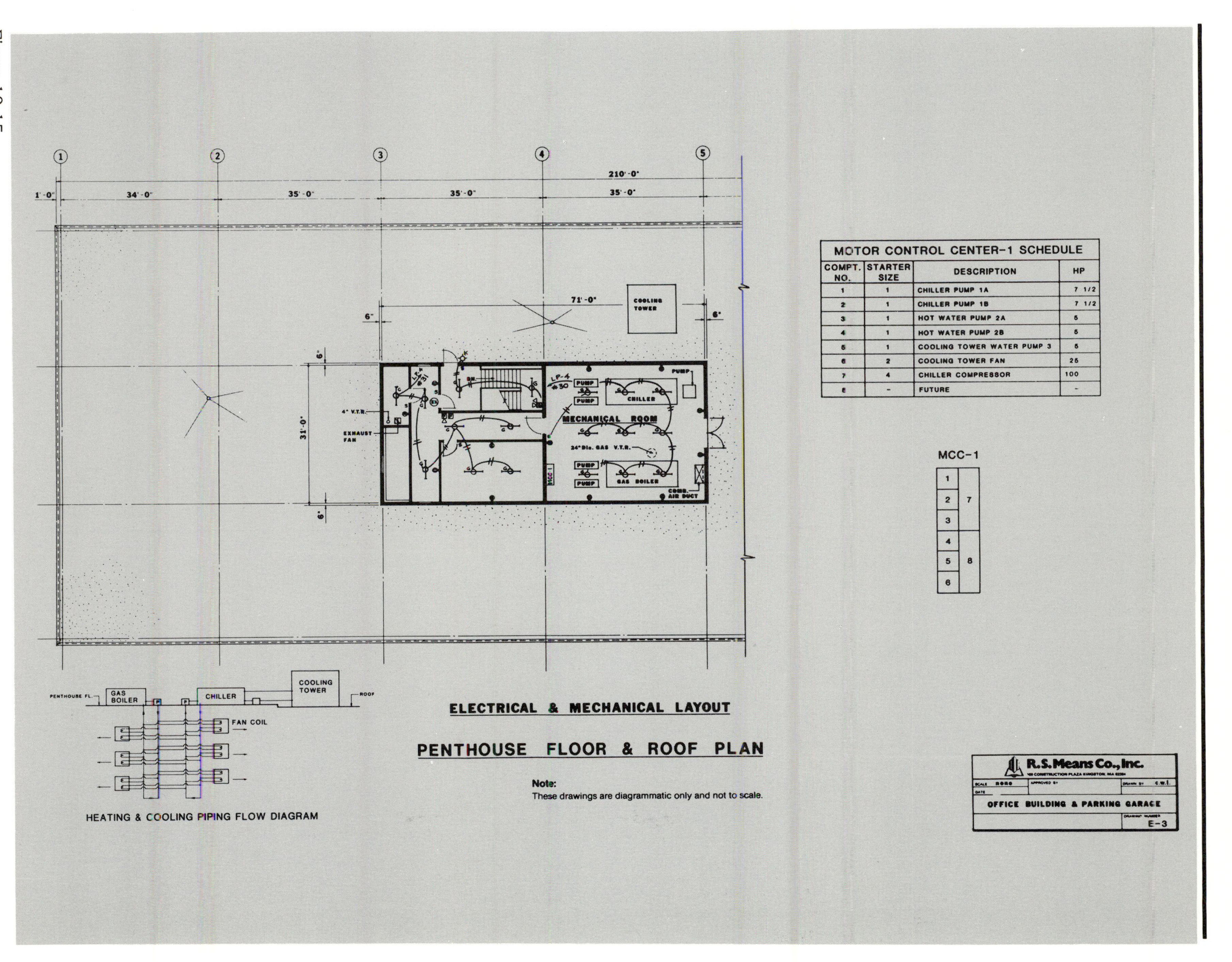
MOTOR CONTROL CENTER-1 SCHEDULE
COMPT. NO.
STARTER SIZE
DESCRIPTION
HP
1 1 CHILLER PUMP 1A 7 1/2
2 1 CHILLER PUMP 1B 7 1/2
3 1 HOT WATER PUMP 2A 5
4 1 HOT WATER PUMP 2B 5
5 1 COOLING TOWER WATER PUMP 3 5
6 2 COOLING TOWER FAN 25
7 4 CHILLER COMPRESSOR 100
8 - FUTURE -
MCC-1
1
2
3
4
5
6
7
8
210'-0"
1'-0"
34'-0"
35'-0"
35'-0"
35'-0"
71'-0"
31'-0"
6"
COOLING TOWER
MECHANICAL ROOM
PUMP
CHILLER
GAS BOILER
24" Dia. GAS V.T.R.
COMB. AIR DUCT
LP-4 #30
4" V.T.R.
EXHAUST FAN
PENTHOUSE FL.
ROOF
GAS BOILER
CHILLER
COOLING TOWER
FAN COIL
HEATING & COOLING PIPING FLOW DIAGRAM
ELECTRICAL & MECHANICAL LAYOUT
PENTHOUSE FLOOR & ROOF PLAN
Note:
These drawings are diagrammatic only and not to scale.
R.S. Means Co., Inc.
OFFICE BUILDING & PARKING GARAGE
E-3

Figure 10.15

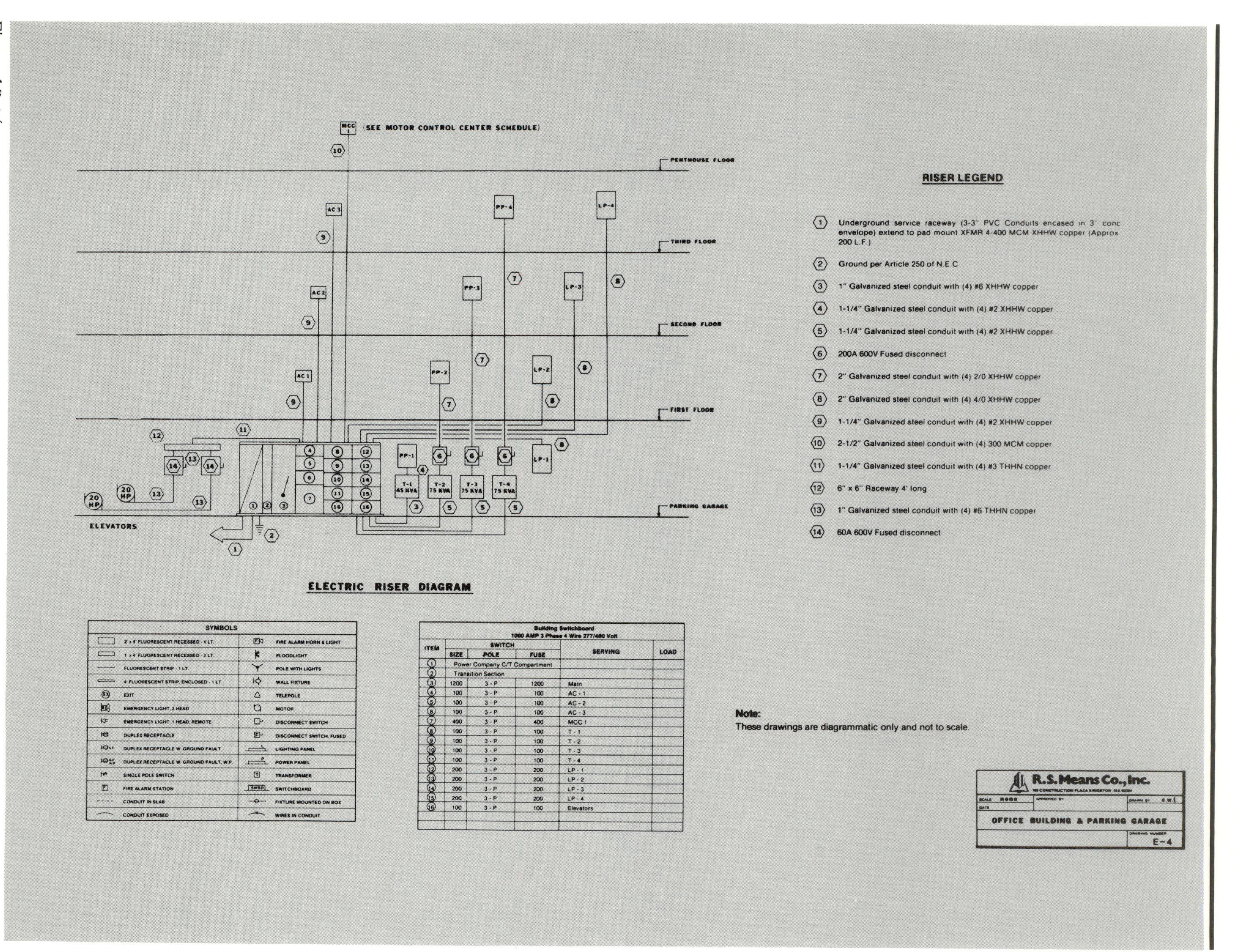

Building Switchboard
1000 AMP 3 Phase 4 Wire 277/480 Volt

ITEM	SWITCH SIZE	SWITCH POLE	FUSE	SERVING	LOAD
1	Power Company C/T Compartment				
2	Transition Section				
3	1200	3 - P	1200	Main	
4	100	3 - P	100	AC - 1	
5	100	3 - P	100	AC - 2	
6	100	3 - P	100	AC - 3	
7	400	3 - P	400	MCC 1	
8	100	3 - P	100	T - 1	
9	100	3 - P	100	T - 2	
10	100	3 - P	100	T - 3	
11	100	3 - P	100	T - 4	
12	200	3 - P	200	LP - 1	
13	200	3 - P	200	LP - 2	
14	200	3 - P	200	LP - 3	
15	200	3 - P	200	LP - 4	
16	100	3 - P	100	Elevators	

Figure 10.16

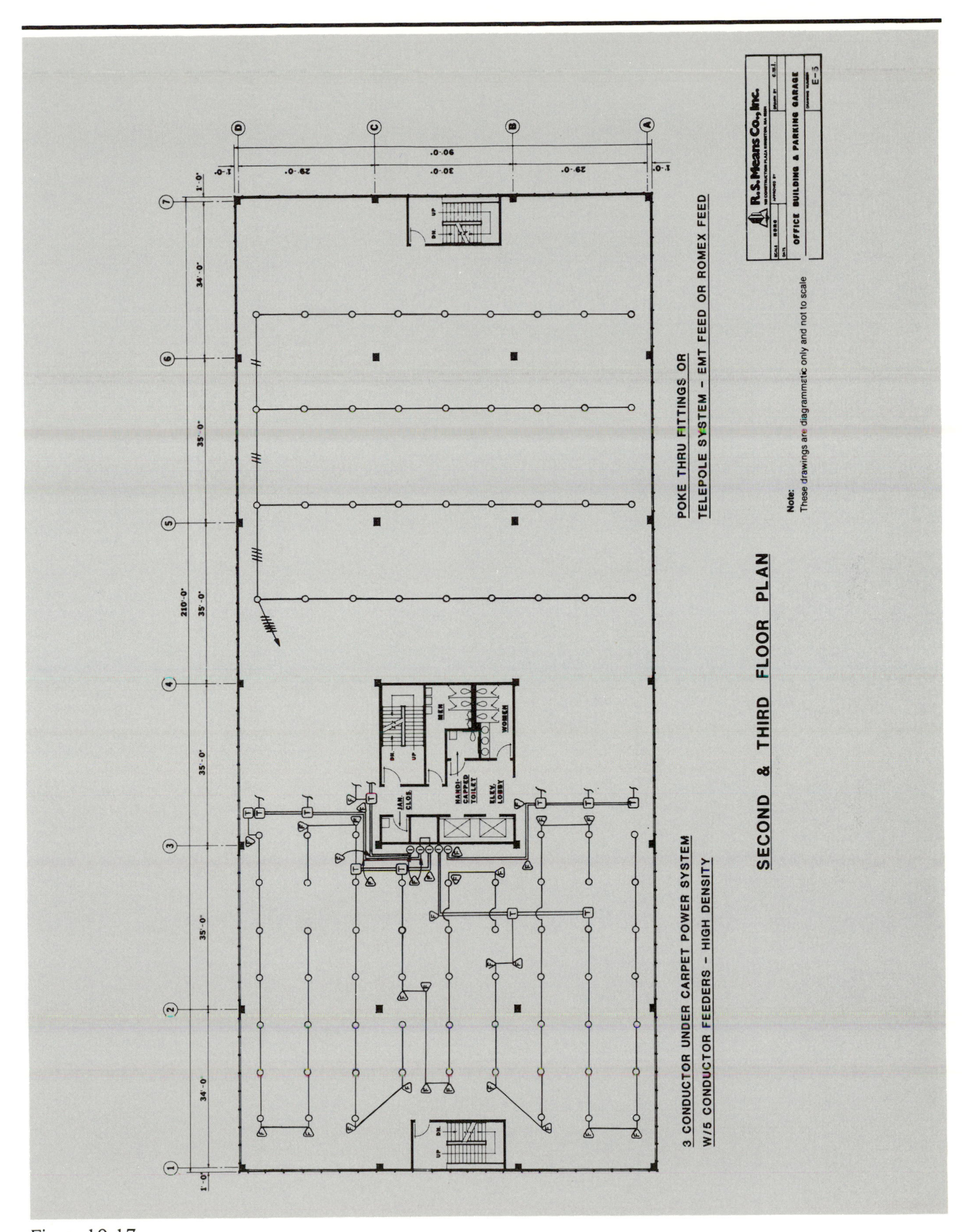
SECOND & THIRD FLOOR PLAN
3 CONDUCTOR UNDER CARPET POWER SYSTEM
W/5 CONDUCTOR FEEDERS - HIGH DENSITY
POKE THRU FITTINGS OR
TELEPOLE SYSTEM - EMT FEED OR ROMEX FEED
Note:
These drawings are diagrammatic only and not to scale
R.S. Means Co., Inc.
OFFICE BUILDING & PARKING GARAGE
E-5
210'-0"
90'-0"
JAN. CLOS.
HANDI-CAPPED TOILET
ELEV. LOBBY
MEN
WOMEN

Figure 10.17

ASSEMBLY NUMBER	DESCRIPTION	QTY	UNIT	TOTAL COST UNIT	TOTAL COST TOTAL	COST PER S.F.
8.0	**Mechanical System**					
9.0	**Electrical**					
9.1-210-0480	1200 Amp Service	1	ea	8725	8725	
9.1-410-0320	✓ Switchgear	✓	ea	14,850	14850	
9.2-212-0840+	Strip Fluor. @ bsmnt & PH	21,101	sf	1.04	21945	
9.2-213-0280	Recessed @ tenant	56,700	sf	3.85	218295	
9.2-522-0200+	Wall recept - @ bsmnt	18,900	sf	.08	1512	
9.2-522-0360	✓ ✓ - @ tenant & PH	58,901	sf	1.10	64791	
9.2-524-0880	U/Carpet - hi density - @ tenant	56,700	sf	1.67	94689	
9.2-542-0200	Switches - assume 1/1000 sf	77,801	sf	.13	10114	
C9.0-110	Equip. Conn.	68,351	sf	.51	34859	
✓	Fire Alarm & Detec.	✓	sf	.27	18455	
	Total				488235	

Page 4 of 6

Figure 10.18

Using the data for Type A fixtures, strip fluorescent fixtures, from Figure 10.10, the line with the density providing the closest match to the garage density is determined by dividing the number of fixtures by the square feet given for each line number. For example, the price in line 9.2-212-0840 is for a density of 64 fixtures per 5000 square feet. The number of fixtures per square foot is 64/5000, or .0128 fixtures per square foot. This density is much higher than is required for the Chambers garage. The price must therefore be adjusted for the Chambers estimate. The adjustment is performed as follows:

$$\frac{\$3.41/\text{S.F.} \times .0039}{.0128} = \$1.04/\text{S.F.}$$

This price is multiplied by the floor area to determine fixture costs.

The cost of the fixtures for the first, second, and third floors, and for the wall and under-carpet receptacles is estimated in the same way. Since no light switches are shown, a density of one switch per 1000 square feet is assumed. Once the prices for electrical service and switchgear, lighting, and devices is estimated, square foot costs from Figures 10.7a and b are used to estimate the cost of equipment connections and the fire alarm and detection system.

As with the plumbing and HVAC estimates, the total electrical cost can be transferred to the systems summary sheet to be included in the total project estimate, or it can be used as a check estimate by making adjustments for sales tax, bid conditions, location, and inflation.

Summary

The special qualities of electrical systems make estimating for this work somewhat different from plumbing and HVAC estimating. Using the square foot method for some of the electrical system components (equipment connections and special systems) helps to keep the procedure within the capabilities of the layman. The assemblies method can be used to estimate the costs for electrical service and switchgear, lighting, receptacles, and switches. Although some accuracy is lost, the method is still accurate enough for the purpose of cost control.

11

IMPROVING ESTIMATE ACCURACY

11

IMPROVING ESTIMATE ACCURACY

The design cost control system presented in the preceding chapters consists of a series of estimates beginning in the Need Identification phase and continuing through project construction. First, the square foot estimating method is used to provide Order of Magnitude estimates during the Need Identification phase. A combination of the assemblies and square foot methods is then used to provide a modeling estimate during the Need Analysis phase, and again for cost control estimates during the Conceptual Design and Early Design Development phases. The assemblies method is used to provide both the final cost control estimate at the end of the Design Development phase and change order estimates during construction.

The cost data sources used for these various estimates may include historical cost data from past projects, or proprietary cost data bases. Trend analysis is used to forecast inflation rates, and a leading indicator "rule of thumb" is used to forecast bid conditions.

The author has successfully used this estimating system in conjunction with Means cost data to control the costs of projects ranging from a $30,000 seed storage building to a $15,000,000 university classroom/laboratory building. The system proved particularly accurate for new construction projects costing in excess of $2,500,000. The average of the final cost control estimates for these projects fell within a range of ±5 % of the low bid. The final cost control estimates of smaller projects averaged around ±10% of the low bid.

The system does have certain limitations and requires some effort in developing and maintaining accuracy. This chapter addresses some of those limitations and demonstrates a method for improving accuracy.

System Limitations

The degree of accuracy that can be attained with the system depends on *project size* and the *type of construction*. Assemblies estimates, made in "broad strokes", provide sufficient detail for large projects. The effects of relatively minor errors in the quantity survey or pricing tend to cancel each other out in these cases. An example can be seen in the Project Report for the Chambers office building at the end of this chapter. The quantities of materials and systems in smaller projects are insufficient for this "balancing" to occur, a limitation that must be recognized.

The type of construction also affects estimate accuracy, depending on the applicability of the cost data and the experience of the estimator. The Means data is based on nationwide averages of the costs of standard building assemblies and the reported costs of the most common building types. This data is sufficient to provide accurate estimates for most projects. For certain applications, however, adjustments must be made or expert help obtained. For example, the square foot cost tables in *Means Assemblies Cost Data* provide costs for college science and engineering laboratories. These prices are reasonably accurate for most laboratory projects. However, chemistry laboratory HVAC and plumbing costs are likely to be much higher than those listed for general science and engineering laboratories. To obtain accurate Need Identification, Need Analysis, and Early Cost Control estimates, the Means square foot prices must be adjusted, or other data sources used. In these cases, historical square foot HVAC and plumbing costs are an appropriate information source. When an unusual building type (such as a sports arena with an air supported roof) is completely outside the scope of the cost data used, or outside the experience of the estimator, then it is appropriate to consult a specialist in that type of construction.

Experience has shown two types of projects that seem to present a particular challenge to the estimator. These are projects using pre-engineered metal building components and those involving a significant amount of demolition. The pre-engineered metal building projects are difficult because of the volatility of that segment of the construction market. The price of the components can change dramatically from week to week depending on the number of projects out for bid and the quantity of material that the manufacturer has in inventory. The best approach seems to be to obtain costs from local pre-engineered building contractors, while keeping in mind the fact that subsequent changes in the market could have a significant effect on the bid price. Projects with extensive demolition can be challenging since it is often very difficult to find an accurate source for demolition costs. The prices given in proprietary data bases may bear little relation to those quoted by salvage/demolition contractors. The estimator may find the use of historical square foot cost data slightly more accurate. Demolition involving the removal of hazardous materials is an entirely different issue. This very specialized field cannot be estimated by a non-expert.

Developing and Maintaining Accuracy

While our cost control system does have some accuracy limitations, a little effort can develop it into a reliable cost control tool. Once the estimating and forecasting procedures are understood, accuracy is developed by practicing the system on completed projects. The practice estimates are compared with actual project costs to suggest modifications and refinements to the individual's application of the system. Accuracy is improved and maintained by constantly checking estimate assumptions, cost data, forecasts, and estimate quantities and prices against actual project conditions.

The Project Report

The basic tool in the comparison/refinement process is the project report. This document provides a brief summary of the assumptions, estimates, forecasts, actual design, and actual costs of a construction project. It details differences between estimates and actual project conditions and the reasons for these differences, and suggests procedural modifications needed to eliminate these discrepancies. The project report also becomes the basis for a historical cost data base. It includes:

1. The date of the report.
2. The project name, location, and owner's name.
3. The names of the designers.
4. The assumptions, cost range, cost data source, date, inflation forecast, and bid conditions forecast for the Need Identification estimate.
5. The assumptions, amount, cost data source, date, inflation forecast, and bid conditions forecast for the Need Analysis estimate.
6. A brief description of the assemblies actually used in the project.
7. The cost data source, amount, date, inflation forecast, and bid conditions forecast for the final Cost Control estimate.
8. The project bid date, the actual number of bidders, and the actual inflation that occurred during the planning/design process.
9. A Cost Analysis table showing the actual cost of the different components of the project, and the square foot estimated costs of those components. Final check estimate costs for the components are also listed, along with the percent difference between the check estimate costs and actual costs.
10. An analysis of the differences between the *estimated* amounts, assumptions, and forecasts and *actual* project conditions. This analysis should identify the reasons for the differences and suggest refinements to eliminate them.

The date, project name, project location, owner's name, and designers' names serve as basic information to identify the project. The estimate assumptions and the design description provide a basis for determining estimate accuracy. A comparison of the estimate assumptions with each other and with the design description shows developmental changes and components that were omitted. The record of developmental changes and omitted components is used in reconciling differences between estimated and actual project costs, and suggests items to consider in future planning estimates. The inflation forecast versus actual inflation, and the bid conditions forecast versus the actual number of bidders are recorded as a basis for checking forecast accuracy and reconciling the differences between the estimates and actual project costs. The Cost Analysis table is used to provide a detailed check of the accuracy of the final cost control estimate, and to provide detailed historical cost data for use in future estimates. The source of the cost data is included so that source accuracy and reliability can be evaluated over a period of time.

The dates of the estimates provide a measure of the duration of the planning/design process for use in future planning. The bid date provides the effective date of the historical cost data, and is used to determine the amount to add for inflation when using the historical data in future estimates. Historical data should be inflated no more than five years beyond the effective dates from the records maintained for each estimate.

Most of the information required for the report is taken from the records maintained for each estimate. The information on actual project conditions must, however, be obtained elsewhere. The design description information is obtained from the project plans and specifications. The number of bidders is the number of companies actually submitting bids for the project. This number is given in the bid analysis and award recommendation that the designer provides as part of his service. The actual inflation data for the project location comes from proprietary construction cost indexes, such as *Means Construction Cost Index*. The actual project cost information is derived from the contractor's schedule of values, submitted with each request for payment. It is acknowledged that the schedule of values may not be precisely accurate regarding the costs of the building systems. For example, it is not unusual for the contractor to "front end load" the schedule of values. This tactic involves "padding" the costs of the systems completed in the early phases of construction, and reducing the costs of systems completed later. The effect is an increase in the amount of the early payments. Component costs may also be adjusted to "hide" the actual costs of proprietary systems from competitors. Although these practices prohibit the accurate checking of unit price estimates, the schedule of values cost information is usually accurate enough to allow a rough check of a "broad stroke" assemblies estimate.

The Chambers Office Building Project Report

An example project report for the Chambers office building project is shown in Figures 11.1 through 11.3. First is the project identification information. This includes the date of the report, and the names and addresses of the project, the owner, and the designer. Next, the date, assumptions, cost range, cost data source, and forecasts for the Need Identification estimate are provided. This information is covered in Chapter 4. The next segment involves the date, assumptions, amount, cost data source, and forecasts for the Need Analysis estimate. This information is obtained from the demonstration in Chapter 6.

The next section of the project report begins with the date of the final Cost Control estimate, the design description, the bid date, number of bidders, and actual inflation values. The final cost control estimate is shown in Figures 11.4 through 11.8. (The bid date, number of bidders, and inflation values are assumed for the purpose of illustration.) The Cost Analysis table shows the actual project cost, the actual square foot costs, the estimated costs, and the percent difference between the actual and estimated costs for the major building assemblies. The actual cost data is obtained from the schedule of values shown in Figure 11.9, and the estimated costs are obtained from the final Cost Control estimate in Figures 11.4 through 11.8. The square foot costs are provided for use as historical cost data in future estimates. The *Percent Difference* column is included to show the accuracy of the estimated costs for each assembly.

Finally, the Estimate Analysis is given. The reasons for the differences in estimated and actual cost amounts are listed, together with suggestions for correcting the differences in future estimates.

Preparing the Cost Analysis table is, perhaps, the most tedious part of the project report process. The first step is to disperse the costs for the adjustment items (sales tax, overhead, profit, bond, competition, location factor, and inflation) into the other cost entries. This is accomplished by multiplying the values of the cost entries by a proration factor.

The proration factor increases the value of each entry by the percentage of the total cost that is attributable to the adjustment items. For the final Cost Control estimate, this factor is obtained by dividing the *Total Cost* by the *Building Subtotal* from Figure 11.8, as follows.

Total Cost/Building Subtotal = Proration Factor
$4,607,197/3,496,171 = 1.3178

The cost entries are then multiplied by the proration factor to complete the distribution. The results of the process are shown in Figure 11.10.

Most of the adjustment costs have already been distributed in the contractor's schedule of values. The one item remaining to be distributed is "Start-up and Move-on". This entry includes the costs of project office, storage facilities, temporary roads and utilities, and other physical plant facilities that will be needed to support project construction. "Start-up and Move-on" also includes the costs for permits, fees, insurance, and bonds. The proration factor to distribute this cost is obtained by dividing the *Total Cost* by the *Total Cost less "Start-up and Move-on"*. The factors for calculation and cost dispersal are shown in Figure 11.11.

Next, cost entries must be combined into categories that reflect the costs of the major building assemblies. A typical list of the major assemblies for a commercial project is shown in the Cost Analysis table. The cost entries that were combined to obtain the values shown in the Cost Analysis table are detailed in Figures 11.12 through 11.14.

Finally, the square foot costs and the difference in percentages are calculated. The square foot costs are calculated by dividing the *actual cost* for each assembly by the *gross building size* obtained from the design description. The percentages are calculated by dividing the *difference between the estimated cost* and the *actual cost* by the *actual cost*. The values are shown in the Cost Analysis table, Figure 11.2.

Summary

Although the Cost Control system outlined in the preceding chapters can serve as an effective control tool, there are limitations to its accuracy. Some of the limiting factors include: the size of the project, the applicability of the cost data, the estimator's experience, the use of pre-engineered metal building components, and a requirement for a significant amount of demolition. Accuracy can be improved, however, by analyzing the differences between estimate assumptions, forecasts, and amounts in the estimate, and the actual project conditions. This analysis may lead to procedural changes that improve accuracy. The tool for making and recording this analysis is the Project Report.

Project Report
10-1-89

Project: Chambers Office Building
200 Chambers Blvd.
Chambers, GA

Owner: Chambers Development Company
488 Chambers Blvd., Suite 240
Chambers, GA

Designers: Chambers Design, Architects & Engineers
488 Chambers Blvd., Suite 120
Chambers, GA

Need identification: 4-1-88
Unfinished shell containing 56,700 G.S.F. & site development;
Total Project Cost Range: $3,360,000 to $2,483,000; '88 *Means Assemblies Cost Data*;
Inflation from 1-88 to assumed bid date 6-89 @ 4.0%/yr;
Bid Conditions: low competition, adj. @ 8%.

Need Analysis: 6-1-88
Design Assumptions:
Structure: 66,150 G.S.F.; conc. fill/mtl. deck/joists/beams @ 2, 3, roof; multispan joist/conc. wall/CIP cols./strip & spread ftgs. @ 1st, bsmnt.; slab on grade @ bsmnt.
Exterior Closure: brick/drywall/alum. windows & doors.
Roofing: Single ply on 3" polystyrene.
Interior Construction: CMU walls @ bsmnt.; drywall @ 1, 2, 3; hollow metal doors; vinyl wall covering @ lobbies; ceramic tile @ rest room, paint elsewhere; quarry tile @ lobbies, ceramic tile @ rest room; acoustical ceiling @ tenant, and lobbies, drywall @ rest room.
Elevators: 2 @ 3000 lbs.
Site: water, sanitary, and storm, HID lighting, asphalt parking for 236 cars.
Estimate: '88 *Means Assemblies Cost Data & Building Construction Cost Data*, Construction Contract: $3,590,827.
Total Project Cost: $3,906,000;
Inflation: 1-88 to 6-89 @ 4%/yr;
Bid Conditions: low competition, adj. @ 8%.

Figure 11.1

Project Report
10-1-89 (continued)

Final Cost Control: 5-1-89

Design: Structure: 68,351 G.S.F.; deck/joists @ penthouse roof; deck/joists/beams/cols. @ main roof; CIP composite slab/beams/cols. @ 2, 3, penthouse; multispan joist slab/CIP walls & cols./strip & spread ftgs. @ 1, bsmnt.; slab on grade @ bsmnt.

Exterior Closure: Alum. frame curtain wall w/tinted, 1/2" insulating glass, insulated spandrel w/drywall backup, alum. entries, hollow metal doors @ stairs, rolling Grilles @ bsmnt. entry.

Interior Construction: CMU @ bsmnt., Penthouse; drywall 1, 2, 3; ceramic tile @ restroom, painted elsewhere; quarry tile @ lobbies, ceramic tile @ rest room; acoustical tile @ tenant & lobbies, drywall @ rest room.

Elevators: 2@ 2500 #.

Plumbing: rest room @ 1, 2, 3; roof drains; perimeter drain; stand pipe w/FHC; Fire Cycle sprinkler system.

HVAC: gas-fired boiler heat; chilled water/cooling tower/fan coil w/ductwork cooling.

Electrical: 1200 amp. service, 100 L.F. to power source, fluor. recessed lighting @ lobbies & tenant, strip @ bsmnt. & penthouse, under-carpet receptacles @ tenant.

Site: Water, sanitary, storm, 1200 amp. feeder, HID lighting; asphalt parking for 205 cars, conc. sidewalk, steps, ramp, retaining wall.

Estimate: '88 *Means Assemblies Cost Data* & *Building Construction Cost Data*;

Construction: $4,607,198;

Inflation: 1-88 to 6-89 @ 4%/yr;

Bid Conditions: low competition, adj. @ 8%.

Bid: 6-1-89, 8 bidders

Inflation: 1988 = 3.9%, 1-89 to 6-89 = 1.5%.

Figure 11.1 (*continued*)

Project Report—Cost Analysis Table

Item:	Actual Cost	Actual Cost/S.F.	Final Check Est.	% Difference
Excavation	$ 44,618	$ .65	$ 45,297	+ 1.52
Paving & Sidewalks	171,997	2.52	138,571	− 19.4
Site Utilities	15,672	.23	40,717	+ 159.81
Structural	1,135,360	16.61	1,215,586	+ 7.07
Masonry	48,198	.71	49,421	+ 2.54
Carpentry	7,728	.11	omitted	− 100.00
Waterproofing	7,976	.11	9,757	+ 22.33
Roofing & Insul.	77,673	1.14	69,419	− 10.63
Doors & Frames	46,762	.68	48,064	+ 2.78
Curtain Wall, Ent.	1,109,749	16.23	803,287	− 27.62
Finishes	223,368	3.27	229,803	+ 2.88
Specialties	15,346	.22	16,635	+ 8.40
Elevator	185,309	2.71	186,598	+ 0.70
Plumbing	125,229	1.83	145.849	+ 16.47
Fire Protection	116,099	1.70	146,361	+ 26.07
HVAC	736,793	10.78	777,527	+ 5.53
Electrical	649,562	9.50	684,305	+ 5.35
	$4,717,439	$69.02	$4,607,197	− 2.34

Figure 11.2

Project Report—Estimate Analysis

Inflation Forecast: Actual inflation = 5.5%.
Forecast inflation = 6.1%.

Bid Conditions Forecast: 8 bidders = moderate competition @ 4%; forecast low competition @ 8%. Business conditions weakened 1st and 2nd Quarters, 1988. Market responds more quickly to decrease than increase?

Need Identification Estimate: Low because of need for parking basement and penthouse, no interior finishes.

Need Analysis Estimate: Low because of need for penthouse mechanical space, curtain wall in lieu of brick veneer potentially more attractive to tenants, estimate left out column fireproofing.

Final Cost Control: Significant differences in cost for site utilities (error in use of trenching costs), omitted costs for carpentry, waterproofing (?), curtain wall (failed to add for anodized finish for frame), plumbing (pipe and fittings and quality/complexity factors too high), and fire protection (used fire cycle in lieu of wet pipe system cost).

Figure 11.3

ASSEMBLY NUMBER	DESCRIPTION	QTY	UNIT	TOTAL COST UNIT	TOTAL COST TOTAL	COST PER S.F.
1.0	**Foundations**					
1.1-120-7750+	Ftg F1 11^{5}□ x 2^{0}	6	ea	1442	8652	
1.1-120-7750+	〃 F2 9^{5}□ x 2^{0}	4	ea	984	3936	
1.1-120-7810+	〃 F3 8^{0}□ x 2^{0}	18	ea	766	13788	
1.1-140-2700	〃 AA 2^{0} x 1^{0}	580^{42}	LF	25^{55}	14830	
1.1-140-2100+	〃 BB 1^{5} x 0^{67}	215^{0}	LF	15^{41}	3313	
1.1-210-3000	Elev Pit Walls 6' x 6"	41^{0}	LF	44^{30}	1816	
1.1-210-7220	Bsmnt Walls 8" x 12'	300	LF	100^{0}	30000	
1.1-210-5020	〃 〃 8" x 8'	186	LF	65^{0}	12090	
1.1-292-6460	Waterproofing perimeter 8' avg ht.	600	LF	12^{34}	7404	
1.9-100-4620+	Bldg Exc & BF 6' avg	18,900	SF	1^{19}	22491	
1.1-294-1000	4" PVC Perimeter Drain	600	LF	4^{0}	2400	
	Total				120720	
2.0	**Slab on Grade**					
2.1-200-4520	Elev. Pit Slab - 6" reinf.	137^{29}	SF	3^{77}	518	
2.1-200-2280	Bsmnt Slab - 4" reinf.	18,228	SF	3^{06}	55778	
	Total				56296	
3.0	**Superstructure**					
3.1-112-1490	Int. Cols 28" ⌀	130	LF	63^{0}	8190	
3.1-114-4000+	Pilasters @ ext. 24" x 12"	234	LF	43^{0}	10062	
3.5-160-7000	CIP Multispan Joist Slab @ 1st	18,900	SF	8^{54}	161406	
3.5-520-6000	Composit Bm & CIP Slab @ M & PH	37,800	SF	11^{15}	421470	
3.1-130-4800	Steel Cols	1,176	LF	29^{78}	35021	
3.5-520-6250	Composit Bm & CIP Slab @ M & PH	2,201	SF	13^{50}	29714	
3.7-420-4700	SH Joists & Bms on Cols @ Roof	16,699	SF	3^{17}	52936	
3.7-430-1700	SH Joists on Walls @ PH	2,201	SF	1^{92}	4226	
3.9-100-0780	Stairs - Conc in pans	10	ea	5470	54700	
	Total				777725	
4.0	**Exterior Closure**					
4.1-211-3300	6" Lt. Wt. CMU w/ perlite	2,244	SF	4^{77}	10704	
4.6-100-7150	Entry	2	opng	3825	7650	
4.6-100-4200	Dbl Dr @ Bsmnt & PH	2	opng	1940	3880	
4.6-100-3950	Sgl Dr 〃 〃	7	opng	975	6825	
4.7-582-1850	Curtain Wall Frame	25,800	SF	10^{75}	277350	
4.7-584-1100	1/2" insul tinted glass	9,840	SF	13^{10}	128904	
4.7-584-3100	1/4" Pl w/ 2" insul spandrel	15,960	SF	12^{26}	195670	
4.6-100-6150	O.H. Rolling Grilles - motor op.	2	opng	4325	8650	
					639633	

Page 2 of 6

Final Cost Estimate Sheet

Figure 11.4

ASSEMBLY NUMBER	DESCRIPTION	QTY	UNIT	TOTAL COST UNIT	TOTAL COST TOTAL	COST PER S.F.
5.0	**Roofing**					
5.1-220-2100	EPDM – .045, Balasted	18,900	sf	.86	16254	
5.1-520-1200	Alum. Roof edge	804	LF	13.50	10854	
5.1-510-1100	Base Flashing	204	LF	15.10	3080	
5.7-101-2350	3" Polystyrene Insul.	18,900	sf	1.19	22491	
	Total				52679	
6.0	**Interior Construction**					
6.1-210-6000	8" CMU @ Bsmnt	3,406	sf	4.92	16758	
6.1-210-5500	6" ✓ @ PH	2,282	sf	4.40	10041	
6.1-510-5400	Drywall @ Core 1 thru 3	15,162	sf	2.54	38511	
6.1-510-5300	✓ @ ext walls	13,620	sf	1.57	21383	
6.1-870-0420	T. Partition	15	ea	465	6975	
6.1-870-1300	U. Screens	3	ea	216	648	
6.4-260-6000	1 hr flush birch	27	ea	384	10368	
	Hdwr - allow	✓	ea	250	6750	
	Specialties - allow	1	LS	5000	5000	
6.5-100-0080	Paint DW	14,391	sf	.56	8059	
6.5-100-0320/40	Paint & Fill CMU	8,532	sf	1.01	8617	
6.6-100-1720	CT Floors	768	sf	4.30	3302	
6.6-100-1820	QT Floors	1625	sf	9.40	15275	
6.7-100-5800	Acou. Ceiling	55,803	sf	1.42	79240	
	Total				230927	
7.0	**Conveying**					
7.1-100-2000t	Elev 2500#, 100 FPM, 4 floors	2	ea	70,800	141600	

Page 3 of 6

Final Cost Estimate Sheet

Figure 11.5

ASSEMBLY NUMBER	DESCRIPTION	QTY	UNIT	TOTAL COST UNIT	TOTAL COST TOTAL	COST PER S.F.
8.0	**Mechanical System**					
8.1-160-1840	WH - 80 gal.	1	ea	2435	2435	
8.1-310-4120+	3" CI Roof Drain	12	ea	581	6972	
8.1-310-4160+	✓ ✓ ✓ addl floors	12	ea	456	5472	
8.1-433-2040	Lav. - Wall - PE on CI 18"x15"	12	ea	735	8820	
8.1-433-1840	✓ - @ Van. - 19" ϕ	9	ea	480	4320	
8.1-434-4340	Serv. Sink - Wall - PE on CI 24"x30"	3	ea	1095	3285	
8.1-450-2000	Urinals - wall	6	ea	755	4530	
8.1-460-1880	EWC - Wall - dual	3	ea	840	2520	
8.1-510-3000	WC - 1st pr	3	ea	1315	3945	
8.1-510-3100	✓ - addl pr	3	ea	1290	3870	
8.1-510-1800	✓ - addl sgl	6	ea	795	4770	
8.2-310-1080	Std. Pipe - 2 1/2" - 1st Flr	1	ea	1420	1420	
8.2-310-1100	✓ ✓ ✓ - addl flr	4	ea	470	1900	
8.2-390-0800	FHC	6	ea	200	1200	
151-125-0440	Auto Deck Drain	6	ea	240	1440	
	Subtotal "A"				56899	
	Water Control @ 15%				8535	
	Pipe & Fittings @ 50%				28450	
8.2-150-0620	Fire Cycle Sprinkler - 1st flr	18,900	sf	1.84	34776	
8.2-150-0740	✓ ✓ ✓ - addl flr	58,901	sf	1.05	61846	
8.3-141-1480	Boiler Assembly - CI, gas	56,700	sf	2.11	119637	
8.4-120-4040	Chilled Water, Cooling Tower	✓	sf	6.76	383292	
8.5-410-0060	Boiler Vent - 6" galv.	15	LF	12.67	190	
~~9.0~~ 8.5-410-1140	~~Electrical~~ ✓ ✓ 90° elbow	1	ea	34	34	
8.5-410-2440	✓ ✓ Roof flash'g	1	ea	26	26	
157-250-0560	Exhaust Duct	2000	lbs	3.53	7060	
157-290-7140	Roof Exhaust Fan	1	ea	960	960	
157-460-1020	Registers	23	ea	20	460	
155-408-0100	Duct Heater	1	ea	300	300	
155-420-1020	Stair Heaters	3	ea	445	1335	
	Subtotal "B"				703800	
	O/C @ 15%				105570	
	TOTAL				809370	
9.0	ELECTRICAL					
9.1-210-0480	1200 Amp Service	1	ea	8725	8725	
9.1-410-0320	✓ Switchgear	✓	ea	14,850	14850	
9.2-212-0840+	Strip Fluor. @ bsmnt & PH	21,101	sf	1.04	21945	
9.2-213-0280	Recessed @ Tenant	56,700	sf	3.85	218295	
9.2-522-0200+	Wall recept. - @ bsmnt	18,900	sf	.08	1512	
9.2-522-0360	✓ ✓ - @ tenant & PH	58,901	sf	1.10	64791	
9.2-524-0320	U/Carpet - hi density - @ tenant	56,700	sf	1.67	94689	
9.2-542-0200	Switches - assume 1/1000 sf	77,801	sf	.13	10114	
C 9.0-110	Equip - Conn.	68,351	sf	.51	34859	
✓	Fire Alarm & Detec.	✓	sf	.27	18455	
					488235	
	Page Total					

Final Cost Estimate Sheet

Figure 11.6

ASSEMBLY NUMBER	DESCRIPTION	QTY	UNIT	TOTAL COST UNIT	TOTAL COST TOTAL	COST PER S.F.
10.0	**General Conditions**					
	See Summary					
11.0	**Special Construction**					
12.0	**Site Work**					
12.5-510-1500	Parking	205	Cars	365	74825	
12.7-140-1600	4' SW	260	LF	7.26	1888	
12.7-610-2000	Curb & Gutter	3120	LF	6.40	19968	
12.3-710-5860	Catch Basins	3	ea	1405	4215	
12.3-510-4650	SS - 24" RCP	310	LF	24	7440	
12.3-110-3900	Trenching	570	LF	29.20	16644	
12.3-510-9060	Sanitary	100	LF	8.65	865	
12.3-540-2160	H_2O	100	LF	17.35	1735	
12.7-500-5820	Mtl Halide Lighting	10	ea	1825	18250	
	Site Exc. - Allow	11,883	sy	1.00	11883	
~~13.0~~	~~Miscellaneous~~					
9.1-310-0480	1200 Amp Feeder	100	LF	128	12800	
	Conc Steps - Allow	1	LS	3000	3000	
1.1-120-2100+	Ftg - BB @ ramp & steps	120	LF	15.41	1849	
1.1-210-1500	Ret'g Wall 4' x 6"	✓	LF	30.20	3624	
	Total				178986	

Page 5 of 6

Final Cost Estimate Sheet

Figure 11.7

PRELIMINARY
ESTIMATE (Cost Summary)

PROJECT Chambers Office Bldg — TOTAL AREA — SHEET NO.

LOCATION 103 Chambers Blvd. - Chambers, Ga. — TOTAL VOLUME — ESTIMATE NO.

ARCHITECT — COST PER S.F. — DATE

OWNER Chambers Investment Inc. — COST PER C.F. — NO. OF STORIES

QUANTITIES BY: RK — PRICES BY: RK — EXTENSIONS BY: RK — CHECKED BY:

NO.	DESCRIPTION	SUB TOTAL COST	COST/S.F.	%
1.0	**Foundation**	120720		
2.0	**Substructure**	56296		
3.0	**Superstructure**	777725		
4.0	**Exterior Closure**	639633		
5.0	**Roofing**	52679		
6.0	**Interior Construction**	230927		
7.0	**Conveying**	141600		
8.0	**Mechanical System**	809370		
9.0	**Electrical**	488235		
10.0	**General Conditions (Breakdown)**	—		
11.0	**Special Construction**	—		
12.0	**Site Work**	178986		
	Building Sub Total	$ 3,496,171		

→ $ 3,496,171

Sales Tax 4 % × Sub Total $ ______ /2 = $ 69,923

General Conditions (%) 16 % × Sub Total $ ______ = ______

General Conditions $ 559,387

Sub Total "A" $ 4,125,481

Overhead & Profit 10 % × Sub Total "A" $ ______ = $ 412,548

Sub Total "B" $ ______

~~Profit~~ Bid Conditions-low @ 8 % × Sub Total "B" $ ______ = $ 330,038

Sub Total "C" $ 4,868,067

Location Factor Atlanta @ 89.2 % × Sub Total "C" $ ______ =

Adjusted Building Cost $ 4,342,316

Architects Fee ______ % × Adjusted Building Cost $ ______ = $ ______

~~Contingency~~ Inflation 1-1-88 to 6-89 – 4 % × Adjusted Building Cost $ ______ = $ x 1.061

Total Cost 4,607,197

Square Foot Cost $ ______ / ______ S.F. = ______ $/S.F.

Cubic Foot Cost $ ______ / ______ C.F. = ______ $/C.F.

Page 6 of 6

Final Cost Estimate Sheet

Figure 11.8

Schedule of Values Chambers Office Building	
Item	Amount
Start-up & Move-on	$ 89,170
Excavation	43,775
Site Utilities	15,376
Asphalt Paving	130,712
Site Concrete	38,034
Concrete, Forms & Finish	536,840
Masonry	47,287
Structural Steel	368,952
Joists & Deck	112,547
Misc. Metals	59,191
Carpentry & Millwork	7,582
Waterproofing	7,825
Insulation	10,035
Roofing & Sheet Metal	66,170
Doors & Frames	13,386
Finish Hardware	15,600
Curtain Wall/Alum. Ent.	1,088,772
Overhead Doors & Grilles	16,892
Drywall	70,849
Fireproofing	36,370
Ceramic Tile	14,352
Acoustical Ceiling	103,822
Flooring	17,645
Painting	12,478
Specialties	15,056
Elevator	181,806
HVAC–Ductwork	231,919
–A/C Equipment	315,791
–Heating Equip.	40,704
–Pipe & Fittings	71,773
–Insulation	17,656
–Balancing	7,812
–Controls	37,211
Plumbing–Fixtures	44,491
–Pipe & Fittings	73,538
–Insulation	3,718
–Testing	1,115
Fire Protection	113,904
Elec.–Lighting	318,541
–Serv. & Dist.	157,095
–Wiring	107,702
–Motors	14,499
–Fire Alarm	39,447
TOTAL	**$4,717,439**

Figure 11.9

Final Cost Control Estimate
Proration Adjustment—Chambers Office Building

Item	Amount		Proration Factor	Adjusted Amount
Ftg. F1	$8,652	×	1.3178	$11,401
F2	3,936	"	"	5,187
F3	13,788	"	"	18,170
AA	14,830	"	"	19,543
BB	3,313	"	"	4,366
Elev. Pit Walls	1,816	"	"	2,393
Bsmnt. Walls 12″ × 12′	30,000	"	"	39,534
Bsmnt. Walls 6″ × 8′	12,090	"	"	15,932
Waterproofing	7,404	"	"	9,757
Bldg. Exc. & Backfill	22,491	"	"	29,638
Perimeter Drain	2,400	"	"	3,163
Elev. Pit. Slab	518	"	"	683
Bsmnt. Slab	55,778	"	"	73,503
Int. Cols.	8,190	"	"	10,793
Ext. Cols.	10,062	"	"	13,260
CIP Multispan Joist Slab	161,406	"	"	212,698
Composite Beam & CIP Slab 2 & 3	421,470	"	"	555,406
Steel Cols.	35,021	"	"	46,150
Composite Beam & CIP Slab @ Mech. & Roof	29,714	"	"	39,157
Stl. Joists & Beams on Cols. @ Roof	52,936	"	"	69,758
Stl. Joists on Walls @ PH	4,226	"	"	5,569
Stairs	54,700	"	"	72,083
6″ Lt. Wt. CMU w/perlite	10,704	"	"	14,106
Entry	7,650	"	"	10,081
Dbl. Dr. @ Bsmnt. Lobby & Mech.	3,880	"	"	5,113
Sgl. Dr. @ Bsmnt. Lobby & Mech.	6,825	"	"	8,994
Curtain Wall Framing	277,350	"	"	365,487
Insul. Tinted Glass	128,904	"	"	169,868
Spandrel	195,670	"	"	257,851
Overhead Rolling Doors	8,650	"	"	11,399
EPDM Roof	16,254	"	"	21,419
Alum. Roof Edge	10,854	"	"	14,303
Base Flashing	3,080	"	"	4,059
3″ Polystyrene Insul.	22,491	"	"	29,638
8″ CMU	16,758	"	"	22,083
6″ CMU	10,041	"	"	13,232
Drywall @ Core	38,511	"	"	50,749
Drywall @ Ext. Walls	21,383	"	"	28,178
Toilet Partition	6,975	"	"	9,192
Urinal Screens	648	"	"	854
1 Flush Birch Door	10,368	"	"	13,663
Hdwr.–Allow	6,750	"	"	8,895
Specialties	5,000	"	"	6,589
Paint DW	8,059	"	"	10,620
Paint & Fill CMU	8,617	"	"	11,355
Ceramic Tile Floors	3,302	"	"	4,351
Quarry Tile Floors	15,275	"	"	20,129
Acou. Ceiling	79,240	"	"	104,421
Elev.	141,600	"	"	186,598

Figure 11.10

Item	Amount		Proration Factor	Adjusted Amount
Plumbing Fixtures Subtotal	56,899	×	1.3178	$74,981
Control	8,535	"	"	11,247
Pipe & Fittings	28,450	"	"	37,491
Fire Cycle Sprinkler 1st Flr.	34,776	"	"	45,827
Fire Cycle Sprinkler Addl. Flr.	61,846	"	"	81,500
Boiler Assembly	119,637	"	"	157,656
Ch. W.C. Tower	383,292	"	"	505,096
Boiler Vent 6"	190	"	"	250
Boiler Vent 90 el.	34	"	"	45
Boiler Vent Roof Flashing	26	"	"	34
Ex. Duct	7,060	"	"	9,304
Roof Exhaust Fan	960	"	"	1,265
Registers	460	"	"	606
Duct Heater	300	"	"	395
Stair Heaters	1,335	"	"	1,759
Mech. Qual./Complex Fctr. @ 15%	105,570	"	"	139,118
1200 Amp. Service	8,725	"	"	11,498
1200 Amp. Switchgear	14,850	"	"	19,569
Strip Fluor. @ Bsmnt.	21,945	"	"	28,919
Recessed @ Tenant	218,295	"	"	287,666
Wall recept. @ Bsmnt.	1,512	"	"	1,992
Wall recept. @ Tenant & PH	64,791	"	"	85,381
Under-carpet @ Tenant	94,689	"	"	124,777
Switches	10,114	"	"	13,328
Equip. Conn.	34,859	"	"	45,937
Fire Alarm	18,455	"	"	24,320
Parking	74,825	"	"	98,603
4' Sidewalk	1,888	"	"	2,488
Curb & Gutters	19,968	"	"	26,314
Catch Basin	4,215	"	"	5,554
Storm 24" RCP	7,440	"	"	9,804
Trenching	16,644	"	"	21,933
Sewer	865	"	"	1,140
Water	1,735	"	"	2,286
Mtl. Halide Lighting	18,250	"	"	24,050
Site Excavation–Allow	11,883	"	"	15,659
1200 Amp. Feeder	12,800	"	"	16,868
Conc. Steps	3,000	"	"	3,953
Ftg. @ Ramp & Steps	1,849	"	"	2,437
Retaining Wall	3,624	"	"	4,776
TOTAL				$4,607,197

Final Cost Control Estimate
Proration Adjustment—Chambers Office Building

Figure 11.10 *(continued)*

Schedule of Values—Proration Adjustment

Item	Amount		Proration Factor	Adjusted Amount
Start-up & Move-on	$89,170			
Excavation	43,775	×	1.019266	$44,618
Site Utilities	15,376	"	"	15,672
Asphalt Paving	130,712	"	"	133,230
Site Concrete	38,034	"	"	38,767
Concrete, Forms & Finish	536,840	"	"	547,183
Masonry	47,287	"	"	48,198
Structural Steel	368,952	"	"	376,060
Joists & Deck	112,547	"	"	114,715
Misc. Metals	59,191	"	"	60,331
Carpentry & Millwork	7,582	"	"	7,728
Waterproofing	7,825	"	"	7,976
Insulation	10,035	"	"	10,228
Roofing & Sheet Metal	66,170	"	"	67,445
Doors & Frames	13,386	"	"	13,644
Finish Hardware	15,600	"	"	15,901
Curtain Wall/Alum. Ent.	1,088,772	"	"	1,109,749
Overhead Doors & Grilles	16,892	"	"	17,217
Drywall	70,849	"	"	72,214
Fireproofing	36,370	"	"	37,071
Ceramic Tile	14,352	"	"	14,629
Acoustical Ceiling	103,822	"	"	105,822
Flooring	17,645	"	"	17,985
Painting	12,478	"	"	12,718
Specialties	15,056	"	"	15,346
Elevator	181,806	"	"	185,309
HVAC –Ductwork	231,919	"	"	236,387
–A/C Equipment	315,791	"	"	321,875
–Heating Equip.	40,704	"	"	41,488
–Pipe & Fittings	71,773	"	"	73,156
–Insulation	17,656	"	"	17,996
–Balancing	7,812	"	"	7,963
–Controls	37,211	"	"	37,928
Plumbing –Fixtures	44,491	"	"	45,348
–Pipe & Fittings	73,538	"	"	74,955
–Insulation	3,718	"	"	3,790
–Testing	1,115	"	"	1,136
Fire Protection	113,904	"	"	116,099
Elect. –Lighting	318,541	"	"	324,678
–Serv. & Dist.	157,095	"	"	160,122
–Wiring	107,702	"	"	109,777
–Motors	14,499	"	"	14,778
–Fire Alarm	39,447	"	"	40,207
TOTAL	$4,717,440			$4,717,439
Total less Start-up	$4,628,270			
Proration Factor	1.0193			

Figure 11.11

Cost Breakdowns for Chambers Office Building

Cost Table Item	Schedule of Values Item	Schedule of Values Amount	Final Cost Estimate Item	Final Cost Estimate Amount
EXCAVATION	Excavation	$44,618	Bldg. Excavation	$29,638
			Site Excavation	15,659
				$45,297
PAVING & SIDEWALK	Asphalt Paving	$133,230	Parking	$98,603
	Site Concrete	38,767	4' Sidewalk	2,488
		$171,997	Curb & Gutter	26,314
			Ftg. BB @ Ramp	2,437
			Retaining Wall	4,776
			Concrete Steps	3,953
				$138,571
SITE UTILITIES	Site Utilities	$15,672	Catch Basins	$5,554
			24" Storm Sewer	9,804
			Trenching	21,933
			Sanitary Sewer	1,140
			Water Line	2,286
				$40,717
STRUCTURAL	Concrete, Forms		Ftg. F1	$11,401
	& Finishes	$547,183	F2	5,187
	Structural Stl.	376,060	F3	18,170
	Joists & Deck	114,715	AA	19,543
	Misc. Metals	60,331	BB	4,366
	Fireproofing	37,071	Elev. Pit Walls	2,393
		$1,135,360	Bsmnt. Walls 12"	39,534
			" " 6"	15,932
			Elev. Pit Slab	683
			Basement Slab	73,503
			Interior Columns	10,793
			Exterior Columns	13,260
			CIP Joist Slab	212,698
			Comp. Slab/Bms	555,406
			Steel Columns	46,150
			Comp. Slab/Bms PH	39,157
			Stl. Deck/Joists Roof	69,758
			Stl. Deck/Joists PH	5,569
			Stairs	72,083
				$1,215,586

Figure 11.12

Cost Table Item	Schedule of Values Item	Schedule of Values Amount	Final Cost Estimate Item	Final Cost Estimate Amount
MASONRY	Masonry	$48,198	6″ Lt. Wt.	$14,106
			8″ CMU	22,083
			6″ CMU	13,232
				$49,421
CARPENTRY & MILLWORK	Carpentry & Millwork	$7,728		
WATERPROOFING	Waterproofing	$7,976	Waterproofing	$9,757
ROOFING	Insulation	$10,228	EPDM Roof	$21,419
	Roofing	67,445	Alum. Edging	14,303
		$77,673	Base Flashing	4,059
			3″ Poly. Insul.	29,638
				$69,419
DOORS & FRAMES	Doors & Frames	$13,644	Dbl. Door	$5,113
	Finish Hdwr.	15,901	Sgl. Doors	8,994
	Overhead Doors & Grilles	17,217	1 Wood Door, HM Frame	13,663
		$46,762	Finish Hdwr.	8,895
			Overhead Rolling Grille	11,399
				$48,064
ALUMINUM FRMS & GLAZING	Curtain Wall & Entries	$1,109,749	Entry	$10,081
			Curtain Wall	365,487
			Insul. Glass	169,868
			Spandrel Panels	257,851
				$803,287
FINISHES	Drywall	72,214	Drywall @ Core	$50,749
	Ceramic Tile	14,629	@ Ext. Walls	28,178
	Ceiling	105,822	Paint Drywall	10,620
	Flooring	17,985	Paint & Fill CMU	11,355
	Painting	12,718	Ceramic Tile	4,351
		$223,368	Quarry Tile Floors	20,129
			Ceiling	104,421
				$229,803
SPECIALTIES	Specialties	$15,346	Toilet Partitions	$9,192
			Urinal Screens	854
			Misc. Specialties	6,589
				$16,635
ELEVATORS	Elevators	$185,309	Elevators	$186,598

Figure 11.13

Cost Table Item	Cost Breakdowns: Schedule of Values — Item	Amount	Final Cost Estimate — Item	Amount
PLUMBING	Fixtures	$45,348	Perimeter Drain	$3,163
	Pipe & Fittings	74,955	Plumbing Fixtures	74,981
	Insulation	3,790	Control	11,247
	Testing	1,136	Pipe/Ftgs.	37,491
		$125,229	Quality/Complex.	18,967
				$145,849
FIRE PROTECTION		$116,099	Fire Cycle Sys.	$45,827
			Addl. Flr.	81,500
			Quality/Complex.	19,034
				$146,361
HVAC	HVAC–Ductwk	$236,387	Boiler Assembly	$157,656
	–A/C Eq.	321,875	Chilled Water	505,096
	–Heat Eq.	41,488	Boiler Vent	250
	–Pipe & Ftgs.	73,156	El.	45
	–Insul.	17,996	Flashing	34
	–Bal.	7,963	Exhaust Duct	9,304
	–Controls	37,928	Roof Exhaust Fan	1,265
		$736,793	Exhaust Registers	606
			Duct Heaters	395
			Stair Heaters	1,759
			Quality/Complex.	101,117
				$777,527
ELECTRICAL	Elec–Lights	$324,678	1200 Amp. Serv.	$11,498
	–Serv. & D	160,122	Switch	19,569
	–Wiring	109,777	Strip Fluorescent	28,919
	–Motors	14,778	Recessed Fluor.	287,666
	–Fire Alarm	40,207	Wall Recep. Bsmnt.	1,992
		$649,562	Tenant/PH	85,381
			Under-Carpet Recept.	124,777
			Switches	13,328
			Equip. Connections	45,937
			Fire Alarm	24,320
			Metal Halide	24,050
			1200 Amp. Feeder	16,868
				$684,305

Figure 11.14

12

COMPUTERS IN DESIGN COST CONTROL

12

COMPUTERS IN DESIGN COST CONTROL

In the past few years, the use of computers in architecture and construction has increased dramatically. Formerly, it was limited to organizations that could afford mini or mainframe computers, and specially written software. The advent of powerful, inexpensive microcomputers, together with full-featured, reasonably-priced software, has brought the benefits of computers within the reach of nearly every company. Beyond word processing and accounting, microcomputer software is available for design and drafting, surveying and mapping, project cost control, contract management, scheduling, and estimating.

Computers can offer a significant advantage to those who prepare conceptual and cost control estimates. The associated calculations can be more quickly and accurately performed by computer. The effects of changes in costs and assemblies choices can also be more quickly assessed using a computer. Some estimating software packages offer increased takeoff productivity and accuracy through the use of a digitizer board and stylus or "mouse" to enter information directly from project drawings.

At the time of this writing, three basic types of estimating software are available. These are general application programs such as spreadsheet and data base packages, estimating programs with pricing and reporting capabilities such as Means ASTRO system, and estimating programs that incorporate digitizer/mouse takeoff capabilities such as Means GALAXY system. In this chapter we will discuss the capabilities and use of these software packages, trends in estimating software, and considerations in choosing a computer estimating system.

General Application Programs

The general application programs that can be used in conceptual and cost control estimating include spreadsheets, such as Lotus 1-2-3 (Lotus Development Corporation) and data base programs, such as dBase III+ (Ashton-Tate). Each of these programs offers the capability of entering and pricing assemblies, but each performs the task in a different way.

Spreadsheets

A spreadsheet is a grid of intersecting columns and rows that resembles an accountant's ledger sheet. Like the ledger sheet, the "cells", or intersections of the rows and columns, are used to enter information. Spreadsheet cells may contain numerical data, alphabetical data, symbols, or mathematical formulas. Calculations can be performed using the formulas in one cell and numbers from other cells. A spreadsheet provides flexibility by creating templates, the spreadsheet equivalent of a printed form, for the entry, storage, and manipulation of data in any column/row format that the user finds beneficial. The spreadsheet template offers an advantage over a printed form in that the formulas stored in the cells automatically perform their computations when data is entered or changed.

For conceptual and cost control estimating, spreadsheet templates are created to convert quantities into cost. The data entered includes certain percentages for sales tax, general conditions, overhead and profit, bid conditions, location, and inflation. Takeoff quantities and prices are also entered. Formulas built into the template automatically multiply quantities by prices to obtain costs. Individual costs for each section are totalled, and the section totals added to get a building subtotal. The building subtotal is then adjusted for sales tax, general conditions, overhead and profit, bid conditions, location, and inflation.

An example of a conceptual and cost control estimating template containing the final cost control estimate for the Chambers project is shown in Figures 12.1 through 12.4. The template begins with a record of the project name, location, estimate type, estimate date, and adjustment percentages. Next are takeoff and pricing sections in the UNIFORMAT divisions used in *Means Assemblies Cost Data*. These sections contain spaces for Means cost code numbers, item descriptions, quantities, units, and item costs. The takeoff and pricing sections also contain the formulas to automatically calculate the cost extensions and division totals. The template copies the project name and estimate date from the first page to the following pages, and transfers the adjustment percentages from the first page to page 4 (Figure 12.4), where the *Building Subtotal* adjustments are made. The template also includes a "macro", a simple program to print the estimate in the *page* and *condensed print* format shown in the example.

Spreadsheets are relatively inexpensive to use since calculations are performed more quickly and accurately using this method than they are by hand. The effects of changes in quantities and/or prices can be seen easily, allowing the designer to quickly perceive the effect that different assemblies have on the total project cost. Time is required, however, to learn to use the spreadsheet and to design the template.

Data Base Programs

Data base programs provide another way in which to enter, store, and manipulate data. Rather than forcing the data into a column and row format, the user is, in this case, free to design the entry and report forms in a way that more closely meets his needs. Many data base programs have the further capability of creating specific applications with an internal programming language. For example, a program can be designed to prompt the user to enter project, estimate, and percentage information; and to enter cost code, item description, and quantity data

ASSEMBLIES ESTIMATE

Project:	Chanbers Office Building					Page 1 of 4
Location:	Chambers, Ga.				Sales Tax	4.00%
Est. Type:	Final Cost Control				Gen Conditions	16.00%
Date:	5-1-89				OH & P	10.00%
					Bid Cond.	8.00%
					Location	89.20%
					Inflation	6.10%

System No.	Descritpion	Qty	Unit	Unit Cost	Ext.	Total
1.0	FOUNDATIONS					
1.1-120-7750+	Ftg F1 - 11.5 sq x 2.0	6	ea	1442	8,652	
1.1-120-7750+	F2 - 9.5 sq x 2.0	4		984	3,936	
1.1-120-7810	F3 - 8.0 sq x 2.0	18		766	13,788	
1.1-140-2700	AA - 2.0 x 1.0	580.42	lf	25.55	14,830	
1.1-140-2100+	BB - 1.5 x .67	215		15.41	3,313	
1.1-210-3000	Elev Pit Walls 6' x 6"	41		44.3	1,816	
1.1-210-7220	Bsmnt Walls 12' x 8"	300		100	30,000	
1.1-210-5020	8' x 8"	186		65	12,090	
1.1-292-6400	Wp perimeter 8' avg ht	600		12.34	7,404	
1.9-100-4620	Bldg Exc & BF 8' avg depth	18900	sf	1.19	22,491	
1.1-294-1000	4" PVC Perimeter Drain	600	lf	4	2,400	
					0	
					0	
					0	$120,720
2.0	SLAB ON GRADE					
2.1-200-4520	Elev Pit Slab - 6" reinf	137.29	sf	3.77	518	
2.1-200-2280	Bsmnt Slab - 4" reinf	18228		3.06	55,778	
					0	
					0	
					0	$56,296
3.0	SUPERSTRUCTURE					
3.1-112-1490	Int Cols 28" dia	130	lf	63	8,190	
3.1-114-4000+	Pilasters @ Ext Wall 24" x 12'	234		43	10,062	
3.5-160-7000	CIP Multispan Joist Slab @ 1st	18900	sf	8.54	161,406	
3.5-520-6000	Composit Bm & CIP Slab @ 1st	37800		11.15	421,470	
3.1-130-4800	Stl Cols	1176	lf	29.78	35,021	
3.5-520-6250	Composit Bm & CIP Slab @ PH	2201	sf	13.5	29,714	
3.7-420-4700	Stl Joists & Bms on Cols @ Roof	16699		3.17	52,936	
3.7-430-1700	Stl Joists on Walls @ PH	2201		1.92	4,226	
3.9-100-0780	Stairs - Conc in mtl pans	10	ea	5470	54,700	
					0	
					0	
					0	$777,725
4.0	EXTERIOR CLOSURE					
4.1-211-3300	6" Lt Wt CMU w/ Perlite	2244	sf	4.77	10,704	
4.6-100-7150	Entry	2	opng	3825	7,650	
4.6-100-4200	Dbl Dr @ bsmnt & PH	2		1940	3,880	
4.6-100-3950	Sgl Dr " "	7		975	6,825	
4.7-582-1850	Curtain Wall Frame	25800	sf	10.75	277,350	
4.7-584-1100	1/2" insul, tinted glass	9840		13.1	128,904	
4.7-584-3100	1/2" Pl w/ 2" insul @ spandrel	15960		12.26	195,670	
4.6-100-6150	OH Rolling Grills - motor op	2	opng	4325	8,650	
					0	
					0	
					0	$639,633

Figure 12.1

ASSEMBLIES ESTIMATE

Project: Chanbers Office Building
Date: 5-1-89

5.0	ROOFING	Qty	Unit	Unit Cost	Ext.	Total
5.1-220-2100	EPDM - .045, balasted	18900	sf	0.86	16,254	
5.1-520-1200	Alum Roof Edge	804	lf	13.5	10,854	
5.1-510-1100	Base Flashing	204		15.1	3,080	
5.7-101-2350	3" polystyrene insul	18900	sf	1.19	22,491	
					0	
					0	
					0	
					0	
					0	
					0	
					0	
					0	
					0	
					0	$52,679
6.0	INTERIOR CONSTRUCTION					
6.1-210-6000	8" CMU @ bsmnt	3406	sf	4.92	16,758	
6.1-210-5500	6" CMU @ PH	2282		4.4	10,041	
6.1-510-5400	DW @ Core 1-3	15162		2.54	38,511	
6.1-510-5300	" @ Ext walls	13620		1.57	21,383	
6.1-870-0420	T Partitions	15	ea	465	6,975	
6.1-870-1300	U Screens	3		216	648	
6.4-260-6000	Wd Dr - 1 hr flush birch	27		384	10,368	
	Hdwr - allow	27		250	6,750	
	Specialties - allow	1	LS	5000	5,000	
6.5-100-0080	Paint DW	14391	sf	0.56	8,059	
6.5-100-0320	Paint & Fill CMU	8532		1.01	8,617	
6.6-100-1720	CT Flooring	768		4.3	3,302	
6.6-100-1820	QT Flooring	1625		9.4	15,275	
6.7-100-5800	Acou Ceiling	55802		1.42	79,239	
					0	
					0	
					0	
					0	
					0	
					0	
					0	
					0	
					0	
					0	
					0	
					0	
					0	$230,927
7.0	CONVEYING					
7.1-100-2000+	Elev 2500# 100 fpm 4 flrs	2	ea	70800	141,600	
					0	
					0	
					0	
					0	
					0	
					0	
					0	
					0	
					0	
					0	
					0	$141,600

Figure 12.2

ASSEMBLIES ESTIMATE

Project:	Chanbers Office Building					Page 3 of 4
Date:	5-1-89					
8.0	MECHANICAL SYSTEM	Qty	Unit	Unit Cost	Ext.	Total
8.1-160-1860	WH - 80 gal	1	ea	2435	2,435	
8.1-310-4120+	3" CI Roof Drain	12		581	6,972	
8.1-310-4160+	" " " " addl flrs	12		456	5,472	
8.1-433-2040	Lav - Wall - PE on CI 18x15	12		735	8,820	
8.1-433-1840	- @ van. - 19" dia	9		480	4,320	
8.1-434-4340	Serv Sink - Wall - PE on CI 24x30	3		1095	3,285	
8.1-450-2000	Urinals - Wall	6		755	4,530	
8.1-460-1880	EWC - Wall - dual	3		840	2,520	
8.1-510-3000	WC - 1st pr	3		1315	3,945	
8.1-510-3100	- addl pr	3		1290	3,870	
8.1-510-1800	- addl sngl	6		795	4,770	
8.2-310-1080	Stand Pipe - 2 1/2" - 1st flr	1		1420	1,420	
8.2-310-1100	" " " - addl flr	4		470	1,900	
8.2-390-0800	FHC	6		200	1,200	
151-125-0440	Auto Deck Drain	6		240	1,440	
	Water Control	56899	$	15.00%	8,535	
	Pipe & Fittings	56899	$	50.00%	28,450	
8.2-150-0620	Fire Cycle Sprinkler - 1st flr	18900	sf	1.84	34,776	
8.2-150-0740	" " " - addl flr	58901		1.05	61,846	
8.3-141-1480	Boiler Assembly - CI, gas	56700		2.11	119,637	
8.4-120-4040	Ch W, Cooling Tower system	56700		6.76	383,292	
8.5-410-0060	Boiler Vent 6" galv	15	lf	12.67	190	
8.5-410-1140	" " " 90 L	1	ea	34	34	
8.5-410-2440	" " " Roof Flaching	1		26	26	
157-250-0560	Exhaust Duct	2000	lbs	3.53	7,060	
157-290-7140	Roof Exhaust Fan	1	ea	960	960	
157-460-1020	Registers	23		20	460	
155-408-0100	Duct Heater	1		300	300	
155-420-1020	Stair Heaters	3		445	1,335	
					0	
					0	
					0	
					0	
					0	
					0	
	Quality/Complexity	703799.4	$	15.00%	105,570	$809,370
9.0	ELECTRICAL SYSTEM					
9.1-210-0480	1200 Amp Service	1	ea	8725	8,725	
9.1-410-0320	1200 Amp Switchgear	1		14850	14,850	
9.2-212-0840+	Strip Fluor @ bsmnt	21101	sf	1.04	21,945	
9.2-213-0280	Recessed fluor @ tenant	56700		3.85	218,295	
9.2-522-0200+	Wall recpt @ bsmnt	18900		0.08	1,512	
9.2-522-0360	" " @ tenant & PH	58901		1.1	64,791	
9.2-524-0320	U/carpet - hi density - @ tenant	56700		1.67	94,689	
9.2-542-0200	Switches - assume 1/1000 sf	77801		0.13	10,114	
C 9.0-110	Equipment Connections	68351		0.51	34,859	
C 9.0-110	Fire Alarm & Detection	68351		0.27	18,455	
					0	
					0	
					0	
					0	
					0	
					0	
					0	
					0	
					0	$488,235

Figure 12.3

ASSEMBLIES ESTIMATE

Project: Chanbers Office Building
Date: 5-1-89

11.0	SPECIAL CONSTRUCTION	Qty	Unit	Unit Cost	Ext.	Total
					0	
					0	
					0	
					0	
					0	
					0	$0
12.0	**SITE WORK**					
12.5-510-1500	Parking	205	cars	365	74,825	
12.7-140-1600	4" SW	260	lf	7.26	1,888	
12.7-610-2000	Curb & Gutter	3120		6.4	19,968	
12.3-710-5860	Catch Basins	3	ea	1405	4,215	
12.3-510-4650	Storm	310	lf	24	7,440	
12.3-110-3900	Trenching	570		29.2	16,644	
12.3-510-9060	Sanitary	100		8.65	865	
12.3-540-2160	Water	100		17.35	1,735	
12.7-500-5820	Mtl Halide Lighting	10	ea	1825	18,250	
	Site Excavation - Allow	11883	sy	1	11,883	
9.1-310-0480	1200 Amp Feeder	100	lf	128	12,800	
	Conc Steps - Allow	1	LS	3000	3,000	
1.1-120-2100+	Ftg - BB @ ramp & steps	120	lf	15.41	1,849	
1.1-210-1500	Retg Wall 4'x6"	120		30.2	3,624	
					0	$178,986
13.0	**MISCELLANEOUS**					
					0	
					0	
					0	
					0	
					0	
					0	
					0	
					0	
					0	
					0	
					0	
					0	
					0	
					0	$0
					Bldg Subtotal	$3,496,171
	Sales Tax	3496170.	x	0.04	x .5	$69,923
	General Conditions	3496170.	x	0.16		$559,387
					Subtotal "A"	$4,125,481
	Overhead & Profit	4125480.	x	0.1		$412,548
	Bid conditions	4125480.	x	0.08		$330,038
					Subtotal "B"	$4,868,067
	Location Factor				x	0.892
					Subtotal "C"	$4,342,316
	Inflation				x	1.061
					TOTAL	$4,607,197

Figure 12.4

for each estimate entry. The program can then automatically insert the cost of the costs; calculate extensions, totals, and adjustments; and print an estimate report.

Generally, data base programs cost about the same as spreadsheets. While they do not have all the math capabilities of a spreadsheet, most data base programs can quickly and accurately perform the simple multiplication and addition required for conceptual and cost control estimating. A "what if" capability, the ability to see the effects of changes, can also be programmed into the application. Although the costs must be entered once into a cost data base, the costs do not have to be entered each time an assembly is used. In this way, time is saved and the opportunity for error is reduced. Learning to use and program a data base does, however, tend to be more difficult and may take much longer than learning to use a spreadsheet.

Pricing and Report Programs

In the early 1970's, the following procedures were representative of the best available pricing/report estimating programs. First, the program would probably have been specially written to be used on the company's mainframe computer. The takeoff would then be read into a dictaphone or tape recorder, the tapes transcribed to punch cards, and the cards read into the computer. A print-out of the takeoff was used for checking. Any corrections would be transcribed and read into the computer. When the takeoff and corrections were completed, the quantity summary sheets were returned for manual pricing.

Many advances have been made in estimating software since that time. The current programs are easier to use and far more powerful, and most can be used on microcomputers. Means ASTRO System is an example of a currently available pricing/report estimating software package. It has the ability to extend and summarize quantity takeoff, and can also do pricing, make adjustments, and provide a wide variety of reports.

Using the ASTRO system, takeoff items in the form of a cost code and quantity are entered into the computer through the keyboard. ASTRO supplies the item description and costs from the *Means Building Construction Cost Data*, *Mechanical*, or *Electrical Cost Data* files (or any Means unit price files), or from a user-created cost data file. Each of the cost items includes a description, unit of measure, crew, productivity, bare material unit price, bare installation (labor and equipment) unit price, total bare unit price, and a total price including the installing contractor's overhead and profit. Adjustment percentages may be added to each takeoff entry. Material, labor, and equipment prices may be changed to more accurately reflect actual conditions. Overhead and profit rates may also be customized. Custom crews may be created and substituted for those in the Means cost data files. Cost items may be combined to create assemblies costs.

Once the cost code and quantity are entered, ASTRO calculates the extension. The cost item is then added to a project data file. When the takeoff is complete, the results may be reported in a variety of ways. Among the reports offered are the burdened and unburdened *Itemized Job Reports*. The Itemized Job Report lists all cost lines by MASTERFORMAT division and subdivision. The unburdened report contains the bare costs for each cost line, subdivision total, and division total. In the burdened report, costs including overhead and profit are provided for each cost line, subdivision total, and division total. Burdened and unburdened *Division Summary Reports* providing subdivision and division totals, can also be produced. The burdened and unburdened *Job Report Summary* lists only the division totals. Reports are also available to summarize the crews used, the cost items that were modified, and the adjustment percentages used.

ASTRO has an update feature that allows an old estimate to be adjusted to reflect new prices. The updating may be performed for material costs only, for installation costs only, or for both. ASTRO can be run on a microcomputer with 256K RAM, a 360K disk drive, and a 10 megabyte hard drive.

Pricing/report software is more expensive than most spreadsheet programs, partly due to the hard drive requirement. Nevertheless, the software is quick and accurate and the available cost data files reduce data entry to a minimum, reducing the opportunity for error. Furthermore, the extensive customizing options allow the package to be tailored to reflect the conditions of the local market.

Electronic Data Entry Programs

The data entry programs have the potential to significantly boost estimating productivity by increasing the speed of the quantity takeoff. The takeoff devices include any of a variety of "mice" or digitizer stylus, and digitizer tablets. A stylus or "mouse" can be used to enter dimensions into the computer as it is moved over the plans in the manner of a plan measure. A digitizer tablet is a plan table with an electronic grid. Plan sheets are placed on the table and the stylus enters dimensions by pinpointing locations on the table grid. These dimensions provide accurate readings of the lengths, perimeters, areas, volumes, and numbers needed in the quantity takeoff. One such program is the Means GALAXY System.

The Means GALAXY System uses a digitizer tablet to enter the data into the computer. Just as with ASTRO, a cost code number is supplied by the estimator for the takeoff item. GALAXY can also access the *Means Building Construction Cost Data* file and the *Means Mechanical* and *Electrical Cost Data* files (and any other Means unit price files), as well as user-created data files, to supply item descriptions and costs. GALAXY provides the same customizing and report features as ASTRO. In addition, GALAXY provides an *Estimate Checklist*, a *Trade Report* listing all the trades required for the project, and a *Labor Cost Report* listing the bare and burdened hourly labor and equipment rates. A scheduling module is also available for use with GALAXY or ASTRO.

The data entry programs are more expensive than either the spreadsheet programs or the pricing/report programs. Since a mouse or digitizer tablet must be included in the computer system, the hardware cost is also significantly higher. The major advantage of the data entry program is the increase in productivity and accuracy afforded by entering data directly from the plans into the computer.

Trends in Estimating Software

The wave of the future in estimating software is to eliminate the need for printed documents. Products have been announced, and should be on the market shortly after this book is published, that will allow the estimator to work directly with drawings stored in the computer. Two approaches are being tried. One is to supply an estimating module that will work within the Computer-Aided Drafting and Design (CADD) program to produce the estimate. This method has the potential of providing a quick and accurate cost control system that designers can use without leaving the CADD environment. Such a system could also be used by contractors.

The alternate approach is a program that will interact directly with the CADD drawing files to prepare estimates. Because the designer would have to leave the CADD environment to use this program, this second method would be more appropriate for use by contractors.

Both packages would provide the cost data file, customizing, and report capabilities listed above. The pre-release prices given for these programs fall between the cost of the data entry and pricing/report programs. The hardware requirements will probably include a fast microprocessor, a math coprocessor, extended machine memory, and a high capacity hard drive. These systems will be expensive; it will be interesting to see how well they work.

Computers in Design Cost Control Estimating

Of the currently available computer estimating systems, only the *application* programs provide a cost effective advantage for design cost control. The *price and report* and *data entry* programs are designed to provide the detail and report capabilities needed by contractors and, though most can be modified to provide assemblies estimating capabilities, they are specifically designed for unit price estimating. If the CADD add-on packages live up their advance billing, then a powerful estimating package will soon become available for design cost control.

Summary

The potential gain in both productivity and accuracy certainly makes using computers in conceptual and cost control estimating worthy of consideration. Relatively inexpensive estimating software is currently available for microcomputers. The three major types are general application programs, such as spreadsheets and data base packages, pricing and report packages, such as Means ASTRO System, and data entry programs, such as Means GALAXY System.

The wave of the future appears to be programs that work within a CADD environment or directly with CADD drawing files to produce estimates. While both should be relatively easy to use, a premium will, no doubt, be paid for these new features and improved productivity. In the end, the choice of system must depend on budget limitations and productivity requirements.

INDEX

U

V

W

Z